Pressure Swing Adsorption

Pressure Swing Adsorption

Douglas M. Ruthven
Shamsuzzaman Farooq
Kent S. Knaebel

WILEY-VCH

NEW YORK • CHICHESTER • WEINHEIM • BRISBANE • SINGAPORE • TORONTO

A NOTE TO THE READER:
This book has been electronically reproduced from digital information stored at John Wiley & Sons, Inc. We are pleased that the use of this new technology will enable us to keep works of enduring scholarly value in print as long as there is a reasonable demand for them. The content of this book is identical to previous printings.

Copyright © 1994 by John Wiley & Sons, Inc.

No part of this publication may be reproduced, stored in a retrieval system or transmitted in any form or by any means, electronic, mechanical, photocopying, recording, scanning or otherwise, except as permitted under Sections 107 or 108 of the 1976 United States Copyright Act, without either the prior written permission of the Publisher, or authorization through payment of the appropriate per-copy fee to the Copyright Clearance Center, 222 Rosewood Drive, Danvers, MA 01923, (978) 750-8400, fax (978) 750-4470. Requests to the Publisher for permission should be addressed to the Permissions Department, John Wiley & Sons, Inc., 111 River Street, Hoboken, NJ 07030, (201) 748-6011, fax (201) 748-6008, E-Mail: PERMREQ@WILEY.COM.

To order books or for customer service please, call 1(800)-CALL-WILEY (225-5945).

Library of Congress Cataloging-in-Publication Data

Ruthven, Douglas M. (Douglas Morris), 1938—
    Pressure swing adsorption/Douglas M. Ruthven, Shamsuzzaman Farooq. Kent S. Knaebel.
       p. cm.
    Includes bibliographical references and index.
    ISBN 0-471-18818-2
    1. Adsorption. I. Farooq. Shamsuzzaman. II. Knaebel. Kent S. 1951— . III. Title.
TP156.A35R78   1993
660'.28423—dc20   93-33965
                                                    CIP

Preface

Although pressure swing adsorption (PSA) is not a new process, it is really only during the past decade that such processes have achieved widespread commercial acceptance as the technology of choice for more than a few rather specific applications. Nowadays, however, PSA processes are widely used, on a very large scale, for hydrogen recovery and air separation, and further important applications such as recovery of methane from landfill gas and production of carbon dioxide appear to be imminent. The suggestion for a book on this subject came from Attilio Bisio, to whom we are also indebted for his continuing support and encouragement and for many helpful comments on the draft manuscript.

The authors also wish to acknowledge the seminal contributions of two pioneers of this field, the late Frank B. Hill and Robert L. Pigford. Several of their publications are cited in the present text, but their influence is far broader than the citations alone would suggest. Suffice it to say that much of the book would not have been written without their encouragement and the stimulus provided by their widsom and insight. Several graduate students and post-doctorals have made major contributions, most of which are recognized explicitly by citations. However, they, as well as others whose work may not have been directly referenced, also contributed in a very real way by helping the authors, through discussion and argument, to understand and appreciate some of the subtleties of PSA systems. It would be remiss not to mention by name M. M. Hassan, J. C. Kayser, N. S. Raghavan, and H. S. Shin.

This book is not intended as an exhaustive review of PSA technology, neither is it a design manual. Rather, we have attempted to present a coherent general account of both the technology and the underlying theory. Perhaps more than in other processes the rational design and optimization of a pressure swing adsorption process requires a reasonably detailed mathematical model. The two commonly used approaches to PSA modeling, equilibrium theory, and dynamic numerical simulation are discussed in some detail in Chapters 4 and 5. Inevitably these chapters are somewhat mathematical in approach. The details may be important only to those who are involved in process design and optimization but we hope that the more general reader will still be able to gain some insight concerning the underlying principles and the strengths and limitations of the various approaches.

A three-way collaboration between authors inevitably raises some difficulties since it becomes hard to maintain consistency in style and emphasis and to avoid repetition between different sections of the text. We hope, however, that the advantages of a more authoritative treatment of the subject will more than compensate for any such deficiencies. From our perspective the collaboration has proved interesting and instructive, and we have encountered no serious disagreements amongst ourselves.

UNB, Fredericton, Canada   D. M. Ruthven

National University of Singapore   S. Farooq

Adsorption Research Inc., Dublin, Ohio   K. S. Knaebel
                                                              June 1993

Contents

List of Symbols xi

Greek Symbols xv

Subscripts xvii

Figure Credits xix

1. Introduction 1
   1.1 Historical Development of PSA Processes 4
   1.2 General Features of a PSA Process 5
   1.3 Major Applications of PSA 7
   References 9

2. Fundamentals of Adsorption 11
   2.1 Adsorbents 11
   2.2 Adsorption Equilibrium 23
   2.3 Adsorption Kinetics 34
   2.4 Adsorption Column Dynamics 52
   References 63

3. PSA Cycles: Basic Principles 67
   3.1 Elementary Steps 67
   3.2 Equilibrium-Controlled Separations for the Production of Pure Raffinate Product 71

3.3 Recovery of the More Strongly Adsorbed Species in Equilibrium-Controlled Separations 83
3.4 Cycles for the Recovery of Pure Raffinate Product in Kinetically Controlled Separations 85
3.5 Cycle for Recovery of the Rapidly Diffusing Species 93
References 94

4. Equilibrium Theory of Pressure Swing Adsorption 95
4.1 Background 95
4.2 Mathematical Model 97
4.3 Model Parameters 102
4.4 Cycle Analysis 105
4.5 Experimental Validation 133
4.6 Model Comparison 137
4.7 Design Example 143
4.8 Heat Effects 148
4.9 Pressurization and Blowdown Steps 151
4.10 Conclusions 161
References 163

5. Dynamic Modeling of a PSA System 165
5.1 Summary of the Dynamic Models 166
5.2 Details of Numerical Simulations 184
5.3 Continuous Countercurrent Models 201
5.4 Heat Effects in PSA Systems 207
References 217

6. PSA Processes 221
6.1 Air Drying 221
6.2 Production of Oxygen 226
6.3 Production of Nitrogen 230
6.4 PSA Process for Simultaneous Production of O_2 and N_2 232
6.5 Hydrogen Recovery 235
6.6 Recovery of CO_2 242
6.7 Recovery of Methane from Landfill Gases 244
6.8 Hydrocarbon Separations 246
6.9 Process for Simultaneous Production of H_2 and CO_2 from Reformer Off-Gas 246
6.10 PSA Process for Concentrating a Trace Component 251
6.11 Efficiency of PSA Processes 258
References 263

7. Extensions of the PSA Concept 265
7.1 The Pressure Swing Parametric Pump 265
7.2 Thermally Coupled PSA 270

CONTENTS

7.3 Single-Column Rapid PSA System   278
7.4 Future Prospects   286
References   287

8. Membrane Processes: Comparison with PSA   289

8.1 Permeability and Separation Factor   289
8.2 Membrane Modules   295
8.3 Calculation of Recovery–Purity Profiles   299
8.4 Cascades for Membrane Processes   301
8.5 Comparison of PSA and Membrane Processes for Air Separation   303
8.6 Future Prospects   305
References   306

Appendix A. The Method of Characteristics   307

References   311

Appendix B. Collocation Form of the PSA Model Equations   313

B.1 Dimensionless Form of the LDF Model Equations   313
B.2 Collocation Form of the Dimensionless LDF Model Equations   315
B.3 Dimensionless Form of the Pore Diffusion Model Equations (Table 5.6)   318
B.4 Collocation Form of the Dimensionless Pore Diffusion Model Equations   320

Appendix C. Synopsis of PSA Patent Literature   327

C.1 Introduction   327
C.2 Inventors and Patents   328
C.3 Concluding Remarks   338

Index   345

List of Symbols

a	sorbate activity; external area per unit volume for adsorbent sample (Eq. 2.46)
A	adsorbent surface area per mole (Eq. 2.10); $A_{cs} + A'$ (Table 5.10); membrane area (Eq. 8.1)
A'	cross-sectional area of column wall (Table 5.10)
A_{cs}	internal cross-sectional area of column
A_s	Helmholtz free energy (Eq. 2.11)
$A(k, i)$	collocation coefficient for internal (intraparticle) concentration profile (Appendix B)
$AH(j, i)$	collocation coefficient for velocity profile during pressurization (Appendix B)
$Ax(j, i)$	collocation coefficient for the external (fluid-phase) concentration profile (Appendix B)
b	Langmuir constant
b_o	pre-exponential factor ($b = b_o\, e^{-\Delta U/RT}$)
B	mobility coefficient (Eq. 2.29); constant in Eq. 4.76
$B(k, i)$	collocation coefficient for the intraparticle (internal) phase Laplacian
$Bx(j, i)$	collocation coefficient for the external fluid-phase Laplacian
c	sorbate concentration in gas phase
c_o	sorbate concentration in feed
C	total gas-phase concentration
C_g	volumetric heat capacity of gas (ρC_p)

LIST OF SYMBOLS

C_s	volumetric heat capacity of solid (ρC_p)
$C_{p\,\text{steel}}$	heat capacity of steel wall (mass basis) (Table 5.10)
d	internal diameter of adsorbent column
D	diffusivity
D_c	micropore or intracrystalline diffusivity
D_e	effective diffusivity
D_K	Knudsen diffusivity
D_L	axial dispersion coefficient
D_m	molecular diffusivity
D_p	pore diffusivity
E	diffusional activation energy
E_A	enrichment of heavy component ($\bar{y}/y_{AF}$)
$f_{i,j}$	isotherm function for component i at composition j
f'_j	isotherm slope (dq^*/dc) at composition j
F	total feed volume
F_s	free energy of adsorbed phase (Eq. 2.11)
F_A, F_B	fractions of components A, B desorbed from column during depressurization
G	purge-to-feed velocity ratio
G_s	Gibbs free energy of adsorbed phase (Eq. 2.8)
h	overall heat transfer coefficient
ΔH	enthalpy change on adsorption
J	flux of sorbate
k	overall mass transfer (LDF) rate coefficient based on adsorbed phase concentration
K	adsorption equilibrium constant or isotherm slope; constant in Eq. 7.5
K_c	adsorption equilibrium constant on crystal (microparticle) volume
K'	adsorption equilibrium constant or isotherm slope based on sorbate pressure
K_o, K'_o	pre-exponential factors (Eq. 2.2)
K_L	effective thermal conductivity of steel wall (Table 5.10)
L	adsorbent bed length
L_B, L_A	phenomenological coefficients
M	molecular weight; constant in quadratic isotherm expression
n	exponent in Freundlich isotherm expression
n_a	moles of adsorbable component (Eq. 2.8)
n_s	moles of solid adsorbent (Eq. 2.8)
N	flux relative to fixed frame of reference (Eq. 2.26); total moles (gaseous and adsorbed) in bed at time t
p	partial pressure of sorbate
p_s	saturation vapor pressure
P	absolute pressure (in column)
P'	rate of change of pressure during feed step (Eq. 4.35)

P_H	high pressure (at end of pressurization)
P_F	feed pressure
P_L	low pressure (during purge step)
P_{cH}	high pressure for compressor
P_{cL}	low pressure for compressor
Pe	Peclet number ($v_{oH}L/D_L$)
$\wp$	absolute pressure ratio P_H/P_L
$\wp_1$	pressure ratio P_H/P_F
$\wp_F$	pressure ratio P_F/P_L
$\wp_H$	pressure ratio P_H/P_L (end of pressurization versus end of blowdown)
$\wp_c$	absolute compression ratio P_{cH}/P_{cL}
q	adsorbed phase concentration
q^*	equilibrium value of q
q_o	value of q at equilibrium with feed (concentration c_o)
$\bar{q}$	value of q averaged over an adsorbent particle
q_s	saturation limit
Q	molar gas flow rate
r	radial coordinate in microparticle
r_c	microparticle radius
r_{in}, r_{out}	inner and outer radii of column
R	radial coordinate in a microparticle; gas constant (R_g); product recovery
R_p	macroparticle radius
S_E	equilibrium selectivity K_A/K_B
S_k	kinetic selectivity D_A/D_B
Sh	Sherwood number $2R_p k_f/D_m$
t	time
t^*	adsorption or desorption time
T	temperature
T_o	feed temperature
ΔU	internal energy change or adsorption
v	interstitial gas velocity
v_o	interstitial gas velocity at inlet
$\bar{v}$	dimensionless interstital gas velocity v/v_{oH}
V	volume
w_c	velocity of concentration front
w_t	velocity of temperature front
w'_c	velocity of shock front
x	mole fraction (of component A) in adsorbed phase; dimensionless adsorbed phase concentration averaged over a macroparticle $\bar{q}_i/\bar{q}_{is}$
x_c	dimensionless adsorbed phase concentration averaged over a microparticle (q/q_s)

X	fraction of complete purge
y	mole fraction of A in gas phase
$\bar{y}_{bd}$	average mole fraction (of raffinate product B) in blowdown gas
$\bar{y}_{pdt}$	average mole fraction (of raffinate product B) in high-pressure product stream
z	axial distance
Z	dimensionless axial distance z/L

Greek Symbols

α permeability ratio (intrinsic separation factor); $t_H/(t_H + t_L)$ (in Table 5.9)

α' separation factor $x(1-y)/y(1-x)$ or $y(1-x)/x(1-y)$

α_k kinetic selectivity (effective)—see Eq. 2.46

β parameter characterizing heat effect $[(\Delta H/C_s)(\partial q^*/\partial T)_p]$ in Eq. 2.46; adsorption selectivity parameter β_A/β_B; $b_i C$

β_i ratio of hold-up component i in void space as fraction of total hold-up $[1 + ((1-\varepsilon)/\varepsilon)K_i]^{-1}$

γ ratio of gas heat capacities at constant pressure and constant volume

γ_E ratio of Langmuir constants b_B/b_A

γ_k ratio of micropore diffusivities D_{cB}/D_{cA}

γ_s ratio of saturation capacities q_{Bs}/q_{As}

Γ dimensionless parameter $(r_c^2/D_{Ac})(3k_f/R_p)(C/q_{As})$

Γ' dimensionless parameter $(r_c/D_{Ac})(C/q_{As})k_f$

ε voidage of adsorbent bed

ϵ_p porosity of adsorbent particle

ζ $[1 + \xi P_H y_F \beta_{A_0}(1-\beta)]^{-1}$ (Chapter 4)

η mechanical efficiency of compression; dimensionless radial coordinate R/R_p

θ adsorption selectivity parameter θ_A/θ_B

θ_i dimensionless concentration q_i/q_{is} (Chapters 2 and 5); parameter $\theta_i(P, y_1, y_2) = [1 + ((1-\varepsilon)/\varepsilon)((f_{i2} - f_{i1})/(y_{i2} - y_{i1}))(RT/P_H)]^{-1}$, where 1 and 2 refer to arbitrary states (Chapter 4)

θ_c dimensionless adsorption or desorption time $(\epsilon_p D_p/R_p^2)(c_0/q_0)\, t^*$ (for macropore control) or $D_c t^*/r_c^2$ for micropore control

λ ratio of dead volume to column volume; non-linearity parameter q_o/q_s

μ chemical potential; viscosity; mean residence time in column

Ξ parameter $((1-\varepsilon)/\varepsilon)(M_A/RT)$

ξ $(1-\varepsilon)M_A/\varepsilon R$

ρ density

σ^2 variance of pulse response

τ dimensionless time variable, tv_{OH}/L (LDF model); tD_c/r_c^2 (pore diffusion model)

ϕ parameter $\epsilon A_{cs} L P_L/\beta_A RT$

Φ surface potential

Ω parameter defined by Eq. 5.16; integral function used in determining recovery for pressurization with feed

Subscripts

A, B	components A (more strongly adsorbed) and B (less strongly adsorbed)
B	blowdown step
c	micropore or intracrystalline
ci	component i in microparticle
C	column
DV	dead volume
ei	equivalent value (for component i) in countercurrent flow model
F	feed or feed end
G	purge-to-feed ratio
H, OH, iH	high-pressure step, at inlet during high-pressure step, and for component i during high-pressure step
i	refers to species i (A or B).
ip, is	species i in microparticle, species i at saturation
I	intermediate
L, oL, iL	low-pressure (purge) step, at inlet during low-pressure step, and for component i during low-pressure step
o, oi	limiting or reference value, limiting or reference value for component i
O	outlet or effluent
p	macropore or macroparticle
P	product end or pressurization step
PU	purge step

SUBSCRIPTS

R	rinse step
s	saturation value
S	condition following pressurization step
SH	shock wave
w	at wall
W	combined blowdown and purge step effluent; waste or byproduct
0, 1, 2	initial state, ahead of shock, and behind shock
Superscript	* is sometimes used to denote and emphasize "equilibrium value"

Figure Credits

Chapter 2

Figure 2.3 From H. Jüntgen, K. Knoblauch, and K. Harder, *Fuel* **60**, 817 (1981). Reprinted with permission of the publishers, Butterworth-Heinemann Ltd.

Figure 2.4 From K. Chihara and M. Suzuki, *Caubon* **17**, 339 (1979). Reprinted with permission of Pergamon Press PLC.

Figure 2.12 Adapted from G. A. Sorial, W. H. Granville, and W. O. Daley, *Chem. Eng. Sci.* **38**, 1517 (1983) with permission of Pergamon Press PLC.

Figure 2.16 From H. J. Schröter and H. Jüntgen in *Adsorption Science and Technology*, NATO ASI E158, p. 289, A. E. Rodrigues, M. D. Le Van, and D. Tondeur, eds. Kluwer, Dordrecht (1989). Reprinted with permission of K. Kluwer, Academic Publishers.

Figures 2.24 and 2.25 From *Principles of Adsorption and Adsorption Processes*, by D. M. Ruthven, John Wiley, New York (1984). Reprinted with permission of John Wiley and Sons Inc.

Figure 2.26 From A. I. Liapis and O. K. Crosser, *Chem. Eng. Sci.* **37**, 958 (1982). Reprinted with permission of Pergamon Press PLC.

Chapter 3

Figure 3.2 From G. F. Fernandez and C. N. Kenney, *Chem. Eng. Sci.* **38**, 834 (1983). Reprinted with permission of Pergamon Press PLC.

Figure 3.5 From C. W. Skarstrom in *Recent Developments in Separation Science* Vol 2, p. 95, N. N. Li, ed., CRC Press, Cleveland, OH (1975). Reprinted with permission of the copyright holder, CRC Press, Boca Raton, FL.

Figure 3.10 From J. C. Davis, *Chemical Engineering*, Oct 16th, 1972 p. 88. Excerpted from *Chemical Engineering* by special permission. Copyright (1972) McGraw Hill Inc., New York, NY 10020.

Figure 3.14 Reprinted with permission from R. T. Yang and S. J. Doong, *AIChE Jl* **31**, 1829 (1985). Copyright American Institute of Chemical Engineers.

Figures 3.15 and 3.16 From K. Knoblauch, *Chemical Engineering* **85** (25), 87 (1978). Excerpted from *Chemical Engineering* by special permission. Copyright (1972) McGraw Hill Inc., New York, NY 10020.

Figure 3.19 From A. Kapoor and R. T. Yang, *Chem. Eng. Sci.* **44** 1723 (1989). Reprinted with permission of Pergamon Press PLC.

Chapter 4

Figure 4.24 From A. E. Rodrigues, J. M. Loureiro, and M. D. Le Van, *Gas Separation and Purification* **5**, 115 (1991). Reprinted with permission of the publishers, Butterworth-Heinemann Inc.

Figure 4.25 From Z. P. Lu, J. M. Loureiro, A. E. Rodrigues, and M. D. Le Van, *Chem. Eng. Sci.* **48**, (1993). Reprinted with permission of Pergamon Press PLC.

Figure 4.26 From D. M. Scott, *Chem. Eng. Sci.* **46**, 2977 (1991). Reprinted with permission of Pergamon Press PLC.

Figure 4.27 From J. Hart, M. J. Baltrum, and W. J. Thomas, *Gas Separation and Purification* **4**, 97 (1990). Reprinted with permission of the publishers, Butterworth-Heinemann Inc.

Chapter 5

Figure 5.1 (a) Reprinted from the PhD thesis of P. M. Espitalier-Noel, University of Surrey (1988); (b) Reprinted from the PhD thesis of H. S. Shin, University of Ohio (1988); with kind permission of the authors.

Figure 5.2 From A. Kapoor and R. T. Yang, *Chem. Eng. Sci.* **44**, 1723 (1989). Reprinted with permission of Pergamon Press PLC.

Figure 5.5 Reprinted with permission from P. L. Cen, W. N. Chen, and R. T. Yang, *Ind. Eng. Chem. Process Design Develop.* **24**, 1201 (1985). Copyright 1985, American Chemical Society.

Figures 5.6 and 5.7 From A. Kapoor and R. T. Yang, *Chem. Eng. Sci.* **44**, 1723 (1989). Reprinted with permission of Pergamon Press PLC.

FIGURE CREDITS

Figure 5.11 Reprinted with permission from M. Suzuki, *AIChE Symp. Ser.* **81** (242), 67 (1985). Copyright American Institute of Chemical Engineers.
Figure 5.13 Reprinted with permission from S. J. Doong and R. T. Yang, *AIChE Jl* **32**, 397 (1986). Copyright American Institute of Chemical Engineers; and from P. Cen and R. T. Yang, *Ind. Eng. Chem. Fund* **25**, 758 (1986). Copyright 1986, American Chemical Society.
Figure 5.14 Reprinted from the PhD thesis of P. M. Espitalier-Noel, University of Surrey (1988), with kind permission of the author.

Chapter 6

Figures 6.2 and 6.3 Reprinted with permission from D. H. White and G. Barclay, *Chem. Eng. Prog.* **85** (1), 25 (1989). Copyright American Institute of Chemical Engineers.
Figure 6.4 From C. W. Skarstrom in *Recent Developments in Separation Science*, Vol. 2, N. N. Li, ed., p. 95, CRC Press, Cleveland (1975). Reprinted with permission of the copyright holder, CRC Press Inc., Boca Raton, FL.
Figure 6.7 From J. Smolarek and M. J. Campbell in *Gas Separation Technology*, p. 28, E. F. Vansant and R. Dewolfs, eds., Elsevier, Amsterdam (1990). Reprinted with permission of Elsevier Science Publishers BV.
Figures 6.10 (a) and (c) and 6.11 From S. Sircar in *Adsorption Science and Technology*, NATO ASI E158 p. 285, A. E. Rodrigues, M. D. Le Van, and D. Tondeur eds., Kluwer, Dordrecht (1989). Reprinted with permission of Kluwer, Academic Publishers.
Figure 6.13 From T. Tomita, T. Sakamoto, U. Ohkamo, and M. Suzuki in *Fundamentals of Adsorption II*, p. 569 (1986), A. I. Liapis, eds. Reprinted with permission of the Engineering Foundation.
Figure 6.16 From E. Pilarczyk and K. Knoblauch in *Separation Technology*, p. 522 (1987), N. Li and H. Strathmann, ed. Reprinted with permission of the Engineering Foundation.
Figure 6.17 From H. J. Schröter and H. Juntgen in *Adsorption Science and Technology* NATO ASI E158 p. 281 (1989). Reprinted with permission of Kluwer, Academic Publishers.
Figure 6.18 From E. Pilarczyk and H. J. Schröter in *Gas Separation Technology*, p. 271 (1990), E. F. Vansant and R. Dewolfs, eds. Reprinted with permission of Elsevier Science Publishers BV.
Figure 6.19 Reprinted with permission from R. T. Cassidy and E. S. Holmes, *AIChE Symp. Ser.* **80** (233), 74 (1984). Copyright American Institute of Chemical Engineers.
Figure 6.20 From S. Sircar in *Adsorption and Technology*, p. 285, NATO ASI 158, A. E. Rodrigues, M. D. Le Van, and D. Tondeur, eds., (1989). Reprinted with permission of Kluwer, Academic Publishers; and from R. Kumar et al.,

paper presented at AIChE National Meeting, Houston, April 1991, with permission of the authors.

Figure 6.24 From S. Sircar, Fourth International Conference on Adsorption, Kyoto, May 1992 (plenary lecture). Reprinted with permission of the author.

Figures 6.25 and 6.26 From R. Banerjee, K. G. Narayankhedkar, and S. P. Sukhatine, *Chem. Eng. Sci.* **45**, 467 (1990). Reprinted with permission of Pergamon Press PLC.

Figure 6.27 From R. Banerjee and K. G. Narayankhedkar, *Chem. Eng. Sci.* **47**, 1307 (1992). Reprinted with permission of Pergamon Press PLC.

Chapter 7

Figure 7.1 Reprinted with permission from N. H. Sweed, *AIChE Symp. Ser.* **80** (233), 44 (1984). Copyright American Institute of Chemical Engineers.

Figures 7.2 and 7.3 Reprinted with permission from U.S. Patent 4,354,854 (1982), with kind permission of George Keller II.

Figures 7.5, 7.6, 7.7 and 7.8 Reprinted from reports of Highquest Engineering Inc., with kind permission of Bowie Keefer, Highquest Engineering Inc.

Figure 7.10 Reprinted with permission from C. W. Kenney, Proceedings of 5th Priestley Conference on Gas Separations, Birmingham (1989) p. 273-286. Copyright Royal Society of Chemistry.

Figure 7.11 Reprinted with permission from P. H. Turnock and R. A. Kadlec, *AIChE Jl* **17**, 335 (1971). Copyright American Institute of Chemical Engineers.

Figures 7.12 and 7.13 Reprinted with permission from D. E. Kowler and R. H. Kadlec, *AIChE Jl* **18**, 1207 (1972). Copyright American Institute of Chemical Engineers.

Figure 7.14 Reprinted with permission from S. J. Doong and R. T. Yang, *AIChE Symp. Ser.* **84** (264), 145 (1988). Copyright American Institute of Chemical Engineers.

Figure 7.15 Reprinted from a hitherto unpublished manuscript with kind permission of the authors D. M. Scott, E. Alpay, and C. N. Kenney.

Chapter 8

Figure 8.3 From K. Haraya, T. Hakuta, K. Obuta, Y. Shindo, N. Itoh, K. Wakabayshi, and H. Yoshitome, *Gas Separation and Purification* **1**, 4 (1987). Reprinted with permission of the publishers, Butterworth-Heinemann Ltd.

Figure 8.4 From W. J. Koros, G. K. Fleming, S. M. Jordan, T. H. Kim, and H. H. Hoehn, *Progress in Polymer Sci.* **13**, 339 (1988). Reprinted with permission of Pergamon Press PLC.

Figure Figure 8.6 (c) Reprinted with permission of Dow-Generon Inc.

Figure 8.10 From R. M. Thorogood, *Gas Separation and Purification* **5**, 83 (1991). Reprinted with permission of the publishers, Butterworth-Heinemann Ltd.

Figure 8.11 Reprinted with permission from R. W. Spillman, *Chemical Engineering Progress* **85** (1), 41 (1989). Copyright American Institute of Chemical Engineers.

CHAPTER

1

Introduction

Pressure swing adsorption (PSA) is not a new process and, like most good inventions, with the advantage of hindsight the principle appears obvious. As in all adsorption separation processes, the essential requirement is an adsorbent that preferentially adsorbs one component (or one family of related components) from a mixed feed. This selectivity may depend on a difference in adsorption equilibrium or on a difference in sorption rates (kinetic selectivity). In certain cases the difference in rates may be so great that the slower-diffusing species is in effect totally excluded from the adsorbent (size-selective sieving), and in this situation a very efficient separation can obviously be achieved.

All adsorption separation processes involve two principal steps: (1) *adsorption*, during which the preferentially adsorbed species are picked up from the feed; (2) *regeneration* or *desorption*, during which these species are removed from the adsorbent, thus "regenerating" the adsorbent for use in the next cycle. The general concept is shown in Figure 1.1. It is possible to obtain useful products from either the adsorption or regeneration steps or from both steps. The effluent during the adsorption step is purified "raffinate" product from which the preferentially adsorbed species have been removed. The desorbate that is recovered during the regeneration step contains the more strongly adsorbed species in concentrated form (relative to the feed) and is sometimes called the "extract" product.

The essential feature of a PSA process is that, during the regeneration step, the preferentially adsorbed species are removed by reducing the total pressure, rather than by raising the temperature or purging with a displacing

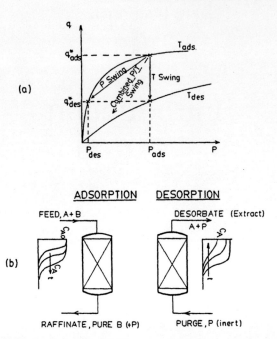

Figure 1.1 The concept of a PSA process. (a) Change in equilibrium loading with pressure. (b) Idealized sketch showing movement of the adsorbed phase concentration profile for the more strongly adsorbed species in a simple two-bed PSA process.

agent (although a low-pressure purge step is commonly included in the cycle). The process operates under approximately isothermal conditions so that the useful capacity is the difference in loading between two points, corresponding to the feed and regeneration pressures, on the same isotherm [Figure 1.1(a)]. Figure 1.1(b) shows schematically the movement of the concentration profiles during the high-pressure feed and low-pressure regeneration steps. The feed step is normally terminated before the more strongly adsorbed component breaks through the bed, while the regeneration step is generally terminated before the bed is fully desorbed. At cyclic steady state the profile therefore oscillates about a mean position in the bed.

A major advantage of PSA, relative to other types of adsorption process such as thermal swing, is that the pressure can be changed much more rapidly than the temperature, thus making it possible to operate a PSA process on a much faster cycle, thereby increasing the throughout per unit of adsorbent bed volume. The major limitation is that PSA processes are restricted to components that are not too strongly adsorbed. If the preferentially adsorbed species is too strongly adsorbed, an uneconomically high vacuum is required to effect desorption during the regeneration step. Thus, for very strongly adsorbed components thermal swing is generally the preferred option since a modest change of temperature produces, in general, a relatively large change in the gas–solid adsorption equilibrium constant.

PSA processes are no more complex than most of the more conventional separation processes, but they are different in one essential feature: the process operates under transient conditions, whereas most processes such as absorption, extraction, and distillation operate under steady-state conditions.

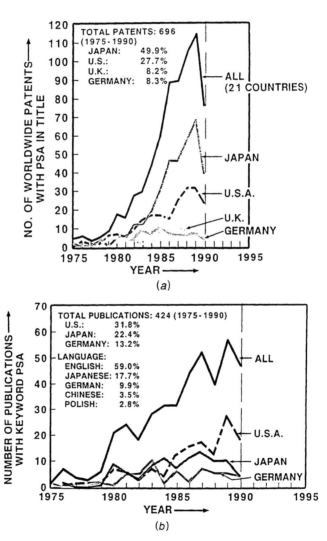

Figure 1.2 The growth of PSA technology from 1975 to 1990 as shown by (a) numbers of patents and (b) numbers of publications. (Courtesy of Dr. S. Sircar, Air Products and Chemicals, Inc.)

As a result, both the conceptual framework and the design procedures are quite different. This difference can best be explained in mathematical terms. A steady-state process can be described mathematically by an ordinary differential equation (or a set of ordinary differential equations), and to obtain the relationship between the operating variables and the process performance requires only the integration of this set of equations. By contrast, a transient process is described by a set of partial differential equations and this requires a more complex solution procedure. As a result the relationship between the process performance and the operational variables is generally less obvious. Procedures for the design and scaleup of PSA units are for the most part available in the open literature. However, they have not yet been generally accepted as part of the normal chemical engineering curriculum and, as a result, a certain air of mystery persists.

Despite their early inception, it was really only during the 1980s that PSA processes gained widespread commercial acceptance. This is illustrated in Figure 1.2, which shows a plot of the annual numbers of publications and U.S. patents relating to PSA processes against the year. The reasons for this unusually long delay between the invention and commercialization of such processes are not entirely clear, but it seems likely that the opposition of entrenched interests in the cryogenic gas industry and the lack of familiarity with the underlying principles among practicing engineers were probably significant factors. During the 1970s interest in alternative separation processes was stimulated by the escalation of energy costs associated with the rising price of crude oil. Although energy costs fell during the 1980s, the impetus to examine alternative processes and to match the technology to the product specifications has continued.

1.1 Historical Development of PSA Processes

The introduction of PSA processes is commonly attributed to Skarstrom[1] and Guerin de Montgareuil and Domine[2] in 1957–1958. However, many of the essential features of this type of process were delineated much earlier in the papers of Kahle[3,4] and in the pioneering patents of Hasche and Dargan,[5] Perley,[6] and Finlayson and Sharp,[7]* which were filed between 1927 and 1930 but have been largely overlooked by more recent authors. The Air Liquide process, developed by Guerin de Montgareuil and Domine, utilized a vacuum swing, whereas the Esso process, pioneered by Skarstrom, used a low-pressure purge to clean the adsorbent bed following the blowdown step. Details of both cycles, which are still in common use, are given in Chapter 3. Some other key dates in the chronological development of PSA technology are

---

* The authors are grateful to Dr. Norman Kirkby of the University of Surrey for pointing out this reference.

INTRODUCTION

Table 1.1. Milestones in the Historical Development of PSA Processes[a]

Date	
1930–1933	First PSA patents issued to Finlayson and Sharp (U.K. 365,092), Hasche and Dargan (U.S. 1,794,377), and Perley (U.S. 1,896,916)
1953–1954	Papers by H. Kahle[3,4] outlining the principle of PSA (including heat storage) and giving details of a PSA process for removal of CO_2, hydrocarbons, and water vapor from air
1955–1956	Synthetic zeolites produced commercially
1957–1958	French patent 1,223,261, P. Guerin de Montgareuil and D. Domine (Air Liquide)[2]; the "vacuum swing" PSA cycle is described. U.S. Patent 2,944,627, C. W. Skarstrom (Esso Research and Engineering)[1]; the low-pressure purge step is introduced, and the importance containing the thermal wave is emphasized
1960–1965	Development and commercialization of the "Heatless Drier" for small-scale air drying and early versions of the "Isosiv" process for separation of linear hydrocarbons
1965–1970	Development and commercialization of PSA hydrogen purification
1970–1972	First large-scale O_2 PSA processes
1972–1973	O_2 selective carbon sieves produced commercially
1976	PSA nitrogen process using CMS adsorbent
1976–1980	Small-scale medical oxygen units
1982	Large-scale vacuum swing processes for air separation
1988	Second generation zeolite adsorbents for air separation by vacuum swing, making VSA competitive with cryogenic distillation up to 100 tons/day

[a] See also R. T. Cassidy and E. S. Holmes, *AIChE. Symp. Series* **80**(233) 68–75 (1984).

summarized in Table 1.1. The patents mentioned are discussed in greater detail in Appendix C.

1.2 General Features of a PSA Process

There are five general features of a PSA system that to a large extent explain both the advantages and limitations of the technology and hence determine the suitability for a given application:

1. Product purity. The raffinate product (the less strongly adsorbed or slower-diffusing species) can be recovered in very pure form, whereas the extract product (the more strongly adsorbed or faster-diffusing species) is generally discharged in impure form as a byproduct. Various modifications to the cycle are possible to allow recovery of the preferentially adsorbed species. However, these all add complexity to the cycle; so the process fits best where a pure raffinate product is required.
2. Yield or fractional recovery. In a PSA process, the fractional recovery (i.e., the fraction of the feed stream that is recovered as pure product) is generally relatively low compared with processes such as distillation,

absorption, or extraction. The recovery can be increased by including additional steps in the cycle and by increasing the number of adsorbent beds, but both these modifications increase the capital cost. A PSA process therefore fits best when the feed is relatively cheap so that a high product yield is not a matter of primary concern.

3. Concentration of trace impurities. Where a highly selective adsorbent is available a PSA process can provide a valuable means of concentrating trace impurities, but this application has not yet been developed to any significant extent.

4. Energy requirements. Like most separation processes, the energy efficiency of a PSA process is relatively low. The First Law efficiency (separation work relative to energy consumed) is in fact comparable with that of processes such as distillation or extraction, but a PSA system uses mechanical energy, which is in general more expensive than heat. The power cost is the major component of the operating cost for a PSA system.

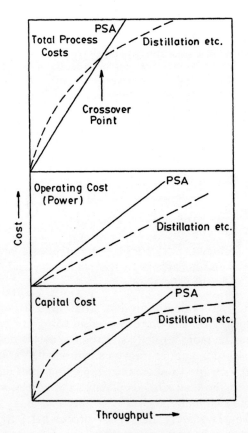

Figure 1.3 Variation of cost with throughput for PSA and cryogenic air separation processes.

INTRODUCTION

However, if the feed is already available at high pressure, these costs may be greatly reduced, since not only are the capital costs and power requirement reduced, but the cost of product recompression will generally be much lower than the cost of compressing the feed to the higher operating pressure. A PSA system is therefore especially useful where the feed is available at elevated pressure.

5. Scaling characteristics. The operating costs of most separation processes increase approximately linearly with throughput. The capital cost of a PSA process is also approximately linear with throughput, but for most other processes the capital cost curve is highly nonlinear, with the incremental cost being smaller for the larger units (Figure 1.3). As a result, when the overall costs are considered, the economics tend to favor PSA at low to moderate throughputs and to favor other processes such as cryogenic distillation for very large-scale operations. Of course the actual costs and the crossover point vary considerably depending on the particular separation and the process configuration, but the form of the cost versus throughput curve is generally similar.

6. Pressure range. The term *vacuum swing adsorption* (VSA) is often used to denote a PSA cycle with desorption at subatmospheric pressure. This is a semantic choice. The performance of any PSA process is governed by the ratio of absolute (rather than gauge) pressures. That desorption at subatmospheric pressure often leads to improved performance is due to the form of the equilibrium isotherm rather than to any intrinsic effect of a vacuum.

1.3 Major Applications of PSA

Some of the major commercial PSA processes are listed in Table 1.2, and a summary of the chronology is given in Table 1.1. The first three applications (air separation, air drying, and hydrogen purification) were in fact foreseen and demonstrated by Skarstrom.[1,8] These remain the most important practical applications for this technology, although newer processes such as carbon dioxide recovery and natural gas purification are gaining increased acceptance. In all three of the major processes the feed is relatively cheap, so that the relatively low recovery is not an overriding economic factor. In both air drying and hydrogen recovery a pure raffinate product is required, and in hydrogen recovery the impure hydrogen is often available at elevated pressure. Purity of the product is important in nitrogen production, but generally somewhat less so in oxygen production. In a typical hydrogen purification process the product purity is commonly 99.995% or even higher. For nitrogen production a purity of 99.9% is easily attainable, but it is generally more economic to produce 99.5% N_2 by PSA with final polishing by a "de oxo" unit. The commonly quoted oxygen product purity of 93–95% is somewhat

Table 1.2. Some Major PSA Processes

Process	Product	Adsorbent	Type of System
H_2 recovery from fuel gas	Ultrapure H_2	Act. C or zeolite	Multiple-bed system
Heatless drier	Dry air (for instruments)	Act. Al_2O_3	Two-bed Skarstrom cycle (or vacuum–pressure swing cycle
Air separation	O_2 (+Ar)	5A Zeolite	Two-bed Skarstrom cycle
Air separation	N_2 (+Ar)	CMS	Two-bed self-purging cycle
Air separation	N_2 and O_2	5A Zeolite or CaX	Vacuum swing system
Isosiv	Linear/Branched hydrocarbons	5A Zeolite	Molecular sieve separation with vacuum swing
Landfill gas separation	CO_2 and CH_4	CMS	Vacuum swing

misleading since the impurity is almost entirely argon—which is adsorbed with the same affinity as oxygen on most adsorbents.

The largest-scale PSA processes are generally to be found in petroleum refinery operations—hydrogen purification and hydrocarbon separation processes such as Isosiv. In such processes product rates up to 10^6 SCFH (> 100 tons/day) are not uncommon. In the other main areas of application (drying and air separation) PSA units are generally economic only at rather smaller scales. For example, for large-scale oxygen or nitrogen production (> 100 tons/day) it is difficult to compete economically with cryogenic distillation. However, there are many small-scale uses for both oxygen and nitrogen (e.g., home oxygen units for asthmatic patients and nitrogen units for purging the fuel tanks of fighter aircraft or for purging the interiors of trucks and warehouses to prolong the shelf life of fruit and vegetables). For such applications the robustness and portability of a PSA system provide additional advantages that reinforce the economic considerations. In these applications the most direct competition comes from small-scale membrane systems, which offer many of the same advantages as a PSA system. A brief comparison of these two classes of process is included in Chapter 8.

To understand the process options and the factors involved in design and optimization of PSA systems, some background in the fundamentals of adsorption and the dynamic behavior of adsorption columns is required. These aspects are considered in Chapter 2. A wide variety of different cycles have been developed in order to increase energy efficiency, improve product purity, and improve the flexibility of the operation. The basic cycles and a few of the more advanced cycles are reviewed in Chapter 3, while more detailed aspects of process modeling are discussed in Chapters 4 and 5. Chapter 6 is devoted to a detailed description of some current PSA processes, while some of the future trends in process development are discussed in Chapter 7.

References

1. C. W. Skarstrom, U.S. Patent 2,944,627 (Feb. 1958) to Esso Research and Engineering Company.
2. P. Guerin de Montgareuil and D. Domine, French Patent 1,223,261 (Dec. 1957) to Air Liquide. See also U.S. Patent 3,155,468 (1964) to Air Liquide.
3. H. Kahle, *Chemie Ing. Technik* **23**, 144 (1953).
4. H. Kahle, *Chemie Ing. Technik* **26**, 75 (1954).
5. R. L. Hasche and W. N. Dargan, U.S. Patent 1,794,377 (1931).
6. G. A. Perley, U.S. Patent 1,896,916 (1933).
7. D. Finlayson and A. J. Sharp, U.K. Patent 365,092 (Oct. 15, 1930) to British Celanese Corp.
8. C. W. Skarstrom, "Heatless Fractionation of Gases over Solid Adsorbents," In *Recent Developments in Separation Science*, Vol. II, pp. 95–106, N. Li ed., CRC Press, Cleveland (1972).

CHAPTER

2

Fundamentals of Adsorption

To understand the design and operation of PSA process requires at least an elementary knowledge of the principles of adsorption and the dynamic behavior of an adsorption column. A brief review of these subjects is therefore included in this chapter. More detailed information can be found in the books of Ruthven,[1] Yang,[2] and Suzuki.[3]

The overall performance of a PSA process depends on both equilibrium and kinetic factors, but the relative importance of these factors varies greatly for different systems. The majority of PSA processes are "equilibrium driven" in the sense that the selectivity depends on differences in the equilibrium affinities. In such processes mass transfer resistance generally has a deleterious effect and reduces the performance relative to an ideal (equilibrium) system. There are, however, several processes in which the selectivity is entirely kinetic (i.e., the separation depends on differences in adsorption rate rather than on differences in equilibrium affinity). In such systems the role played by mass transfer resistance is clearly pivotal, and a more fundamental understanding of kinetic effects is needed in order to understand and model this class of process.

2.1 Adsorbents

2.1.1 Forces of Adsorption

A gas molecule near a solid surface experiences a reduction in potential energy as a consequence of interaction with the atoms (or molecules) in the solid. The result is that gas molecules tend to concentrate in this region so that the molecular density in the vicinity of the surface is substantially greater than in the free-gas phase. The strength of the surface forces depends on the nature of both the solid and the sorbate. If the forces are relatively weak, involving only van der Waals interactions supplemented in the case of polar or quadrupolar species by electrostatic forces (dipole or quadrupole interactions), we have what is called "physical adsorption" or "physisorption." By contrast, if the interaction forces are strong, involving a significant degree of electron transfer, we have "chemisorption." Chemisorption is limited to a monolayer, whereas, in physical adsorption, multiple molecular layers can form. Most practical adsorption separation processes (including PSA) depend on physical adsorption rather than on chemisorption, since, except for a few rather specialized applications, the capacities achievable in chemisorption systems are too small for an economic process. Since the adsorption forces depend on the nature of the adsorbing molecule as well as on the nature of the surface, different substances are adsorbed with different affinities. It is this "selectivity" that provides the basis for adsorption separation processes.

The role of the adsorbent is to provide the surface area required for selective sorption of the preferentially adsorbed species. A high selectivity is the primary requirement, but a high capacity is also desirable since the capacity determines the size and therefore the cost of the adsorbent beds. To achieve a high capacity commercial adsorbents are made from microporous materials. As a result the rate of adsorption or desorption is generally controlled by diffusion through the pore network, and such factors must be considered in the selection of an adsorbent and the choice of operating conditions. Certain materials (zeolites and carbon molecular sieves) that have very fine and uniformly sized micropores show significant differences in sorption rates as a result of steric hindrance to diffusion within the micropores. Such adsorbents offer the possibility of achieving an efficient kinetic separation based on differences in sorption rate rather than on differences in sorption equilibrium.

2.1.2 Hydrophilic and Hydrophobic Behavior

For equilibrium-controlled adsorbents, the primary classification is between "hydrophilic" and "hydrophobic" surfaces. If the surface is polar, generally as a result of the presence of ions in the structure but possibly also as a result of the presence of ions or polar molecules strongly bound to the solid surface, it will preferentially attract polar molecules—in particular water. This is because the field-dipole and/or field gradient-quadrupole interactions provide additional contributions to the energy of adsorption. This additional energy will arise only when both conditions are fulfilled (i.e., a polar or quadrupolar molecule and a polar adsorbent). If either of these is lacking there can be no significant electrostatic contribution to the energy of sorption. Thus, on highly polar adsorbents such as zeolites or activated alumina, water (a small polar molecule) is strongly adsorbed while methane

Table 2.1. Limiting Heats of Sorption for CH_4 and H_2O (kcal / mole)

	Act. carbon (nonpolar)	4A Zeolite (polar)
CH_4 (nonpolar)	4.3	4.5
H_2O (polar)	6.0	18.0

(a small nonpolar molecule of similar molecular weight and therefore with comparable van der Waals interaction energy) is only weakly adsorbed. In contrast, on a clean activated carbon (a nonpolar surface) both these compounds are adsorbed to a comparable extent. Furthermore, while the affinity of the zeolite surface for water is much higher than that of the carbon surface, methane is retained with comparable affinity on both these adsorbents (see Table 2.1). Clearly the polar zeolite surface is "hydrophilic" and, by comparison, the nonpolar carbon surface is "hydrophobic."

Ionic adsorbents such as the zeolites owe their hydrophilic nature to the polarity of the heterogeneous surface. However, when the surface contains hydroxyl groups (e.g., silica gel, alumina, or some polymeric resins) molecules such as water can also interact strongly by hydrogen bond formation. As with polar adsorbents, water is therefore preferentially adsorbed, but in this case the hydrophilic selectivity is attributable mainly to the hydrogen bond energy rather than to surface polarity.

It should be noted that hydrophobic surfaces do not actually repel water. In general water will be adsorbed on any surface with at least the affinity dictated by the van der Waals forces. The point is that on a hydrophilic surface water (and other polar molecules) will be adsorbed much more strongly than would be expected simply from the van der Waals forces alone. Furthermore, while hydrophilic adsorbents generally also show selectivity for other polar molecules relative to similar nonpolar species, this is not always true. Where the hydrophilic selectivity comes from hydrogen bonding, polar molecules with no "active" hydrogens will be held only with an affinity comparable to nonpolar sorbates.

The possibility of creating polar selectivity by pretreatment of the surface is well illustrated by activated carbon adsorbents (see Figure 2.1). On a clean carbon surface n-hexane is adsorbed much more strongly than sulfur dioxide (a polar sorbate), but on an oxidized surface this selectivity is reversed. Control and modification of surface polarity is indeed the most important practical tool in the tailoring of equilibrium selectivity.

2.1.3 Pore Size Distribution

According to the IUPAC classification, pores are divided into three categories by size:

Micropores < 20Å; Mescopores 20–500 Å; Macropores > 500 Å

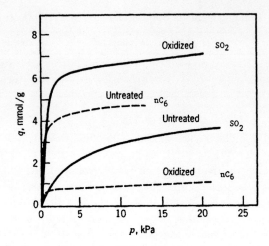

Figure 2.1 Equilibrium isotherms for SO_2 and n-hexane on activated carbon showing the effect of surface modification. (Data from Mastsumura.[4])

In a micropore the guest molecule never escapes from the force field of the solid surface, even at the center of the pore. It is therefore reasonable to consider all molecules within a micropore to be in the "adsorbed" phase. By contrast, in mesopores and macropores, the molecules in the central region of the pore are essentially free from the force field of the surface; so it becomes physically reasonable to consider the pore as a two-phase system containing both adsorbed molecules at the surface and free gaseous molecules in the central region. Of course the IUPAC classification is arbitrary, and it is clear from the description presented that the distinction between a microp-ore and mesopore really depends on the ratio of pore diameter to molecular diameter rather than on absolute pore size. Nevertheless, for PSA processes that deal in general with relatively small molecules, the arbitrary figure of 20 Å is a reasonable choice.

Macropores contain very little surface area relative to the pore volume and so contribute little to the adsorptive capacity. Their main role is to facilitate transport (diffusion) within the particle by providing a network of super highways to allow molecules to penetrate rapidly into the interior of the adsorbent particle.

Representative pore size distributions for several different adsorbents are shown in Figure 2.2. Many commercial adsorbents (e.g., most zeolitic adsor-bents and carbon molecular sieves) (see Table 2.2) consists of composite particles crystals (or char particles) aggregated together and formed into a macroporous pellet, often with the aid of a binder. Such particles have a well-defined bimodal pore size distribution in which the first peak represents the micropores within the microparticles and the second peak represents the large intraparticle pores resulting from the pelletization process. The impli-cations for mass transfer are discussed in Section 2.3.

FUNDAMENTALS OF ADSORPTION

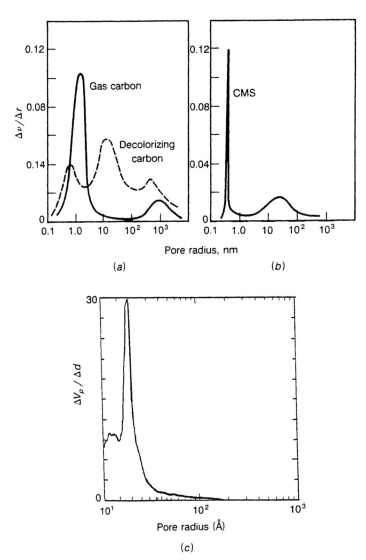

Figure 2.2 Pore size distributions for (a) typical activated carbons; (b) carbon molecular sieve; (c) typical activated alumina.

2.1.4 Kinetically Selective Adsorbents

While most adsorbents have a relatively wide distribution of pore size, kinetic selectivity depends on steric hindrance and therefore requires a very narrow distribution of pore size. This is a characteristic feature of zeolitic adsorbents since these materials are crystalline and the dimensions of the micropores are

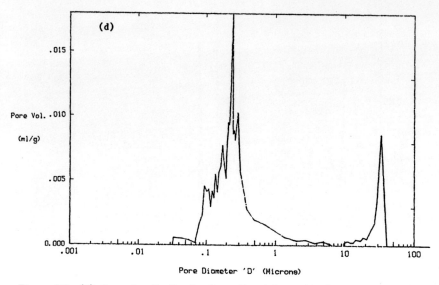

Figure 2.2 (d). Pore size distribution for pelleted 5A zeolite (only the extracrystalline pores are shown).

determined by the crystal structure. Some control of pore size can be achieved by procedures such as silanation and by ion exchange, since, in many zeolites, the cations partially (or even totally) obstruct the intracrystalline micropores.* By contrast, the carbon molecular sieves are amorphous materials similar to high-area activated carbons but with a much narrower

Table 2.2. Classification of Commercial Adsorbents

Equilibrium selective		Kinetically selective	
Hydrophilic	Hydrophobic	Amorphous	Crystalline
Activated alumina	Activated carbon	Carbon molecular sieves (CMS)	Small-pore zeolites and zeolite analogs
Silica gel	Microporous silica		
Al-rich zeolites	Silicalite, dealuminated mordenite, and other silica-rich zeolites		
Polymeric resins containing –OH groups or cations	Other polymeric resins		

---

* For a detailed discussion of this topic, see: E. F. Vansant, *Pore Size Engineering in Zeolites*, Wiley Chichester, U.K. (1990).

distribution of pore size. This uniformity of pore size is achieved in two ways: by careful control of the conditions during the activation step and by controlled deposition of easily crackable or polymerizable hydrocarbons such as acetylene. Control of these processes provides the means by which the pore size can be adjusted.[5,6] In this respect there is somewhat greater flexibility than with crystalline microporous materials in which the pore dimensions are fixed by the crystal structure. In kinetically selective adsorbents the primary parameters determining the selectivity are the pore size and pore size distribution. The nature of the material is generally of secondary importance. Thus, despite the difference in chemical nature, small-pore zeolites and molecular sieve carbons exhibit very similar kinetic selectivities.

2.1.5 Physical Strength

Repeated pressurization and depressurization of an adsorbent bed tends to cause attrition of the adsorbent particles. Physical strength is therefore a prime consideration in the choice of an adsorbent for a PSA process. Such considerations may indeed preclude the use of an otherwise desirable adsorbent in favor of a material that, from kinetic and equilibrium considerations alone, may appear to have inferior properties. Both the "crush strength" and the "abrasion resistance" are strongly dependent on the way in which the adsorbent particles are manufactured, including such factors as the nature of the binder and the pretreatment conditions, but only very limited information is available in the open literature.*

2.1.6 Activated Carbon and Carbon Molecular Sieves

Activated carbon is produced in many different forms that differ mainly in pore size distribution and surface polarity. The nature of the final product depends on both the starting material and the activation procedure. For liquid-phase adsorption a relatively large pore size is required, and such materials can be made by both thermal and chemical activation procedures from a wide range of carbonaceous starting materials. The activated carbons used in gas adsorption generally have much smaller pores, with a substantial fraction of the total porosity in the micropore range. These adsorbents are generally made by thermal activation from a relatively dense form of carbon such as bituminous coal. High-area small-pore carbons may also be made from sources such as coconut shells, but the product generally has insufficient physical strength for PSA applications.

---

* A useful reference is: C. W. Roberts, "Molecular Sieves for Industrial Applications." In *Properties and Applications of Zeolites*, R. P. Townsend, ed., Special Publ. No. 33, The Chemical Society, London (1980).

The thermal activation procedure is a two-step process in which volatile material is first driven off by controlled pyrolysis followed by a controlled "burnout" of the pores using oxidizing gases such as steam or CO_2 at 800°C (or even higher temperatures).[7] The surface of such activated carbons is partially oxidized; so where a nonpolar surface is required, a further step is often included, involving either evacuation or purging with an inert gas at elevated temperature. This eliminates most of the oxides as CO or CO_2.

In many liquid-phase applications activated carbon is used in powder form, but for gas-phase applications larger particles are needed. These are made either directly by crushing and screening or more commonly by granulation of the powder using binders such as pitch, which can be activated to some extent during the final thermal treatment. The preparation of activated carbon in fiber form is a relatively new development which holds considerable promise for the future. The diameter of the fibers is small (~ 10 μm) so diffusional resistance is reduced to an insignificant level. To date such materials do not appear to have been used in PSA processes, but the rapid kinetics make this an intriguing possibility.

The preparation of carbon molecular sieves (Figure 2.3) is broadly similar but often includes an additional treatment with species such as benzene or

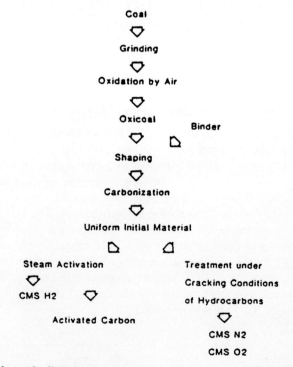

Figure 2.3 Schematic diagram showing the processes involved in the manufacture of carbon molecular sieve adsorbents. (From Jüntgen et al.,[7] with permission.)

FUNDAMENTALS OF ADSORPTION

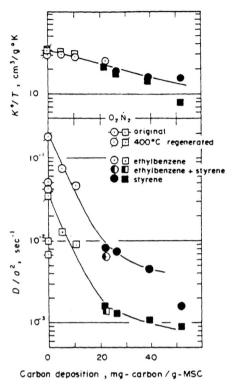

Figure 2.4 Effect of controlled carbon deposition on sorption rates for oxygen and nitrogen in a carbon molecular sieve. (From Chihara and Suzuki,[5] with permission.)

actelylene that are easily polymerized or cracked on the surface (Figure 2.4). By careful control of the conditions a very uniform pore size is achieved. It appears that such control is more easily achieved by carbon deposition than in the burnout step. Brief details of some representative carbon adsorbents are included in Table 2.3.

2.1.7 Silica Gel

A pure silica surface is inactive and "hydrophobic," but if hydroxyl groups are present the surface becomes hydrophilic as a result of the possibilities for hydrogen bond formation. Silica "gel" is formed as a colloidal precipitate when a soluble silicate is neutralized by sulfuric acid. The size of the collidal particles and the nature of their surface are strongly influenced by trace components present in the solution. When water is removed from the "gel," an amorphous microporous solid is formed, but the size of the silica particles and therefore the pore size depend on the conditions during the water

Table 2.3. Physical Properties of Some Common Adsorbents

Adsorbent	Sp. pore vol. (cm³ g⁻¹)	Av. pore diam. (Å)	Pore size distrib.	Sp. area (m² g⁻¹)	Particle density (g cm⁻³)
Silica gel (1)	0.43	22	Unimodal	800	1.09
Silica gel (2)	1.15	140	Unimodal	340	0.62
Act. alumina	0.50	30–3000	Unimodal	320	1.28
Act. carbon	0.15–0.5	Wide range	Bimodal	200–2000	0.6–0.9
CMS	0.25	—	Bimodal	400	0.98

removal step. Brief details of two representative materials are included in Table 2.3. The large-pore material is used in many liquid-phase applications, while the small-pore material is widely used as a desiccant in vapor-phase systems.

Adsorption isotherms for water vapor on silica gel, activated alumina, and 4A zeolite are compared in Figure 2.5. Silica gel does not retain water vapor as strongly as the other adsorbents, but it has a higher ultimate capacity. Furthermore, it can be regenerated at moderate temperatures (150–200°C). It is therefore a useful desiccant where the moisture load is high and the dew point required is not too low. If silica gel is heated above about 300°C, most of the hydroxyls are removed. The adsorbent loses surface area and the

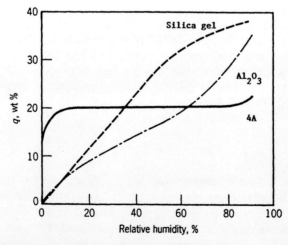

Figure 2.5 Comparative isotherms showing the adsorption of water vapor on silica gel, activated alumina, and 4A zeolite. (When plotted in terms of relative humidity, the isotherms are approximately independent of temperature.)

resulting surface is no longer hydrophilic. Despite its widespread use as a desiccant silica gel is not commonly used in PSA processes as its physical strength is inferior to that of alumina or zeolite based desiccants.

2.1.8 Activated Alumina

Activated alumina is essentially a microporous (amorphous) form of Al_2O_3 and is made by several different methods. The most common route is by controlled dehydration of the trihydrate $Al_2O_3 \cdot 3H_2O$ formed in the Bayer process but some aluminas are made by precipitation from a soluble salt in a manner similar to the production of silica gel.

2.1.9 Zeolites

In contrast to the other adsorbents so far considered, the zeolites are crystalline rather than amorphous, and the micropores are actually intracrystalline channels with dimensions precisely determined by the crystal structure. There is therefore virtually no distribution of micropore size, and these adsorbents show well-defined size-selective molecular sieve properties—exclusion of molecules larger than a certain critical size and strong steric restriction of diffusion for molecules with dimensions approaching this limit. The framework structures of three of the most important zeolites are shown schematically in Figure 2.6. The frameworks consist of tetrahedrally connected assemblages of SiO_2 and AlO_2 units. To translate the schematic diagrams into actual structures one must consider that the lines represent the diameters of oxygen atoms (or ions), while the much smaller Si or Al atoms are located at the apices of the polyhedra. Within rather broad limits Si and Al atoms are interchangeable in the lattice, but each Al introduces a net negative charge that must be balanced by an exchangeable cation. In many structures, notably zeolite A, the exchangeable cations partially (or totally) obstruct the micropores. The equilibrium distribution of the exchangeable cations among the various possible cation "sites" has been extensively studied and is well established for most of the common zeolites.[9] For example, in zeolite A there are three types of site, as indicated in Figure 2.6(a). The most favorable are the type I sites (eight per cage) so in the Ca^{2+} form (six cations per cage) all cations can be accommodated in the type I sites where they do not obstruct the channels. The effective dimension of the channel is then limited by the aperture of the eight-membered oxygen ring (window), which has a free diameter of about 4.3 Å. Since molecules with diameters up to about 5.0 Å can penetrate these windows, this is referred to as a "5A" sieve.

The Na^+ form contains 12 cations per cage; so not only are all eight type I sites filled, but all window sites (3.0 per cage) are also filled. (The twelfth Na^+ cation is accommodated in the relatively unfavorable type III site.) The Na^+ cation partially obstructs the windows, reducing the effective size cutoff

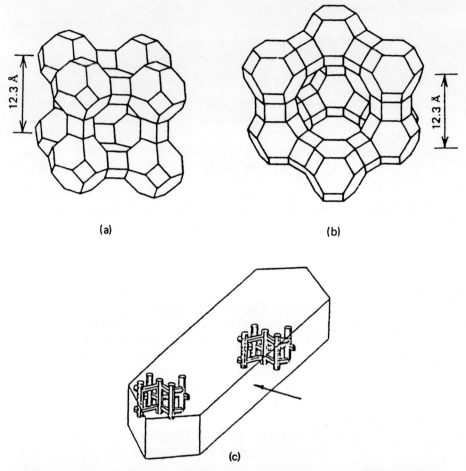

Figure 2.6 Schematic diagrams showing the framework structures of three common zeolites. (a) Zeolite A (the three exchangeable cation sites are indicated), (b) Zeolite X or Y, (c) silicalite or ZSM-5. More detailed descriptions of these structures are given by Breck[8] as well as in more recent reviews.

to about 4 Å—hence the term 4A sieve. Replacement of Na^+ by the larger K^+ cation reduces the dimensions even further so that only water and other very small molecules such as NH_3 can penetrate at an appreciable rate (3A).

The framework structures of X and Y zeolites are the same, and these materials differ only in the Si-to-Al ratio—and therefore in the number of exchangeable cations. The pore structure is very open, the constructions being twelve-membered oxygen rings with free diameter ~7.5 Å. Molecules with diameters up to about 8.5 Å can penetrate these channels with little steric hindrance, and this includes all common gaseous species. Size-selective sieving is observed for larger molecules, but such effects are not relevant to

FUNDAMENTALS OF ADSORPTION

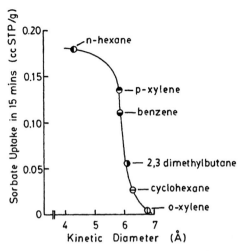

Figure 2.7 Size-selective sorption in silicalite at half-saturation vapor pressure, 298 K. (Data of Harrison et al.[10])

PSA processes. The nature of the cation can have a profound effect on the adsorption equilibria in these materials, but channel-blocking effects are much less important than in the A zeolites.

Silicalite and HZSM-5 are essentially the same material. They are high silica structures. HZSM-5 normally contains measurable aluminum (Si-to-Al ratio ~30–100) with a corresponding proportion of cations. This is important since partial obstruction of the pores as well as strong modification of the adsorption equilibria can result from even a small concentration of cations. "Silicalite" typically has a Si-to-Al ratio of 1000; so the Al may be regarded as an adventitious impurity rather than a true component. The pore network is three dimensional, and the dimensions of the channels are limited by ten-membered oxygen rings having a free aperture of about 6.0 Å. Size-selective sieving is therefore observed for molecules such as the C_8 aromatics, as illustrated in Figure 2.7. In contrast to most Al-rich zeolites, silicalite (and even HZSM-5) are "hydrophobic," but this property appears to be associated with the very high Si-to-Al ratio rather than with the nature of the channel structure, since at high Si-to-Al ratios zeolites of the Y or mordenite type also become hydrophobic.

2.2 Adsorption Equilibrium

2.1.1 Henry's Law

The adsorbed layer at the surface of a solid may be regarded as a distinct "phase" in the thermodynamic sense. Equilibrium with the surrounding gas (or liquid) is governed by the ordinary laws of thermodynamics. Physical adsorption from the gas phase is an exothermic process; so equilibrium favors adsorption at lower temperatures and desorption at higher temperatures. At sufficiently low concentration the equilibrium relationship generally approaches a linear form (Henry's Law):

$$q = K'p \quad \text{or} \quad q = Kc \tag{2.1}$$

and the constant of proportionality (K' or K) is referred to as the "Henry's Law" constant or simply the Henry constant. It is evident that the Henry constant is simply the adsorption equilibrium constant, and the temperature dependence can be expected to follow the usual vant Hoff relations:

$$K' = K_0' e^{-\Delta H/RT}; \qquad K = K_0 e^{-\Delta U/RT} \tag{2.2}$$

where $\Delta H = \Delta U - RT$ is the enthalpy change on adsorption. (For an exothermic process ΔH and ΔU are negative, and the Henry constant therefore decreases with increasing temperature.) Representative plots showing conformity with Eq. 2.2 (for oxygen, nitrogen, and methane in zeolite A) are shown in Figure 2.8.

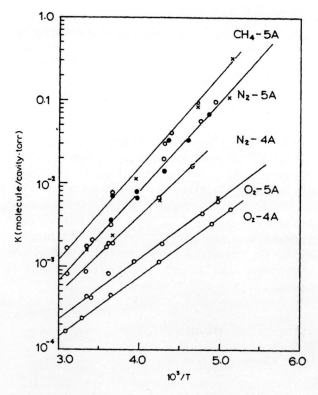

Figure 2.8 Temperature dependence of Henry constants for oxygen, nitrogen, and methane on type A zeolites.[13]

FUNDAMENTALS OF ADSORPTION

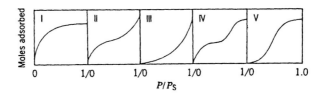

Figure 2.9 The Brunauer classification of isotherms.

2.2.2 Brunauer's Classification

At higher concentrations the equilibrium relationship becomes curved. Brunauer classified the commonly observed forms of isotherm into the five types illustrated in Figure 2.9. Reference to the isotherm for water vapor (Figure 2.5) shows that H_2O-4A is type I, H_2O-alumina is type II, while H_2O-silica gel is type IV. Type I is characteristic of chemisorption, where the saturation limit corresponds to occupation of all surface sites, or to physical adsorption in a microporous material where the saturation limit corresponds to complete filling of the micropores. Type III behavior corresponds to the situation where the sorbate-surface interaction is weaker then the sorbate-sorbate interaction, as, for example, in the adsorption of water vapor on a carbon surface. In a PSA system the isotherms are generally of type I or type II form, and further discussion is therefore restricted to these cases.

2.2.3 "Favorable" and "Unfavorable" Equilibria

In the analysis of adsorption column dynamics it is convenient to classify adsorption equilibria as "favorable," "linear," or "unfavorable" depending on the shape of the dimensionless $(x-y)$ equilibrium diagram. The meaning of these terms is evident from Figure 2.10. (In the "favorable" case the dimensionless adsorbed phase concentration is always greater than the dimensionless fluid phase concentration.) This classification assumes that the direction of mass transfer is from fluid phase to adsorbed phase (i.e., an adsorption process). Since for desorption the initial and final states are reversed, an isotherm that is "favorable" for adsorption will be "unfavorable" for desorption and vice versa.

2.2.4 Langmuir Isotherm

For microporous adsorbents the isotherm can often be represented, at least approximately, by the ideal Langmuir model:

$$\frac{q}{q_s} = \frac{bc}{1 + bc} \tag{2.3}$$

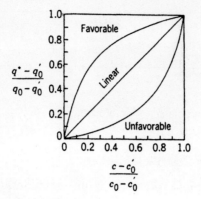

Figure 2.10 Dimensionless equilibrium isotherm showing the meaning of the terms "favorable," "linear," and "unfavorable."

This form may be derived from simple mass action considerations by considering the balance between occupied and unoccupied sites. Equation 2.3 clearly shows the correct asymptotic behavior since it approaches Henry's Law in the low-concentration region and the saturation limit ($q \to q_s$) at high concentrations. In the original Langmuir formulation the saturation limit was assumed to coincide with saturation of a fixed number of identical surface sites and, as such, it should be independent of temperature. In fact a modest decrease of q_s with temperature is generally observed and is indeed to be expected if the saturation limit corresponds with filling of the micropore volume, rather than with the saturation of a set of surface sites. b is an equilibrium constant that is directly related to the Henry constant ($K = bq_s$). Since adsorption is exothermic, it follows that b, like K, will decrease with temperature so at higher temperature the isotherms become less sharply curved, as illustrated in Figure 2.11.

The isosteric enthalpy of sorption is given by:

$$\left(\frac{\partial \ln p}{\partial T}\right)_q = \frac{\Delta H}{RT^2} \tag{2.4}$$

and it follows from Eqs. 2.3 and 2.4 that if q_s is independent of temperature, the isosteric heat will be independent of concentration—a well-known feature of ideal Langmuir behavior.

Although there are relatively few systems that conform accurately to the Langmuir model, there are a great many systems that show approximate conformity, and this model has the further advantage that it reduces to Henry's Law in the low-concentration limit, which is a requirement for thermodynamic consistency in any physical adsorption system. For these reasons the Langmuir model has become widely accepted as the basis for most qualitative or semiquantitative studies of PSA systems.

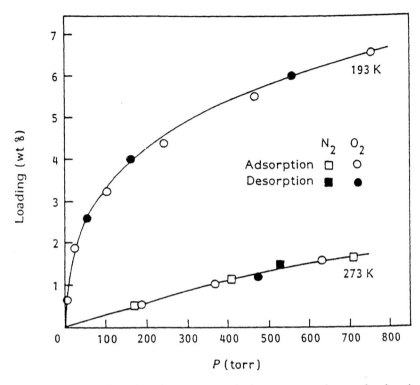

Figure 2.11 Equilibrium data for oxygen and nitrogen on carbon molecular sieve showing the similarity between the isotherms and the effect of temperature on isotherm shape.[14]

2.2.5 Freundlich and Langmuir – Freundlich Isotherms

An alternative expression that is sometimes used to represent a favorable (type I) isotherm is the Freundlich equation:

$$q = bc^{1/n}, \quad n > 1.0 \tag{2.5}$$

This form of expression can be derived from plausible theoretical arguments based on a distribution of affinity among the surface adsorption sites, but it is probably better regarded simply as an empirical expression. Both the Freundlich and Langmuir equations contain two parameters, but, unlike the Langmuir expression, the Freundlich form does not reduce to Henry's Law in the low-concentration limit. Nevertheless, Eq. 2.5 can represent the behavior of several systems over a wide range of conditions. To obtain greater flexibility as an empirical correlation the Langmuir and Freundlich forms are sometimes combined:

$$\frac{q}{q_s} = \frac{bc^{1/n}}{1 + bc^{1/n}} \tag{2.6}$$

Equation 2.6 contains three constants (b, q_s, and n), but it should be stressed that this form is purely empirical and has no sound theoretical basis.

2.2.6 BET Isotherm

Both the Langmuir and Freundlich isotherms are of type I form (in Brunauer's classification). This is the most commonly observed form of isotherm, particularly for microporous adsorbents. However, materials such as activated alumina and silica gel commonly show type II behavior. This form is commonly represented by the BET equation[11]:

$$\frac{q}{q_s} = \frac{b(p/p_s)}{(1 - p/p_s)(1 - p/p_s + bp/p_s)} \tag{2.7}$$

where p_s is the saturation vapor pressure, although the physical model from which this expression was originally derived is probably not realistic, particularly for microporous solids. The BET model is most commonly encountered in connection with the experimental measurement of surface area by nitrogen adsorption at cryogenic temperatures, but it has also been used to represent the isotherms for moisture on activated alumina, where the isotherms are of the well-defined type II form.[12]

2.2.7 Spreading Pressure and the Gibbs Adsorption Isotherm

To understand the Gibbs adsorption isotherm requires a short digression into the formal thermodynamics of adsorption and an introduction to the concept of "spreading pressure." It is convenient to adopt the Gibbsian formulation and consider the adsorbent simply as an inert framework that provides a force field that alters the free energy (and other thermodynamic properties) of the sorbate–sorbent system. The changes in the thermodynamic properties are ascribed entirely to the sorbate. Since the adsorbed layer is a condensed phase, its thermodynamic properties are relatively insensitive to the ambient pressure.

If we consider n_a moles of adsorbent and n_s moles of sorbate, the chemical potential of the adsorbed phase is given by:

$$\mu_s = \left(\frac{\partial G_s}{\partial n_s}\right)_{T, n_a} \tag{2.8}$$

just as for a binary bulk system containing n_s moles of component s and n_a moles of component a. We may also define a specific energy Φ by the partial derivative:

$$\Phi = -\left(\frac{\partial G_s}{\partial n_a}\right)_{T, n_s} \tag{2.9}$$

This quantity has no direct analog for a bulk phase. For example, for a vapor-phase system, the differentiation would have to be performed at constant total pressure, maintaining the same number of moles of s and changing the number of moles of a by adding or removing an inert component to maintain the total moles (and total pressure) constant. This would yield a measure of the partial molar interaction energy between components a and s, which would be very small. For an adsorbed phase Φ can be regarded as the change in internal energy, per unit of adsorbent, due to the spreading of sorbate over the surface. This change in energy may be regarded as a work term—the product of a force and a displacement. Thus, depending on whether one chooses to regard the adsorbed phase as a two-dimensional fluid (area A per mole) or a three-dimensional fluid contained within the pore volume (V per mole):

$$\Phi \, dn_a = \pi \, dA = \phi \, dV \tag{2.10}$$

where π is the "spreading pressure" and ϕ the three-dimensional analog. It is evident that ϕ (or π) fulfills the role of the pressure in a bulk system and the relevant free energy quantity for an adsorbed phase (F_s) is given by:

$$F_s \equiv A_s + \Phi n_a = A_s + \pi A \approx G_s + \pi A \tag{2.11}$$

(since $G_s \approx A_s$). The similarity with the definition of Gibbs free energy, for a bulk phase ($G = A + PV$) is obvious.

Following essentially the same logic as in the derivation of the Gibbs–Duhem equation leads directly to the Gibbs adsorption isotherm:

$$\left(\frac{\partial \Phi}{\partial p}\right)_T = \frac{RT}{p} \frac{n_s}{n_a} \quad \text{or} \quad \left(\frac{\partial \pi}{\partial p}\right)_T = \frac{RT}{p} \frac{n_s}{A} \tag{2.12}$$

By inserting different equations of state for the adsorbed phase [$\pi(n_s, A, T)$], corresponding forms for the equilibrium adsorption isotherm $q(p)$ may therefore be found.

2.2.8 Binary and Multicomponent Sorption

The Langmuir model (Eq. 2.3) yields a simple extension to binary (and multicomponent) systems, reflecting the competition between species for the adsorption sites:

$$\frac{q_A}{q_s} = \frac{b_A p_A}{1 + b_A p_A + b_B p_B + \cdots} \tag{2.13}$$

It is clear that at a given temperature (which determines the value of b) and at given partial pressures the quantity of component 1 adsorbed will be lower than for a single-component system at the same partial pressure. Like the single-component Langmuir equation, Eq. 2.13 provides a useful approximation to the behavior of many systems, but it is quantitatively accurate only for a few systems. It is however widely used in the modeling of PSA systems largely because of its simplicity but also because many PSA systems operate

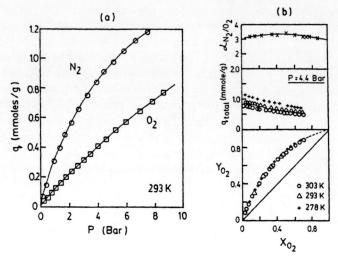

Figure 2.12 Equilibrium isotherms for oxygen, nitrogen, and binary O_2–N_2 mixtures on 5A zeolite showing (a) single-component isotherms and (b) variation of separation factor with loading and X–Y diagram for the binary mixture from Sorial et al.[24] with permission.

under conditions where the loading is relatively low ($q/q_s < 0.5$). Under these conditions, as a first-order deviation from Henry's Law, the Langmuir model is relatively accurate.

It follows from Eq. 2.13 that the equilibrium separation factor (α') corresponds simply to the ratio of the equilibrium constants:

$$\alpha'_{AB} \equiv \left(\frac{x_A}{y_a}\right)\left(\frac{1-y_A}{1-x_A}\right) = \frac{b_A}{b_B} \qquad (2.14)$$

This is evidently independent of composition and the ideal Langmuir model is therefore often referred to as the *constant separation factor* model.

As an example of the applicability of the Langmuir model, Figure 2.12 shows equilibrium data for N_2, O_2, and the N_2–O_2 binary on a 5A molecular sieve. It is evident that the separation factor is almost independent of loading, showing that for this system the Langmuir model provides a reasonably accurate representation.

When the Langmuir model fails, the multicomponent extension of the Langmuir–Freundlich or Sipps equation (Eq. 2.6) is sometimes used:

$$\frac{q_A}{q_s} = \frac{b_A p_A^{1/n_A}}{1 + b_A p_A^{1/n_A} + b_B p_B^{1/n_B} + b_C p_C^{1/n_C} + \cdots} \qquad (2.15)$$

with similar expressions for components B and C. This has the advantage of providing an explicit expression for the adsorbed phase but suffers from the

FUNDAMENTALS OF ADSORPTION 31

disadvantage that it is essentially an empirical data fit with little theoretical basis.

2.2.9 Ideal Adsorbed Solution Theory[15]

A more sophisticated way of predicting binary and multicomponent equilibria from single-component isotherms is the ideal adsorbed solution theory. For a single-component system the relationship between spreading pressure and loading can be found directly by integration of the Gibbs isotherm (Eq. 2.12):

$$\frac{\pi A}{RT} = \int_0^{P_i^0} q_i^0(p) \frac{dp}{p} \tag{2.16}$$

where A is now expressed on a molar basis. The Gibbs isotherm for a binary system may be written as:

$$\frac{A\,d\pi}{RT} = q_A\,d\ln p_A + q_B\,d\ln p_B \tag{2.17}$$

or, at constant total pressure (P):

$$\frac{A\,d\pi}{RT} = q_A\,d\ln y_A + q_B\,d\ln y_B \tag{2.18}$$

where y_i is the mole fraction in the vapor phase.

If the adsorbed phase is thermodynamically ideal, the partial pressure p_i at a specified spreading pressure (π) is given by:

$$p_i = p_i^0(\pi) x_i = y_i P \tag{2.19}$$

where x_i is the mole fraction in the adsorbed phase and p_i^0 is the vapor pressure for the single-component system at the same spreading pressure, calculated from Eq. 2.16. In the mixture the spreading pressure must be the same for both components for a binary system; so we have the following set of equations:

$$\pi_A^0 = \psi_A(p_A^0) = \psi_B(p_B^0) = \pi_B^0$$
$$p_A = Py_A = p_A^0 x_A$$
$$p_B = Py_B = p_B^0 x_B$$
$$y_A = y_B = 1.0$$
$$x_A = x_B = 1.0 \tag{2.20}$$

This is a set of seven equations relating the nine variables ($x_A, x_B, y_A, y_B, P, \pi_A^0, \pi_B^0, p_A^0, p_B^0$); so with any two variables (e.g., P and y_A) specified one may calculate all other variables.

The total concentration in the adsorbed phase is given by:

$$\frac{1}{q_{\text{tot}}} = \frac{x_A}{q_A^0} + \frac{x_B}{q_B^0} \tag{2.21}$$

where q_A^0, q_B^0 are the adsorbed phase concentrations of components A and B, at the same spreading pressure, in the single-component systems. To achieve this spreading pressure in the single-component system the actual pressure for the less strongly adsorbed component must be higher (in some cases much higher) than the total pressure in the binary system. The development outlined here is for a binary system, but the extension to a multicomponent system follows naturally.

It should be stressed that the assumption of ideal behavior defined by Eq. 2.20 does not require a linear equilibrium relationship and does not preclude the possibility of interactions between the adsorbed molecules. The implication, however, is that any such interactions in the mixed adsorbed phase are the same as in the single-component systems. Such as assumption is in fact less restrictive than it might at first appear. However, it is difficult to tell a a priori whether or not this approximation is valid for any particular system. To confirm the validity requires at least limited experimental data for the binary system. From the perspective of PSA modeling a more serious disadvantage of the ideal adsorbed solution theory (IAST) approach is that it provides the equilibrium relationship in implicit rather than explicit form. This makes it inconvenient for direct incorporation into a numerical simulation code.

2.2.10 Adsorption of Atmospheric Gases

Since air separation is one of the major applications of pressure swing adsorption, a brief summary of the available equilibrium data for sorption of argon, oxygen, and nitrogen on some of the more commonly used adsorbents is included here. Table 2.4 lists the Henry constants and heats of sorption, while Table 2.5 gives a summary of the available single and multicomponent

Table 2.4. Henry Constants and Heats of Adsorption for Atmospheric Gases on Some Common Adsorbents

Sorbate	Adsorbent	$K_0 \times 10^{7a}$ (mmole/g Torr)	$-\Delta H$ (kcal/mole)
O_2	4A	6.84	3.2
	5A	8.3	3.3
	CMS	10.5	3.8
N_2	4A	3.6	4.35
	5A	2.0	5.0
	CMS	10.5	3.8
Ar	5A	5.82	3.36
	CMS	8.0	4.0

[a] K_0 is expressed per gram of zeolite crystal. To estimate the value for pelleted adsorbent it is necessary to correct for the presence of the binder (assumed inert). Binder content is typically 15–20% by weight. Data are from Derrah et al.[16] and Ruthven and Raghavan.[35] Values are approximate, since, particularly for CMS adsorbents, there is considerable variation between different materials.

Table 2.5. Published Equilibrium Data for Sorption of Atmospheric Gases on Common Adsorbents[a]

Sorbent	Sorbate	Temp. range (K)	Press. range (atm)	Reference
4A Zeolite	Ar	200–300	0–0.8	Ruthven[16]
		200–300	0–0.7	Eagan[17]
		200–300	0–1.0	Springer[18]
		306–363	0–0.1	Kumar[19]
4A Zeolite	O_2	200–300	0–0.8	Ruthven[16]
		300–360	Henry const.	Haq[20]
		123–173	0–1.0	Eagan[17]
		77	$0-P_{sat}$	Stakebake[21]
4A Zeolite	N_2	200–300	0–0.8	Ruthven[16]
		300–360	Henry const.	Haq[20]
		305	0–1.0	Kumar[19]
		195–223	0–1.0	Eagan[17]
5A Zeolite	Ar	200–300	0–0.8	Ruthven[16]
		304–334	0–1.0	Kumar[19]
		203–297	0–4.5	Miller[22]
		195–348	0–11.0	Wakasugi[23]
5A Zeolite	O_2	200–300	0–0.8	Ruthven[16]
		300–350	Henry const.	Haq[26]
		300–394	0–1.0	Kumar[19]
		203–297	0–4.5	Miller[22]
		273–303	0–0.8	Sorial[24]
		144	0–2.1	Danner[25]
		77	$0-P_{sat}$	Stakebake[21]
		298	0–0.8	Huang[30]
5A Zeolite	N_2	200–300	0–0.8	Ruthven[16]
		300–360	Henry const.	Haq[26]
		300–421	0–1.0	Kumar[19]
		144	0–1.0	Danner[25]
		200–300	0–1.0	Springer[18]
		195, 295	0–30	Lederman[27]
		203–297	0–4.5	Miller[22]
		278–303	0–5	Sorial[24]
		76	P_{sat}	Kidnay[28]
		77–348	0–17.5	Wakasugi[23]
		274–348	0–4.2	Verelst[29]
5A Zeolite	O_2–N_2 binary	283–323	1.0	van der Vlist[31]
		144	1.0	Danner[25]
		298, 304	1.0	Kumar[19]
		299, 320	1.0–4.0	Verelst[29]
		278–303	1.7–4.4	Sorial[24]
		298	0.2–4.0	Miller[22]
		144	0–1.8	Danner[25]
		144	0–2.9	Dorfman[32]
		172–273	0–2.1	Nolan[33]
5A Zeolite	N_2	144	0–2.9	Dorfman[32]
		78–273	0–2.1	Nolan[33]
		144	0–1.2	Danner[25]

(Continued)

Table 2.5. (*Continued*)

Sorbent	Sorbate	Temp. range (K)	Press. range (atm)	Reference
5A Zeolite	O_2-N_2	144	1.0	Danner[25]
	binary	172–273	1.0	Nolan[33]
CMS(Takeda)	O_2	195–323	1–11.0	Kawozoe[34]
CMS(BF)		303	0–0.9	Ruthven[35]
		190–273	0–0.9	Ruthven[14]
CMS(Takeda)	N_2	77	0–0.9	Horvath[37]
		77–323	0–0.9	Kawozoe[34]
CMS(BF)		273–333	0–0.9	Ruthven[35]

[a] See also *Adsorption Equilibrium Data Handbook*, D. P. Valenzuela and A. L. Myers, Prentice Hall, Englewood Cliffs, N.J. (1989), which provides a useful summary of the available adsorption equilibrium data for a wide range of systems.

isotherm data with literature references. The molecules of argon, oxygen, and nitrogen are of similar size and polarizability so their van der Waals interactions are similar. As a result nonpolar adsorbents show very little selectivity between these species, as exemplified by the similarity in the isotherms for nitrogen and oxygen on a carbon molecular sieve (Figure 2.11). By contrast, the aluminum-rich zeolites show preferential adsorption of nitrogen as a result of the field gradient quadrupole interaction energy. 5A zeolite is the most commonly used adsorbent for air separation (to produce oxygen) and the separation factor (essentially the same as the ratio of Henry constants) for this adsorbent is about 3.3 at ambient conditions (see Figure 2.12). This value is almost independent of composition in conformity with the Langmuir model. The separation factors for most other commercial zeolites are similar although very much higher separation factors (8–10) have been reported by Coe for well dehydrated Ca·X or Li X as well as for Ca or Li chabazites.[38,39]

The electric field gradient within a zeolite is enhanced by the presence of divalent cation (Ca^{2+}). However, any traces of moisture can lead to cation hydrolysis, leading to the formation of two singly charged ions:

$$Ca^{2+} + 2H_2O = CaOH^+ + H_2O^+ \tag{2.22}$$

with consequent loss of nitrogen selectivity.

2.3 Adsorption Kinetics

The rate of physical adsorption is generally controlled by diffusional limitations rather than by the actual rate of equilibration at a surface, which, for physical adsorption, is normally very rapid. From the perspective of sorption kinetics, adsorbents may be divided into two broad classes: homogeneous and

FUNDAMENTALS OF ADSORPTION

Table 2.6. Pore Structure of Typical Adsorbents

Homogeneous-Unimodal Pore Size Distribution	Composite-Bimodal Pore Size Distribution
Silica Gel	Carbon Molecular Sieves
Activated Alumina	Pelleted Zeolites
Activated Carbon	Macroreticular ion exchange resins
Homogeneous ion exchange resins	

composite (Table 2.6). These are illustrated in Figure 2.13. In the "homogeneous" adsorbents the pore structure persists, on the same scale, throughout the entire particle; so the distribution of pore size is unimodal. By contrast the composite adsorbent particles are formed by aggregation of small microporous microparticles, sometimes with the aid of a binder. As a result the pore size distribution has a well-defined bimodal character with micropores within the microparticles connected through the macropores within the pellet.

In a composite adsorbent there are three distinct resistances to mass transfer, as illustrated in Figure 2.14. Under practical conditions of operation the external film resistance is seldom, if ever, rate limiting; so that the sorption/desorption rate is generally controlled by either macropore or micropore diffusion or by the combined effects of these resistances.

A proper understanding of kinetic effects in PSA systems therefore requires an understanding of the mechanisms of both macropore and micropore diffusion. Only a brief summary is given here; a more detailed account has been given by Kärger and Ruthven.[40]

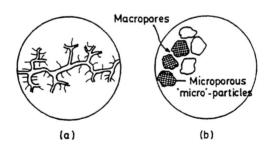

Figure 2.13 Two common types of microporous adsorbent. (a) Homogeneous particle with a wide range of pore size (e.g., activated alumina or silica gel. (b) Composite pellet formed by aggregation of small microporous microparticles (e.g., zeolite or carbon molecular sieve adsorbents).

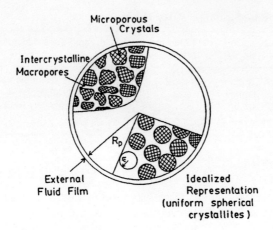

Figure 2.14 The resistances to mass transfer in a composite adsorbent pellet.

2.3.1 Diffusion in Mesopores and Macropores

There are four distinguishable diffusion mechanisms that contribute in varying degrees to transport within macro and mesopores (in which the pore diameter is substantially greater than the diameter of the diffusing sorbate): bulk diffusion, Knudsen flow, Poiseuille flow, and surface diffusion. When the pore diameter is large, relative to the mean free path, bulk or molecular diffusion is dominant. Knudsen diffusion, which depends on collisions between the diffusing molecule and the pore wall, becomes important at low pressures and in small pores when the mean free path is equal to or greater than the pore diameter.

The molecular diffusivity varies approximately according to the relationship:

$$D_m \propto \frac{T^{1.7}}{P\sqrt{M}} \tag{2.23}$$

where M is the mean molecular weight, defined by:

$$\frac{1}{M} = \frac{1}{M_A} + \frac{1}{M_B} \tag{2.24}$$

In a binary system the molecular diffusivity is independent of composition, but this is not precisely true of a multicomponent system. The Knudsen diffusivity is independent of pressure and varies only weakly with temperature:

$$D_K \propto r\sqrt{T/M} \tag{2.25}$$

In the transition region, where both mechanisms are significant, it is easy to show from momentum transfer considerations that the combined diffusivity is given by:[41]

$$\frac{1}{D_A} = \frac{1}{D_{K_A}} + \frac{[1 - y_A(1 + N_A/N_B)]}{D_m} \quad (2.26)$$

where N_A, N_B are the fluxes of components A and B measured relative to a fixed frame of reference. If either $N_A = -N_B$ (equimolar counterdiffusion) or y is small (dilute system), this reduces to the simple reciprocal addition rule:

$$\frac{1}{D} = \frac{1}{D_K} + \frac{1}{D_m} \quad (2.27)$$

It is evident from Eqs. 2.23–2.26 that at high pressures $D \to D_m$ and at low pressures $D \to D_K$.

In addition to molecular and Knudsen diffusion there may be a contribution to the flux from forced flow (Poiseuille flow). The equivalent Poiseuille diffusivity is given by:

$$D = \text{P}r^2/8\mu \quad (2.28)$$

from which it is clear that this contribution is significant only in relatively large pores and at relatively high pressures. It can be important in PSA systems, particularly in the pressurization step. Any such contribution is directly additive to the combined diffusivity from the molecular and Knudsen mechanisms.

In the mechanisms so far considered the flux is through the gas phase in the central region of the pore. Where the adsorbed phase is sufficiently mobile and the concentration sufficiently high, there may be an additional contribution from surface diffusion[42] through the adsorbed layer on the pore wall. Any such contribution is in parallel with the flux from Knudsen and molecular diffusion and is therefore directly additive. Surface diffusion is an activated process and is in many ways similar to micropore diffusion. In particular the patterns of concentration and temperatures dependence are similar to those for micropore diffusion, as discussed in the next section.

2.3.2 Micropore Diffusion

We use here the term *micropore diffusion* to mean diffusion in pores of dimensions comparable with the diameters of the diffusing molecules. In this situation the diffusing molecule never escapes from the force field of the pore wall. The process resembles surface diffusion in that it is an activated process, but steric restrictions are also important and in many instances the diffusional activation energy is in fact largely determined by the size of the diffusing molecule relative to the smallest free diameter of the pore. In such small pores it no longer makes physical sense to distinguish between adsorbed molecules on the pore wall and "gaseous" molecules in the central region of the pore, and it is preferable to regard all sorbate molecules within the micropores as the "adsorbed phase."

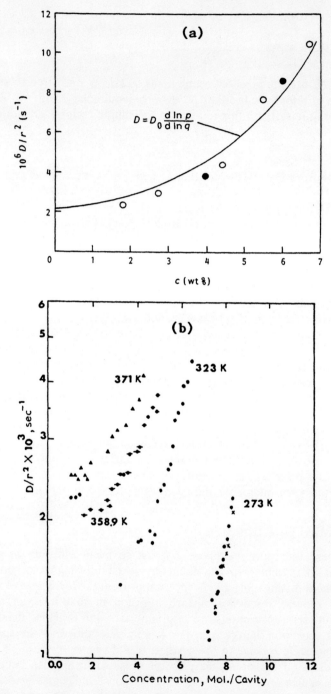

Figure 2.15 Variation of diffusivity with sorbate concentration showing conformity with Eq. 2.31 (a) O_2 in carbon molecular sieve at 193 K[14] and (b) and (c) CO_2 in 4A zeolite[43] showing variation of time constant (D/r_c^2) and constancy of "corrected" time constant (D_0/r_c^2).

FUNDAMENTALS OF ADSORPTION

A strong concentration dependence of the micropore diffusivity is commonly observed, and in many cases this can be accounted for simply by considering the effect of system nonlinearity. The true driving force for any diffusive process is the gradient of chemical potential, rather than the gradient of concentration, as assumed in the Fickian formulation:

$$J = -Bq\frac{d\mu}{dz} \qquad (2.29)$$

where the chemical potential is related to the activity by:

$$\mu = \mu^0 + RT \ln a \qquad (2.30)$$

For an ideal vapor phase the activity is essentially equal to the partial pressure; so Eqs. 2.29 and 2.30 reduce to:

$$D = D_0 \frac{d \ln p}{d \ln q}; \qquad D_0 = BRT \qquad (2.31)$$

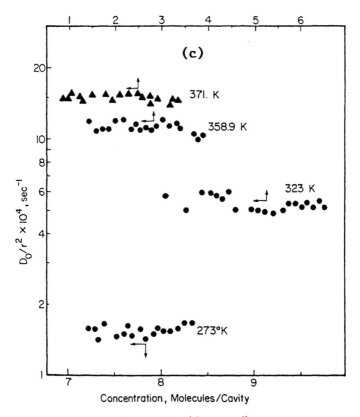

Figure 2.15 (*Continued*).

where D is the Fickian diffusivity, defined in the usual way by:

$$J = -D\frac{dq}{dz} \tag{2.32}$$

In the limit of a linear system (Henry's Law) $d \ln p/d \ln q \to 1.0$ and the Fickian diffusivity becomes independent of concentration. For most microporous adsorbents, however, the isotherm is of type I form; so Eq. 2.31 predicts an increasing trend of diffusivity with concentration. In particular, for the Langmuir isotherm (Eq. 2.3):

$$\frac{d \ln p}{d \ln q} = \frac{1}{1 - q/q_s} = \frac{1}{1 - \theta}; \quad D = \frac{D_0}{1 - q/q_s} \tag{2.33}$$

from which it may be seen that, in the saturation region, the concentration dependence is very strong. Although there is no sound theoretical reason to expect the corrected diffusivity (D_0) to be independent of concentration, this pattern of behavior has been observed experimentally for several sorbates on

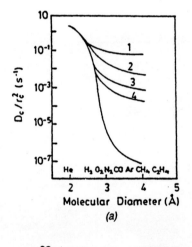

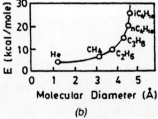

Figure 2.16 Correlation of diffusivity and diffusional activation energy with molecular diameter for several sorbates in 4A and 5A zeolites and carbon molecular sieves. (a) Diffusional time constants for different molecular sieve carbons; (b) and (c) diffusional activation energies for various molecular sieve carbons and 4A and 5A zeolites. (From Schröter and Jüntgen[36] and Ruthven,[1] with permission.)

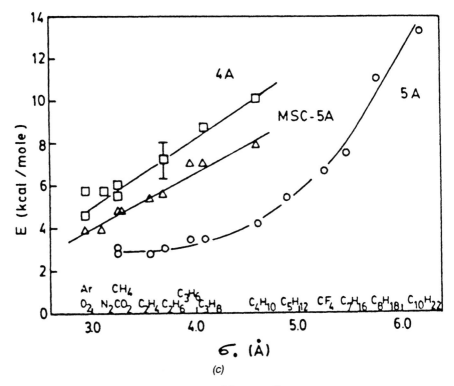

Figure 2.16 (*Continued*).

both small-pore zeolite and carbon molecular sieve adsorbents (see Figure 2.15).

Micropore diffusion is an activated process; so, in contrast to molecular or Knudsen diffusivities, the temperature dependence is strong and generally follows the Arrhenius form:

$$D_0 = D_\infty e^{-E/RT} \tag{2.34}$$

where E is the activation energy. In view of the concentration dependence of D, it is obviously more useful to calculate the activation energy from the temperature dependence of D_0, rather than from that of D. In small-pore zeolites and carbon molecular sieves the major energy barrier is simply the repulsive interactions associated with the molecule passing through constrictions in the pore. As a result there is a well-defined correlation between activation energy and molecular diameter, as illustrated in Figure 2.16.

2.3.3 Uptake Rates in Single Adsorbent Particles

In a packed adsorption column (for example, in a PSA system) the adsorbent particles are subjected to a time-dependent surface concentration, and in such circumstances the sorption/desorption rate depends on both the resistance to mass transfer and the time dependence of the local gas-phase concentration. The modeling of such systems is considered in Section 2.4. However, in order to understand their behavior, it is helpful first to consider the simpler problem of sorption in a single adsorbent particle subjected to a step change in surface concentration. To do this it is necessary to consider in sequence the various possible mass transfer resistances that may control the sorption rate. Of course in practice more than one of these resistances may be significant, but in order to avoid undue complexity we assume here spherical adsorbent particles and a single rate-controlling process. We assume a general expression for the equilibrium isotherm $[q^* = f(c)]$ and in all cases given here the assumed initial and boundary conditions are:

$$t < 0, \quad q = c = 0; t > 0, \quad c = c_0, \quad q|_{R_p} = Kc_0 \qquad (2.35)$$

2.3.4 External Fluid Film Resistance

Sorption rate:

$$\frac{d\bar{q}}{dt} = \frac{3k_f}{R_p}[c_0 - c^*], \quad c^* = f(\bar{q}) \qquad (2.36a)$$

Uptake:

$$\frac{\bar{q}}{q_0} = 1 - \exp\left(\frac{-3k_f t}{R_p}\right), \quad q_0 = f(c_0) \qquad (2.36b)$$

The mass transfer coefficient (k_f) depends in general on the hydrodynamic conditions but in the special case of a stagnant gas (Sh = 2.0)$k_f \approx D_m/R_f$. In practice the external fluid film resistance is normally smaller than the internal (intraparticle or intracrystalline) diffusional resistances; so this process is seldom if ever rate controlling, although in many systems it makes some contribution to the overall resistance.

2.3.5 Solid Surface Resistance

If mass transfer resistance is much higher at the surface than in the interior of the adsorbent particle, for example, as a result of partial closure of the pore mouths, the concentration profile will show a steplike form with a sharp change in concentration at the surface and an essentially constant concentration through the interior region. In this situation the expression for the uptake rate is similar to the case of external film resistance but with the mass transfer coefficient k_s representing the diffusional resistance at the solid surface. Sorption rate:

$$\frac{d\bar{q}}{dt} = \frac{3k_s}{R_p}(q_0 - \bar{q}), \quad q_0 = f(c_0) \qquad (2.37a)$$

Uptake:

$$\frac{\bar{q}}{q_0} = 1 - \exp\left(\frac{-3k_s t}{R_p}\right) \tag{2.37b}$$

2.3.6 Micropore Diffusion

We assume instantaneous equilibration at the external surface with the approach to equilibrium in the interior of the spherical particle controlled by Fickian diffusion with the diffusivity defined on the basis of the gradient of the adsorbed phase concentration. Local sorption rate:

$$\frac{\partial q}{\partial t} = D_c\left(\frac{2}{r}\frac{\partial q}{\partial r} + \frac{\partial^2 q}{\partial r^2}\right) \tag{2.38a}$$

Uptake:

$$\frac{\bar{q}}{q_0} = 1 - \frac{6}{\pi^2}\sum_{n=1}^{\infty}\frac{e^{-n^2\pi^2 D_c t/r_c^2}}{n^2} \tag{2.38b}$$

At short times this expression is approximated by:

$$\frac{\bar{q}_t}{\bar{q}_\infty} = \frac{6}{r_c}\sqrt{\frac{D_c t}{\pi}} - 3\frac{Dt}{r_c^2} \tag{2.39}$$

This expression is accurate to within 1% for $m_t/m_\infty < 0.85$ (or $D_c t/r_c^2 < 0.4$). The first term alone provides an adequate approximation for the initial region ($m_t/m_\infty < 0.15$ or $D_c t/r_c^2 < 0.002$). Conformity with these expressions is illustrated in Figure 2.17. The difference between the forms of the uptake curve derived from the diffusion model and the surface resistance models (Eq. 2.37 or 2.38) is illustrated in Figure 2.20, while the temperature dependence of D_0 is shown in Figure 2.18.

The situation is more complicated in binary or multicomponent systems, since it is then necessary to take account of the effect of component B on the chemical potential of component A. As the simplest realistic example we consider an idealized system in which the cross terms in the flux equation can be neglected and in which the mobility is independent of composition. The detailed analysis has been given by Round, Newton, and Habgood[48] and by Kärger and Bülow.[49] We have for the fluxes:

$$N_A = -D_{0A}\left(\frac{d\ln p_A}{d\ln q_A}\right)\frac{\partial q_A}{\partial z} \tag{2.40}$$

$$N_B = -D_{0B}\left(\frac{d\ln p_B}{d\ln q_B}\right)\frac{\partial q_B}{\partial z}$$

If the equilibrium isotherm is of binary Langmuir form (Eq. (2.13), the

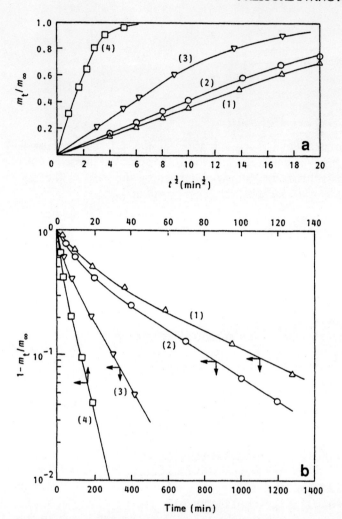

Figure 2.17 Experimental uptake curves (a) and (b) for O_2 in the Bergbau-Forschung carbon molecular sieve at 193 K and (c) and N_2 in three different size fractions of 4A zeolite crystals, showing conformity with the diffusion model. From Ruthven[14] and Yucel and Ruthven.[44]

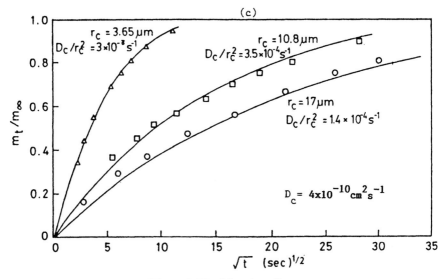

Figure 2.17 (*Continued*).

diffusion equation takes the form:

$$\frac{\partial \Theta_A}{\partial t} = \frac{D_{A0}}{(1 - \theta_A - \theta_B)} \left[(1 - \theta_B) \left(\frac{\partial^2 \theta_A}{\partial r^2} + \frac{2}{r} \frac{\partial \theta_A}{\partial r} \right) \right.$$

$$\left. + \theta_A \left(\frac{\partial^2 \theta_B}{\partial r^2} + \frac{2}{r} \frac{\partial \theta_B}{\partial r} \right) \right] \quad (2.41)$$

$$+ \frac{D_{A0}}{(1 - \theta_A - \theta_B)^2} \left((1 - \theta_B) \frac{\partial \theta_A}{\partial r} + \theta_A \frac{\partial \theta_B}{\partial r} \right) \left(\frac{\partial \theta_A}{\partial r} + \frac{\partial \theta_B}{\partial r} \right)$$

with a similar expression for component *B*. These expressions are used in the modeling the dynamic behavior of the kinetically selective CMS adsorbents (see Section 5.2).

2.3.7 Macropore Diffusion

Local sorption rate:

$$\epsilon_p \frac{\partial c}{\partial t} + (1 - \epsilon_p) \frac{\partial q}{\partial t} = \epsilon_p D_p \left(\frac{2}{R} \frac{\partial c}{\partial R} + \frac{\partial^2 c}{\partial R^2} \right) \quad (2.42)$$

If the equilibrium is linear ($q^* = Kc$), this reduces to:

$$\frac{\partial c}{\partial t} = \frac{\epsilon_p D_p}{\epsilon_p + (1 - \epsilon_p) K} \left(\frac{2}{R} \frac{\partial c}{\partial R} + \frac{\partial^2 c}{\partial R^2} \right) \quad (2.43)$$

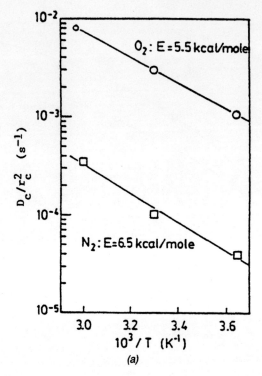

Figure 2.18 Arrhenius plot showing the temperature dependence of micropore diffusivities for (a) O_2 and N_2 in Bergbau carbon molecular sieves[35] and (b) for several light gases in 5A and 13X zeolite crystals.[50]

which has the same form as Eq. 2.38a with the effective diffusivity given by:

$$D_e = \frac{\epsilon_p D_p}{\epsilon_p + (1 - \epsilon_p)K} \quad (2.44a)$$

The sorption curve is then of the same form as Eq. 2.38a but with D replaced by D_e and r replaced by R_p. Since K varies with temperature in accordance with Eq. 2.38b, the uptake behavior gives the appearance of an activated diffusion process with $E \sim -\Delta H$. The case of a nonlinear equilibrium relationship is more complex and corresponds formally with a concentration-dependent effective diffusivity given by:

$$D_e = \frac{\epsilon_p D_p}{\epsilon_p + (1 - \epsilon_p)f'(c)} \quad (2.44b)$$

where $f'(c)$ represents the slope of the equilibrium isotherm (dq^*/dc).

FUNDAMENTALS OF ADSORPTION

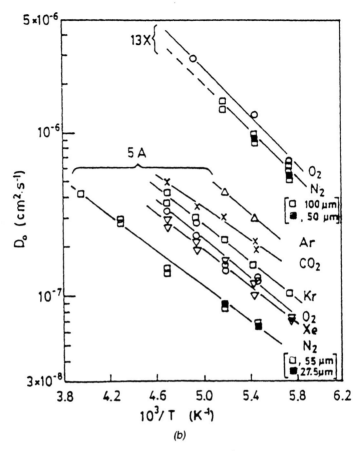

Figure 2.18 (Continued).

2.3.8 Heat Transfer Control

Since adsorption or desorption is generally associated with a significant heat effect (exothermic for adsorption), sorption/desorption rates may be influenced or even controlled by the rate of heat dissipation. Such effects have been investigated both theoretically and experimentally.[45,46] In the limiting situation in which all mass transfer processes are rapid, the sorption rate is controlled entirely by the rate of heat dissipation, and the sorption/desorption curve assumes a very simple form:

$$\frac{m_t}{m_\infty} = 1 - \frac{\beta}{1+\beta}\exp\left(-\frac{ha}{C_s}\frac{t}{(1+\beta)}\right) \quad (2.45)$$

The experimental adsorption/desorption curves for carbon dioxide in 5A zeolite crystals, presented in Figure 2.19, conform to this simple model. As with the diffusion or surface resistance mass transfer models, the approach to

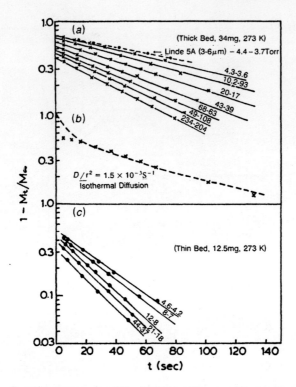

Figure 2.19 Sorption curves for CO_2 in 5A zeolite crystals showing conformity with the heat transfer control model. (From Ruthven et al.[45])

equilibrium in the long-time region is logarithmic. However, in the case of mass transfer control the intercept of a plot of $\log(1 - m_t/m_\infty)$ versus t is invariant, whereas for heat transfer control this intercept $[\beta/(1 + \beta)]$ varies with sorbate concentration because of the nonlinearity of the equilibrium relationship.

2.3.9 Kinetically Selective Adsorbents

The different rate-controlling mechanisms delineated here are clearly illustrated by the sorption kinetics of oxygen and nitrogen in the common PSA adsorbents. The adsorbents used in the PSA production of nitrogen (carbon molecular sieves or 4A zeolite) depend on the difference in sorption rates between oxygen and nitrogen. The oxygen molecule is slightly smaller and therefore diffuses faster in critically sized micropores (~ 4 Å). Representative gravimetric uptake curves for oxygen and nitrogen in 4A zeolite and in carbon molecular sieve showing conformity with the diffusion model are shown in Figure 2.17, and the Arrhenius temperature dependence of the micropore diffusivities is shown in Figure 2.18. A summary of diffusivities and diffusional activation energies is given in Table 2.7. However, not all carbon

FUNDAMENTALS OF ADSORPTION

Table 2.7. Diffusion of Atmospheric Gases in Various Zeolites and Molecular Sieve Carbons[a]

Sorbent	Sorbate	T (K)	D (cm^2 s^{-1})	E (kcal/mole)
13X Zeolite	O_2	200	2.5×10^{-6}	3.4
	N_2	200	2.0×10^{-6}	3.3
5A Zeolite	O_2	200	2.2×10^{-7}	2.7
	N_2	200	1.2×10^{-7}	2.5
	Ar	200	5.5×10^{-7}	—
4A Zeolite	O_2	273	1.6×10^{-8}	4.5
	N_2	273	4×10^{-10}	5.6
	Ar	273	3×10^{-11}	5.7
CMS	O_2	300	3×10^{-11}	5.5
(Bergbau)	N_2	300	1.1×10^{-12}	6.5
CMS	O_2	300	1.9×10^{-9}	6.6
(Takeda)	N_2	300	3.4×10^{-10}	7.4

[a] Data are from Xu et al.,[50] Ruthven and Derrah,[16] Ruthven et al.,[35] and Chihara et al.[51] Intracrystalline diffusivity values are quoted at the specified temperature. For MSC a nominal microparticle radius of 1 μm is assumed. In practice, in 5A and 13X zeolites the controlling resistance is macropore diffusion; so these values do not relate directly to sorption rates.

molecular sieve adsorbents exhibit diffusion control. The data reported by Dominguez et al.[47] (Figure 2.20) show that some carbon sieves conform much more closely to the surface resistance model (Eq. 2.37). Such differences are not unexpected in view of the way in which carbon molecular sieve adsorbents are produced. If in the final deposition process carbon is deposited predominately at the surface, thus partially closing the pore mouths, the kinetics can be expected to follow the surface resistance model, whereas if carbon is deposited more or less uniformly through the particle, diffusion-controlled behavior is to be expected.

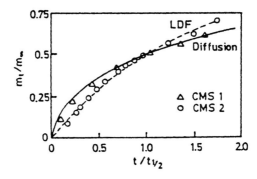

Figure 2.20 Uptake curves for N_2 in two different samples of carbon molecular sieve. CMS 1 obeys the diffusion model; CMS 2 obeys the surface resistance model. (After Dominguez et al.[47])

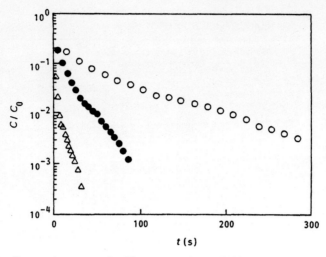

Figure 2.21 Desorption curves for N_2 measured at $-80°C$, under similar conditions, with three different particle sizes of 5A zeolite pellets. See Table 2.8. (From Ruthven.[54])

2.3.10 Equilibrium Selective Adsorbents

The adsorbents used in the PSA oxygen process are generally zeolites (CaA, NaX, or CaX). In these materials diffusion of both oxygen and nitrogen is rapid and the separation depends on the preferential (equilibrium) adsorption of nitrogen. Sorption rates in these adsorbents are controlled by macropore diffusion, as may be clearly seen from measurements with different particle sizes (Figure 2.21 and Table 2.8). The variation of effective diffusivity with temperature is shown in Figure 2.22. At ambient temperature transport within the macropores occurs mainly by molecular diffusion. The effective diffusivity is given by Eq. 2.44 with $\epsilon_p D_p \approx D_m/10$. At lower temperatures the contribution of surface diffusion becomes significant, and, as a result, the Arrhenius plot shows distinct curvature.

Table 2.8. Diffusion Time Constants for N_2 in Different Size Fractions of Commercial 5A Zeolite Adsorbent Particles[a]

T (K)	$R_1 = 1.03$ mm D_e/R_1^2 (s^{-1})	$R_2 = 0.42$ mm D_e/R_2^2 (s^{-1})	Time const. ratio	Ratio $(R_1/R_2)^2$
193	0.0016	0.0083	5.3	6.0
174	0.00064	0.0038	5.9	

[a] The time constant varies with R^2, showing macro diffusion control.

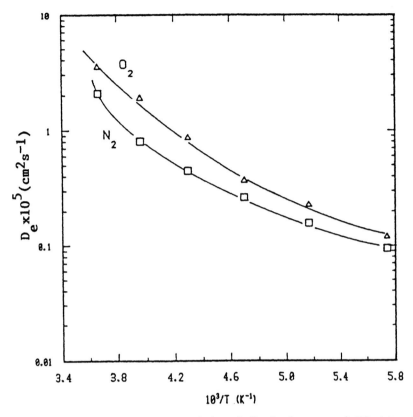

Figure 2.22 Arrhenius plot showing variation of effective (macropore) diffusivity with temperature for O_2 and N_2 in pelleted 5A zeolite. (From Ruthven.[54])

2.3.11 Separation Factor and Selectivity

In an equilibrium based separation the selectivity of the adsorbent is governed by the separation factor, defined in Eq. 2.14. For a Langmuir system this factor is equivalent to the ratio of the Henry's Law constants, so comparison of the Henry constants (or the chromatographic retention volumes which are directly related to the Henry constants through Eq. 2.61) provides a simple and convenient approach for preliminary screening of potential adsorbents.

In a kinetically controlled separation process the situation is somewhat more complicated, since the selectivity then depends on both kinetic and equilibrium effects. In a membrane type of process which operates under steady-state conditions (see Section 8.1), the separation factor, at high pressure ratios, approaches the permeability ratio (Eq. 8.8) i.e., the product of the ratio of diffusivities and equilibrium constants. The reduction in

Table 2.9. Kinetic Selectivities for O_2 / N_2 Separation at 298K[a]

Adsorbent	K_{O_2}/K_{N_2}	D_{O_2}/D_{N_2}	α_K
CMS (Bergbau)	1.0	147	12.1
4A Zeolite	0.28	40	1.75
RS 10 (Modified 4A)	0.5	40	3.2

[a] Kinetic and equilibrium parameters for CMS and 4A zeolite are from Tables 2.4 and 2.7. Values for RS-10 are from S. Farooq, M. N. Rathor, and K. Hidajat, *Chem. Eng. Sci.* (in press).

selectivity which occurs when kinetic and equilibrium selectivities are in opposition is obvious.

A somewhat similar situation arises in kinetically controlled PSA processes, which operate under transient conditions. When the kinetics are controlled by a diffusive process (normally intracrystalline or micropore diffusion), the uptake, following a step change in gas phase concentration, is given by Eq. 2.39. For a linear isotherm this reduces, in the short time region, to:

$$q_t = \frac{6}{r_c}\sqrt{\frac{Dt}{\pi}} \cdot Kc_0 \qquad (2.46a)$$

If two species (A and B) diffuse independently and their isotherms are also independent, the ratio of their uptakes at any time will be given by:

$$\alpha_K = \left(\frac{q_A/c_A}{q_B/c_B}\right) = \frac{K_A}{K_B}\sqrt{\frac{D_A}{D_B}} \qquad (2.46b)$$

This parameter provides a useful approximate measure of the actual kinetic selectivity of the adsorbent. In any real system the assumption that the two species diffuse independently is unlikely to be accurately fulfilled, but Eq. 2.46b is still very useful as a rough guide for initial screening of kinetically selective adsorbents. It shows clearly that the actual selectivity depends on both kinetic and equilibrium effects.

Values of α_K for three kinetically selective adsorbents for O_2/N_2 separation are given in Table 2.9. The superiority of the carbon molecular sieve over the zeolite adsorbents is clearly apparent. Furthermore, it is evident that the advantage of RS-10 compared with regular 4A zeolite stems from a less adverse equilibrium rather than from any difference in the intrinsic diffusivity ratio.

2.4 Adsorption Column Dynamics

Since PSA processes are generally carried out with packed adsorption columns, an elementary understanding of the dynamic behavior of a packed adsorbent bed is an essential prerequisite for process modeling and analysis.

FUNDAMENTALS OF ADSORPTION

The dynamic behavior of an adsorption column depends on the interplay between adsorption kinetics, adsorption equilibrium, and fluid dynamics. However, the overall pattern of the dynamic behavior is generally determined by the form of the equilibrium relationship. This pattern may be strongly modified by kinetic effects (finite resistance to mass transfer), but, in general, kinetic effects do not give rise to qualitative differences in behavior. It is therefore useful to consider first the analysis of the dynamics of an ideal system with infinitely rapid mass transfer (equilibrium theory) and then to show how the ideal patterns of behavior are modified in a real system by the intrusion of finite resistance to mass transfer.

2.4.1 Equilibrium Theory

The formal analysis of adsorption column dynamics starts from the basic differential equation derived from a transient mass balance on an element of the column. If the flow pattern is represented by the axially dispersed plug flow model, this assumes the form:

$$-D_L \frac{\partial^2 c}{\partial z^2} + \frac{\partial}{\partial t}(vc) + \left(\frac{1-\varepsilon}{\varepsilon}\right)\frac{\partial \bar{q}}{\partial t} = 0 \qquad (2.47)$$

If axial dispersion and pressure drop through the column can be neglected and if the concentration of the adsorbable species is small, this expression reduces to:

$$v\frac{\partial c}{\partial z} + \frac{\partial c}{\partial t} + \left(\frac{1-\varepsilon}{\varepsilon}\right)\frac{\partial \bar{q}}{\partial t} = 0 \qquad (2.48)$$

In the absence of mass transfer resistance local equilibrium prevails at all points (i.e., $q = q_i^*$) and if the system is isothermal, $q_i^* = f(c_i)$, where $f(c)$ represents the equilibrium isotherm. Under these conditions Eq. 2.48 becomes:

$$\left[v \Big/ \left(1 + \frac{(1-\varepsilon)}{\varepsilon}\frac{dq_i^*}{dc_i}\right)\right]\frac{\partial c}{\partial z} + \frac{\partial c}{\partial t} = 0 \qquad (2.49)$$

This equation has the form of the kinematic wave equation with the wave velocity given by:

$$w_c = v \Big/ \left[1 + \left(\frac{1-\varepsilon}{\varepsilon}\right)\frac{dq^*}{dc}\right] \qquad (2.50)$$

If the equilibrium relationship is linear ($q = Kc$),

$$w_c = v \Big/ \left[1 + \left(\frac{1-\varepsilon}{\varepsilon}\right)K\right] \qquad (2.51)$$

and it is evident that the wave velocity is independent of concentration. For an unfavorable equilibrium relationship (Figure 2.9) dq^*/dc increases with concentration so w decreases with concentration, leading to a profile that

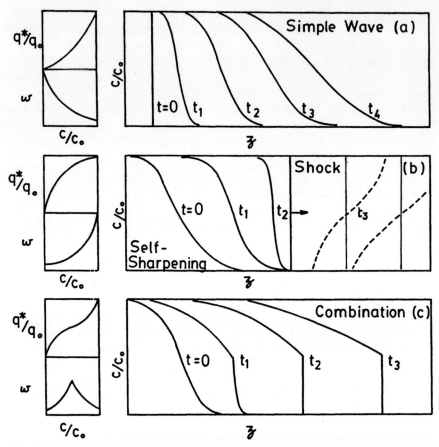

Figure 2.23 Development of the concentration profile in an adsorption column with negligible mass transfer resistance. (a) For an "unfavorable" equilibrium relationship the profile spreads as it propagates, approaching proportionate pattern behavior. (b) For a "favorable" equilibrium relationship an initially dispersed profile is sharpened as it propagates, approaching a shock wave. (c) For a BET-type isotherm the asymptotic form is a combination of a shock and a proportionate pattern wave.

spreads as it propagates [Figure 2.23(a)]. Since the profile spreads in direct proportion to the distance traveled, this is referred to as "proportionate pattern" behavior.

The case of a favorable equilibrium isotherm is slightly more complex. dq^*/dc decreases with concentration; so, according to Eq. 2.49, w will increase with concentration. This leads to what is commonly referred to as "self-sharpening" behavior. An initially dispersed profile will become less and less dispersed as it propagates [Figure 2.23(b)], eventually approaching a shock transition. Equation 2.50 predicts that the sharpening of the profile would continue, even beyond the rectangular shock form, to give the physi-

FUNDAMENTALS OF ADSORPTION

cally unrealistic overhanging profile sketched in the figure. In fact this does not occur; when equilibrium theory predicts an overhanging profile the continuous solution is in fact replaced by the corresponding shock, which travels with a velocity (w') dictated by a mass balance over the transition:

$$w'_c = v \bigg/ \left[1 + \left(\frac{1-\varepsilon}{\varepsilon}\right)\frac{\Delta q^*}{\Delta c}\right] \qquad (2.52)$$

If the isotherm has an inflexion point (e.g., a type II isotherm), it may be regarded as a combination of "favorable" and "unfavorable" segments. Equilibrium theory then predicts that the asymptotic form of the concentration profile will be a composite wave consisting of a shock front with a proportionate pattern wave or a proportionate pattern wave followed by a shock [see Figure 2.23(c)].

Another situation in which a shock solution is obtained arises in bulk separations, where the change in flow rate due to adsorption is relatively large. For a bulk separation we have in place of Eq. 2.48:

$$v\frac{\partial c}{\partial z} + c\frac{\partial v}{\partial z} + \frac{\partial c}{\partial t} + \left(\frac{1-\varepsilon}{\varepsilon}\right)\frac{\partial \bar{q}}{\partial t} = 0 \qquad (2.53)$$

where, for an isobaric system with an adsorbable component in an inert carrier:

$$\frac{v}{v_0} = \frac{1-y_0}{1-y} \qquad (2.54)$$

Expressed in terms of the mole fraction of the adsorbable (or more adsorbable) component, Eq. 2.53 becomes, for a linear equilibrium system:

$$\left\{v_0(1-y_0)/(1-y)^2\left[1 + \left(\frac{1-\varepsilon}{\varepsilon}\right)K\right]\right\}\frac{\partial y}{\partial z} + \frac{\partial y}{\partial t} = 0 \qquad (2.55)$$

which evidently represents a traveling wave with the wave velocity given by:

$$\frac{w}{v_0} = \left\{(1-y_0)/(1-y)^2\left[1 + \left(\frac{1-\varepsilon}{\varepsilon}\right)K\right]\right\} \qquad (2.56)$$

Clearly w increases with increasing y, just as in the case of a trace system with favorable equilibrium, so that, according to equilibrium theory, there will be a shock transition.

2.4.2 Asymptotic Behavior: Effect of Mass Transfer Resistance and Axial Dispersion

When the isotherm is of unfavorable form, mass transfer resistance and axial dispersion have only a relatively minor effect on the asymptotic form of the concentration profile. This may be understood from Figure 2.24, which shows the qualitative form of the concentration profiles in a column following a step change in concentrations at the inlet. Because the isotherm is of unfavorable

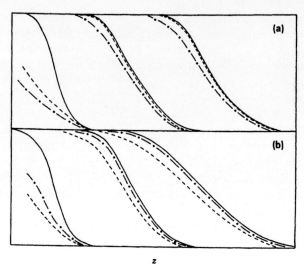

Figure 2.24 Schematic diagram showing (a) approach to constant pattern behavior for a system with a favorable isotherm (b) approach to proportionate pattern behavior for a system with an unfavorable isotherm. y axis: c/c_0,———; q/q_0,---; c^*/c_0,·—·—·.

form, the profile q^*/q_0, representing local equilibrium between fluid and adsorbed phases, lies above the actual adsorbed phase profile (q/q_0). Since mass transfer is from the fluid phase to the adsorbed phase, as the profile propagates, the profiles in the adsorbed and fluid phases tend to approach each other. The asymptotic limit corresponds to local equilibrium at all points in the column (i.e., the profiles q^*/q_0 and c/c_0 are coincident). Thereafter the profile will continue to propagate in the proportionate pattern mode dictated by equilibrium theory. In this situation the effect of mass transfer resistance or axial mixing in the column is simply to increase the distance required to approach the proportionate pattern limit, but the ultimate from of the asymptotic profile is not affected.

When the isotherm is of favorable form, the order of the profiles is reversed [Figure 2.24(a)], and the profile q/q_0 now lies above q^*/q_0. As the profiles propagate and converge, the limiting situation in which $q/q_0 = c/c_0$ is approached with the profile q^*/q_0 still lagging. This represents a stable situation since there remains a finite driving force for mass transfer but this is the same at all concentration levels. As a result this asymptotic profile will propagate without any further change of shape. For obvious reasons this is referred to as "constant pattern" behavior. Thus, where equilibrium theory predicts a shock transition, in practice there will be a constant pattern front. The distance required to approach the constant pattern depends on the extent of mass transfer resistance and the degree of axial mixing in the column. If these effects are small, the constant pattern limit will be approached rapidly, and the resulting profile will be very sharp, approaching shock form. On the other hand, if mass transfer resistance and/or axial mixing effects are large, the distance required to approach the constant pattern limit will be large and the form of the asymptotic profile will be correspondingly dispersed.

As with the unfavorable case, both the form of the asymptotic profile and the distance required to approach this limit are also affected by the curvature of the isotherm. When the isotherm is strongly curved (highly favorable), the asymptotic limit will be approached rapidly and will be correspondingly sharp, whereas where the isotherm is of only slightly favorable form, the asymptotic profile will be reached only in a very long column and will be correspondingly dispersed. It is evident that, in the case of an isotherm with an inflection, where equilibrium theory predicts a composite wave one may expect to see in practice a combination of constant pattern and proportionate pattern profiles as the asymptotic waveform.

2.4.3 Linear Systems

When the isotherm is linear, equilibrium theory predicts that the profile will propagate without change of shape. The effect of mass transfer resistance and axial dispersion is to cause the profile to spread as it propagates. Detailed analysis (see the following) shows that the spread of the profile increases in proportion to the square root of distance (or time). There is no asymptotic limit; such behavior continues indefinitely.

2.4.4 Dynamic Modeling

Knowledge of the asymptotic profile forms is helpful in understanding the dynamic behavior of an adsorption column, but in most PSA systems the column length and cycle time are not sufficiently long or the isotherm sufficiently strongly curved for the asymptotic profiles to be closely approached. To model the dynamic behavior it is therefore necessary to solve simultaneously the differential fluid phase mass balance for the column with the appropriate adsorption rate expression. If heat effects are significant, the problem becomes even more difficult, since it is then necessary also to solve the differential heat balance equation. The general formulation of this problem together with various possible simplifications are summarized in Table 2.10. Even if the equilibria are simple (e.g., linear or Langmuir), the problem is far from trivial and the numerical computations are bulky. It is therefore essential to consider carefully the possibility of introducing appropriate simplifying approximations.

2.4.5 The LDF Rate Expression

In most adsorption systems the kinetics are controlled mainly by intraparticle diffusion, but the use of a diffusion equation to model the kinetics introduces

Table 2.10. Mathematical Model for an Adsorption Column

Differential balance for fluid phase:	$-D_L \dfrac{\partial^2 c}{\partial z^2} + \dfrac{\partial}{\partial z}(vc) + \left(\dfrac{1-\varepsilon}{\varepsilon}\right)\dfrac{\partial \bar{q}}{\partial t} = 0$
	$D_L = 0$ for plug flow; $\dfrac{\partial v}{\partial z} = 0$ for trace system
Heat balance:	$vC_g \dfrac{\partial T}{\partial z} + \left[C_g + \left(\dfrac{1-\varepsilon}{\varepsilon}\right)C_S\right]\dfrac{\partial T}{\partial t}$
	$= (-\Delta H)\left(\dfrac{1-\varepsilon}{\varepsilon}\right)\dfrac{\partial \bar{q}}{\partial t} - \dfrac{4h}{\varepsilon d}(T - T_0)$
Initial conditions:	Adsorption, $\bar{q}(z,0) = 0$, $c(0,t) = 0$; Desorption, $\bar{q}(z,0) = q_0$, $c(0,t) = 0$
Equilibrium:	Linear, $q^* = Kc$; Langmuir, $\dfrac{q^*}{q_S} = \dfrac{bc}{1+bc}$

1. Linear rate models	2. Solid diffusion	3. Pore diffusion
a. Fluid film resistance $\dfrac{\partial \bar{q}}{\partial t} = \dfrac{3k_f}{R_p}(c - c^*)$	$\dfrac{\partial q}{\partial t} = \dfrac{1}{r^2}\dfrac{\partial}{\partial r}\left(D_c r^2 \dfrac{\partial q}{\partial r}\right)$	$\varepsilon_p \dfrac{\partial c}{\partial t} + (1-\varepsilon_p)\dfrac{\partial \bar{q}}{\partial t}$ $= \dfrac{\varepsilon_p D_p}{R^2}\dfrac{\partial}{\partial R}\left(R^2 \dfrac{\partial c}{\partial R}\right)$
	i. D_c = constant ii. $D_c = D_0(1 - q/q_S)^{-1}$	D_p constant
b. Solid film resistance $\dfrac{\partial \bar{\bar{q}}}{\partial t} = k(q^* - \bar{q})$	$q(r,0) = 0$ or q_0 $q(r_c, t - z/v) = q^*(z,t)$ $\dfrac{\partial q}{\partial r}(0, t - z/v) = 0$ $\bar{\bar{q}} = \bar{q} = \dfrac{3}{r_c^3}\int_0^{r_c} qr^2\, dr$	$\bar{q}(r,0) = 0$ or q_0 $\bar{q}(R_p, t - z/v) = q^*(z,t)$ $\dfrac{\partial \bar{q}}{\partial R}(0,t) = 0$ $\bar{\bar{q}} = \dfrac{3}{R_p^3}\int_0^{R_p}[\bar{q}(1-\varepsilon_p)$ $+ \varepsilon_p c]R^2\, dR$

an additional differential equation with associated boundary conditions. For many different boundary conditions diffusion-controlled kinetics may be satisfactorily represented by the so-called "linear driving force" (LDF) expression:

$$\frac{\partial q}{\partial t} = k(q^* - \bar{q}) \qquad (2.57)$$

where $k \approx 15 D_e/R^2$. The validity of this approximation, first introduced by Glueckauf,[52] has been confirmed for many different initial and boundary conditions. Its applicability to a simple Langmuir system is illustrated in Figure 2.25. It is evident that with the time constant defined in an appropriate manner, the LDF approximation provides a reasonable prediction of the breakthrough curves over a wide range of conditions. It is at its best when the

FUNDAMENTALS OF ADSORPTION

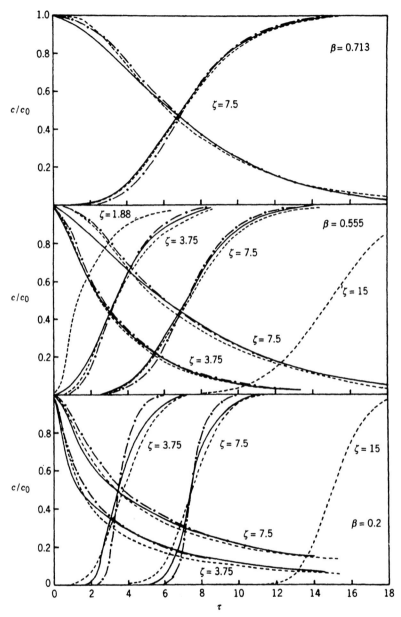

Figure 2.25 Theoretical breakthrough curves calculated for a nonlinear (Langmuir) system showing the comparison between the LDF model (———), the macropore diffusion model (----), and the intracrystalline diffusion model (-·-·), based on the Glueckauf approximation. $k = 15D_e/R^2, \tau = kt, \zeta = kq_0 z(1-\varepsilon)/\varepsilon v c_0$. For intracrystalline diffusion $D_e/R^2 = D_c/r_c^2$; for macropore diffusion $D_e/R^2 = \varepsilon_p D_p/[\varepsilon_p + (1-\varepsilon_p)dq^*/dc]R_p^2$. $\beta = 1 - q_0/q_s$. (From Ruthven,[1] with permission.)

isotherm does not deviate too greatly from linearity, and it tends to break down as the rectangular limit is approached. A more serious defect, from the perspective of modeling PSA systems, is that the Glueckauf approximation does not give a good representation in the initial region of the uptake. This is of little consequence when the column is relatively long ($L/v \gg R^2/D_e$), but it proves to be a serious limitation in certain PSA processes where the cycle time is short relative to the diffusion time.

2.4.6 Combination of Resistances

In a real adsorption systems several different mass transfer resistances may contribute to the overall kinetics. When the equilibrium is linear (or at least not severely nonlinear), it is relatively simple to combine these resistances into a single overall linear driving force mass transfer coefficient based on the reciprocal addition rule:

$$\frac{1}{Kk} = \frac{R_p}{3k_f} + \frac{R_p^2}{15\epsilon_p D_p} + \frac{r_c^2}{15KD_c} \tag{2.58}$$

This rule may be justified in a number of different ways, but the simplest proof rests on an analysis of the moments of the dynamic response.

The first and second moments of the pulse response are defined by:

$$\mu \equiv \frac{\int_0^\infty ct\,dt}{\int_0^\infty c\,dt} \tag{2.59}$$

$$\sigma^2 + \mu^2 = \frac{\int_0^\infty c(t-\mu)^2\,dt}{\int_0^\infty c\,dt} \tag{2.60}$$

For a linear adsorption system it may be shown that the first moment is related to the equilibrium constant by:

$$\mu = \frac{L}{v}\left[1 + \left(\frac{1-\varepsilon}{\varepsilon}\right)K\right] \tag{2.61}$$

where, for a biporous adsorbent, $K = \epsilon_p + (1-\epsilon_p)wK_c$. The reduced second moment is given by:

$$\frac{\sigma^2}{2\mu^2} = \frac{D_L}{vL} + \frac{v}{L}\left(\frac{\varepsilon}{1-\varepsilon}\right)\left(\frac{R_p}{3k_f} + \frac{R_p^2}{15\epsilon_p D_p} + \frac{r_c^2(K_c - \epsilon_p)}{15K^2 D_c}\right) \tag{2.62}$$

$$\times \left(1 - \frac{\varepsilon}{(1-\varepsilon)K}\right)^{-2}$$

FUNDAMENTALS OF ADSORPTION

For $K \gg 1.0$ this simplifies to:

$$\frac{\sigma^2}{2\mu^2} = \frac{D_L}{vL} + \frac{v}{L}\left(\frac{\varepsilon}{1-\varepsilon}\right)\left(\frac{R_p}{3k_f} + \frac{R_p^2}{15\epsilon_p D_p} + \frac{r_c^2}{15KD_c}\right) \qquad (2.63)$$

where the three terms within the final set of large parentheses represent, respectively, the film resistance, macropore resistance, and micropore resistance. For a similar system in which the mass transfer rate is controlled by a linear rate expression (Eq. 2.57) the corresponding expression for the reduced second moment is:

$$\frac{\sigma^2}{2\mu^2} = \frac{D_L}{vL} + \frac{v}{L}\left(\frac{\varepsilon}{1-\varepsilon}\right)\frac{1}{kK} \qquad (2.64)$$

whence it is evident that the equivalence relation is provided by Eq. 2.58.

2.4.7 Multicomponent and Nonisothermal Systems

So far in our discussion of column dynamics we have considered only an isothermal single adsorbable component in an inert (nonadsorbing) carrier. In such a system there is only one mass transfer zone which may approach a constant pattern, proportionate pattern, or a combined form, depending on the shape of the equilibrium isotherm. The situation remains qualitatively similar when there are two adsorbable components (with no inert) since the continuity condition then ensures that there can be only one transition or mass transfer zone with the velocity and shape determined by the binary equilibrium isotherm. The addition of another component, even an inert, however, changes the situation in a rather dramatic way by introducing a second mass transfer zone. The two mass transfer zones will propagate with different velocities so the profile will assume the form sketched in Figure 2.26 with an expanding plateau region between the two transitions. Both transitions may be of proportionate pattern, constant pattern, or combined form, and the plateau concentration may be higher, lower, or intermediate between the initial and final states depending on the precise form of the isotherm. It is evident that even with only three components, the profile may assume a wide range of different forms.

These conclusions, reached here by intuitive arguments, follow directly from the equilibrium theory analysis. For a three-component system there will be two equations of the form of Eq. 2.47 plus the overall continuity equation, which, where pressure drop can be neglected, simply takes the form:

$$c_1 + c_2 + c_3 = c_0 \text{ (constant)} \qquad (2.65)$$

Corresponding to each of the two differential balance equations there will be a wave velocity (from Eq. 2.50 or 2.52), and it is clear that since these velocities depend on the local isotherm slope, they will in general be

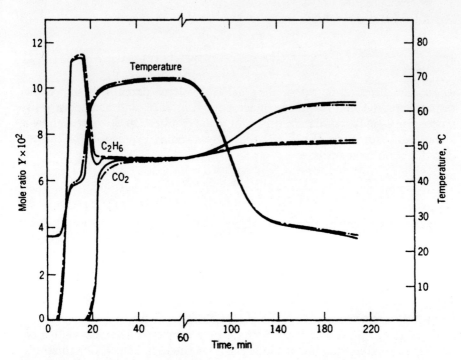

Figure 2.26 Comparison of theoretical (——) and experimental (–·–·) concentration and temperature breakthrough curves for sorption of C_2H_6–CO_2 mixtures from a N_2 carrier on 5A molecular sieve. Feed: 10.5% CO_2, 7.03% C_2H_6 (molar basis) at 24°C, 116.5 kPa (1.15 atm). Column length, 48 cm. Theoretical curves were calculated numerically using the linear driving force model with a Langmuir equilibrium isotherm. (From Liapis and Crosser,[53] with permission.)

different. In general, for an n-component isothermal system, there will be $(n-1)$ transitions and $(n-2)$ intermediate concentration plateaus between the initial and final states.

The effect of nonisothermality is similar. A differential heat balance for an element of the column yields, for a system with negligible axial conduction:

$$vC_g\frac{\partial T}{\partial z} + \left(C_g + \frac{1-\varepsilon}{\varepsilon}C_s\right)\frac{\partial T}{\partial t} = (-\Delta H)\left(\frac{1-\varepsilon}{\varepsilon}\right)\frac{\partial q}{\partial t} - \frac{4h}{\varepsilon d}(T-T_0) \quad (2.66)$$

The temperature and concentration are coupled through the temperature dependence of the adsorption equilibrium constant:

$$\ln\left(\frac{b}{b_0}\right) = \frac{-\Delta H}{R}\left(\frac{1}{T} - \frac{1}{T_0}\right) \quad (2.67)$$

The left-hand side of this equation is clearly of the same form as that of Eq.

2.48; so it is evident that the temperature profile will propagate with a wave velocity given by:

$$w_t = v \bigg/ \left[1 + \left(\frac{1-\varepsilon}{\varepsilon}\right)\frac{C_s}{C_f} - \left(\frac{1-\varepsilon}{\varepsilon}\right)\left(\frac{-\Delta H}{C_f}\right)\frac{dq^*}{dT} \right] \qquad (2.68)$$

where dq^*/dT represents the temperature dependence of the equilibrium loading. In effect heat behaves as an additional component in the system with its own characteristic propagation velocity. Since the adsorption equilibria for all species are temperature dependent, it is evident that a temperature transition will in general cause changes in concentration levels of all species. The only exception arises when the velocity of the temperature front is faster than that of all concentration fronts. In that situation, which is in fact quite common for adsorption of light gases at ambient temperature and pressure, a "pure thermal wave" will be formed and will pass through the column ahead of all concentration changes.

An example showing the form of the effluent concentration and temperature curves for a two-component (plus carrier) adiabatic system is given in Figure 2.26. The least strongly adsorbed species (ethane) passes most rapidly through the column, emerging as a relatively sharp constant pattern front. The ethane concentration rises well above the feed concentration level as a result of displacement by the slower-moving carbon dioxide. The second front (due to adsorption of carbon dioxide) is also sharp and is accompanied by a simultaneous decrease in ethane concentration. The final front is due to the thermal wave, which in this system propagates more slowly than either of the mass transfer fronts. This third front is of proportional pattern form and is accompanied by simultaneous changes in the concentrations of both ethane and carbon dioxide, resulting from the temperature dependence of the equilibrium isotherms. A numerical simulation based on the simultaneous solution of the differential heat and mass balance equations (Eqs. 2.66 and 2.53) with a simple linearized rate expression (Eq. 2.57), and a Langmuir equilibrium isotherm (Eq. 2.13) provides a very good representation of the observed behavior.

References

1. D. M. Ruthven, *Principles of Adsorption and Adsorption Processes*, John Wiley, New York (1984).
2. R. T. Yang, *Gas Separation by Adsorption Processes*, Butterworth, Stoneham, MA (1987).
3. M. Suzuki, *Adsorption Engineering*, Kodansha Elsevier, Tokyo (1990).
4. Y. Matsumura, *Proc. 1st Indian Carbon Conference*, New Delhi, pp. 99–106 (1982).
5. K. Chihara and M. Suzuki, Carbon **17**, 339 (1979).
6. J. Koresh and A. Soffer, *J. Chem. Soc. Faraday Trans. I* **76**, 2457 (1980).

7. H. Jüntgen, K. Knoblauch, and K. Harder, *Fuel* **60**, 817 (1981).
8. D. W. Breck, *Zeolite Molecular Sieves*, Wiley, NY (1974).
9. W. J. Mortier, *Compilation of Extra Framework Cation Sites in Zeolites*, Butterworth, Guildford, U.K. (1982).
10. I. D. Harrison, H. F. Leach, and D. A. Whan, *Proc. Sixth Internat. Zeolite Conf. Reno* (1973), p. 479, Butterworth, Guildford, U.K. (1984).
11. S. Brunauer, P. H. Emmett, and E. Teller, *J. Am. Chem. Soc.* **60**, 309 (1938).
12. R. Desai, M. Hussain, and D. M. Ruthven, *Can. J. Chem. Eng.* **70**, 699 (1992).
13. D. M. Ruthven, *AIChE J.* **22**, 753 (1976).
14. D. M. Ruthven, *Chem. Eng. Sci.* **47**, 4305 (1992).
15. A. L. Myers and J. M. Prausnitz, *AIChE J.* **11**, 121 (1965).
16. D. M. Ruthven and R. I. Derrah, *J. Chem. Soc. Faraday Trans. I* **71**, 2031 (1975).
17. J. D. Eagan and R. B. Anderson, *J. Coll. Interface Sci.* **50**, 419 (1975).
18. C. Springer, Ph.D. Thesis, Iowa State Univ., Ames, Iowa (1964).
19. R. Kumar and D. M. Ruthven, *Ind. Eng. Chem. Fund.* **19**, 27 (1980).
20. N. Haq and D. M. Ruthven, *J. Colloid Interface Sci.* **112**, 154 (1986).
21. J. L. Stakebake and J. Fritz, *J. Coll. Interface Sci.* **105**, 112 (1985).
22. G. W. Miller, K. S. Knaebel, and K. G. Ikels, *AIChE J.* **33**, 194 (1987).
23. J. Wakusugi, S. Ozawa, and Y. Ogino, *J. Colloid Interface Sci.* **79**, 399 (1981).
24. G. A. Sorial, W. H. Granville, and W. O. Daly, *Chem. Eng. Sci.* **38**, 1517 (1983).
25. R. P. Danner and L. A. Wenzel, *AIChE J.* **15**, 515 (1969).
26. N. Haq and D. M. Ruthven, *J. Colloid Interface Sci.* **112**, 164 (1986).
27. P. B. Lederman, Ph.D. Thesis, Univ. of Michigan, Ann Arbor (1961).
28. A. J. Kidnay and M. J. Hiza, *AIChE J.* **12**, 58 (1966).
29. H. Verelst and G. V. Baron, *J. Chem. Eng. Data* **30**, 66 (1985).
30. J. T. Huang, M.Sc. Thesis, Worcester Polytechnic Institute, Worcester, MA (1970).
31. E. van der Vlist and J. van der Meijden, *J. Chromatog.* **79**, 1 (1973).
32. L. R. Dorfman and R. P. Danner, *AIChE Symp Ser.* **71**(152), 30 (1975).
33. J. T. Nolan, T. W. McKeehan, and R. P. Danner, *J. Chem. Eng. Data* **26**, 112 (1981).
34. K. Kawazoe, T. Kawai, Y. Eguchi, and K. Itoga, *J. Chem. Eng. Japan* **7**, 158 (1974).
35. D. M. Ruthven, N. S. Raghavan, and M. M. Hassan, *Chem. Eng. Sci.* **41**, 1325 (1986).
36. H. J. Schröter and H. Jüntgen, in A. E. Rodrigues, M. D. LeVan, and D. Tondeur, eds., *NATO ASI E158, Adsorption Science and Technology*, Kluwer, Dordrecht (1988), p. 269.
37. G. Horvath and K. Kawazoe, *J. Chem. Eng. Japan* **16**, 470 (1983).
38. C. G. Coe, G. B. Parris, R. Srinivasan, and S. Auvil, *Proc. Seventh Internat. Zeolite Conf.*, Tokyo, Aug. 1986, p. 1033. Y. Murakami, A. Lijima, and J. W. Ward, eds., Kodansha Elsevier, Tokyo (1986).

39. C. G. Coe, in *Gas Separation Technology*, pp. 145–59, E. F. Vansant and K. Dewolfs, eds., Elsevier, Amsterdam (1990).
40. J. Kärger and D. M. Ruthven, *Diffusion in Zeolites and Other Microporous Solids*, John Wiley, New York (1992).
41. D. S. Scott and F. A. L. Dullien, *AIChE J.* **8**, 113 (1962).
42. A. Kapoor, R. T. Yang, and C. Wong, *Catal. Revs. Sci. Eng.* **31**, 129 (1989).
43. H. Yucel and D. M. Ruthven, *J. Colloid Interface Sci.* **74**, 186 (1981).
44. H. Yucel and D. M. Ruthven, *J. Chem. Soc. Faraday Trans. I* **76**, 60 (1980).
45. D. M. Ruthven, L. K. Lee, and H. Yucel, *AIChE J.* **26**, 16–23 (1980).
46. D. M. Ruthven and L. K. Lee, *AIChE J.* **27**, 654 (1981).
47. J. A. Dominguez, D. Psaris, and A. I. LaCava, *AIChE Symp. Ser.* **84**(264), 73 (1988).
48. G. F. Round, H. W. Habgood, and R. Newton, *Sep. Sci.* **1**, 219 (1966).
49. J. Kärger and M. Bülow, *Chem. Eng. Sci.* **30**, 893 (1975).
50. Z. Xu, M. Eic, and D. M. Ruthven, *Proc. Ninth Internat. Zeolite Conf. Montreal*, July 1992, R. von Ballmoos, J. B. Higgins, and M. M. J. Treacy, eds., p. 147, Butterworth Heinemann, Stoneham, MA (1993).
51. K. Chihara, M. Suzuki, and K. Kawazoe, *AIChE J.* **24**, 237 (1978).
52. E. Glueckauf and J. E. Coates, *J. Chem. Soc.*, 1315 (1947); and E. Glueckauf, *Trans. Faraday Soc.* **51**, 1540 (1955).
53. A. I. Liapis and O. K. Crosser, *Chem. Eng. Sci.* **37**, 958 (1982).
54. D. M. Ruthven and Z. Xu, *Chem. Eng. Sci.* (in press).

CHAPTER

3

PSA Cycles: Basic Principles

The mode of operation of a PSA separation process was explained in general terms in Chapter 1 without any discussion of the details of the operating cycle. The choice of a suitable operating cycle is in fact critical, and a wide range of different cycles have been proposed to optimize different aspects of the overall process. The underlying principles governing the choice of operating cycle are reviewed in this chapter.

PSA processes may be categorized according to the nature of the adsorption selectivity (equilibrium or kinetic) and whether the less strongly (or less rapidly) adsorbed species (the raffinate product) or the more strongly (or more rapidly) adsorbed species (the extract product) is recovered at high purity. The various possible combinations of these criteria yield four different classes of process. Since the factors that dominate the choice of operating cycle and the mode of operation are somewhat different, these four categories are discussed sequentially.

3.1 Elementary Steps

Any PSA cycle can be considered as a sequence of elementary steps, the most common of which are:
1. Pressurization (with feed or raffinate product);
2. High-pressure feed with raffinate withdrawal;
3. Depressurization or "blowdown" (cocurrent or countercurrent to the feed);

Table 3.1. Summary of the Elementary Steps Used in PSA Cycles

Elementary step	Mode of operation	Principal features
Pressurization	1. Pressurization with feed from the feed end	Enrichment of the less selectively adsorbed species in the gas phase at the product end
	2. Pressurization with raffinate product from the product end prior to feed pressurization	Sharpens the concentration front, which improves the purity and recovery of raffinate product
High-pressure adsorption	1. Product (raffinate) withdrawal at constant column pressure	Raffinate product is delivered at high pressure
	2. The column pressure is allowed to decrease while the raffinate product is drawn from the product end	Very high recovery of the less selectivity adsorbed species may be achieved, but the product is delivered at low pressure
Blowdown	1. Countercurrent blowdown to a low pressure	Used when only raffinate product is required at high purity; prevents contamination of the product end with more strongly adsorbed species
	2. Cocurrent blowdown to an intermediate pressure prior to countercurrent blowdown	Used when extract product is also required in high purity; improves extract product purity and may also increase raffinate recovery
Desorption at low pressure	1. Countercurrent desorption with product purge	Improves raffinate product at the expense of decrease in recovery; purge at subatmospheric pressure reduces raffinate product loss but increases energy cost
	2. Countercurrent desorption without external purge	Recovery enhancement while maintaining high product purity is possible only in certain kinetic separation
	3. Evacuation	High purity of both extract and raffinate products; advantageous over product purge when the adsorbed phase is very strongly held

(Continued)

PSA CYCLES: BASIC PRINCIPLES

Table 3.1. (*Continued*)

Elementary step	Mode of operation	Principal features
Pressure equalization	The high- and low-pressure beds are either connected through their product ends or the feed and product ends of the high-pressure bed are connected to the respective ends of the low-pressure bed	Conserves energy and separative work
Rinse	The bed is purged with the preferentially adsorbed species after high-pressure adsorption at feed pressure in the direction of the feed	Improves extract product purity when the lighter species are coadsorbed in large amount with heavier components

4. Desorption at the lower operating pressure; this may be accomplished by evacuation, purging the bed with the raffinate product or, in a kinetically controlled process, by slow equilibration with consequent evolution of the slower-diffusing sorbate;
5. Pressure equalization (which is used in many cycles, prior to the blowdown step, to conserve energy and separative work);
6. Rinse (purging with the preferentially adsorbed species at high pressure, following the adsorption step).

The processes differ from one another in the sequence of the elementary steps and in the way in which these steps are carried out. Some of the more important variants and the benefits derived from them are summarized in Table 3.1.

To understand a PSA cycle properly it is necessary to know the way in which the concentration profile moves and changes shape during each of the elementary steps. Gas-phase concentration profiles in an adsorption column that undergoes in sequence pressurization, high-pressure adsorption, blowdown, and low-pressure desorption are shown in Figure 3.1 for both equilibrium and kinetic separations. The profiles were calculated for pressurization with feed gas (from the feed end with the product end closed), reverse flow blowdown, and desorption with product purge, also in the reverse flow direction. It is clear that the movements of the concentration wave are similar in both cases. During pressurization the initial gas in the bed is pushed toward the closed product end, where it forms a plateau that is significantly enriched in the less strongly adsorbed species. The region before the plateau shows the penetration of the feed gas. This behavior is in

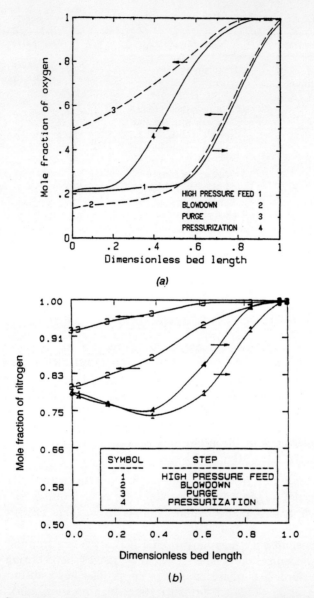

Figure 3.1 Computed steady-state gas-phase concentration profiles at the end of four elementary steps in (a) equilibrium-controlled PSA separation of air on 5A zeolite and (b) kinetically controlled separation of air on a modified 4A zeolite. The arrows indicate the direction of flow, the sequence of operation is 4,1,2,3; the simulation models are described in Section 5.2.

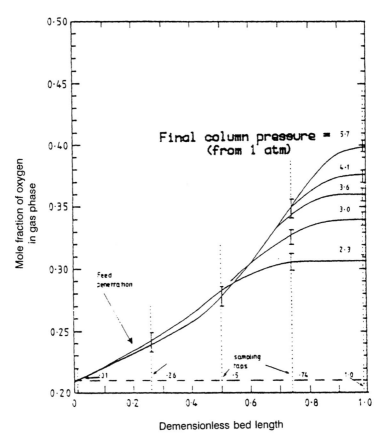

Figure 3.2 Experimental concentration profiles for air pressurization in a bed of 5A zeolite. The lines represent the best curves through the experimental data. (From Ref. 1; reprinted with permission.)

agreement with the profiles measured by Fernandez and Kenney[1] for air pressurization in a bed of 5A zeolite (Figure 3.2). In the high-pressure feed step the concentration wave front travels down the column, and a raffinate product, enriched in the less strongly adsorbed species, is withdrawn at the product end. In the blowdown and purge steps the concentration wavefront is pushed back and a relatively clean initial bed condition is established for the next cycle. The more important features of these steps and their variants will be discussed in relation to various different cycles.

3.2 Equilibrium-Controlled Separations for the Production of Pure Raffinate Product

The early PSA cycles were developed for equilibrium separations, primarily for the recovery of the less strongly adsorbed raffinate product at high purity.

In such a process the purity of the raffinate product and the extent of adsorbent regeneration depend on the partial pressure of the strongly adsorbed species in the void volume of the column at the end of the desorption step. In that step the desorbing gas, which is rich in strongly adsorbed species, occupies the void volume of the column, and, unless adequately removed from the bed, this gas will contaminate the raffinate product. The early PSA cycles were based on two different techniques for regenerating the adsorbent and cleaning the void volume. The cycle developed by Skarstrom[2] (see the following) employed atmospheric desorption with product purge, while the Air Liquide cycle[3] utilized vacuum desorption. Evacuation to a very low absolute pressure may be necessary to achieve reasonable regeneration by vacuum desorption, especially when the isotherm for the more strongly adsorbed component is of favorable (type I) form. However, vacuum desorption has other advantages (such as reduction in the power requirement) and is still widely used, particularly for kinetic separations.

3.2.1 The Skarstrom Cycle

The Skarstrom cycle[2,4] in its basic form utilizes two packed adsorbent beds, as shown schematically in Figure 3.3. The following four steps comprise the cycle:

1. Pressurization;
2. Adsorption;
3. Countercurrent blowdown; and
4. Countercurrent purge.

Both beds undergo these four operations and the sequence, shown in Figure 3.4, is phased in such a way that a continuous flow of product is maintained. In step 1, bed 2 is pressurized to the higher operating pressure, with feed from the feed end, while bed 1 is blown down to the atmospheric pressure in the opposite direction. In step 2, high-pressure feed flows through bed 2. The more strongly adsorbed component is retained in the bed and a gas stream enriched in the less strongly adsorbed component leaves as effluent at a pressure only slightly below that of the feed. A fraction of the effluent stream is withdrawn as product and the rest is used to purge bed 1 at the low operating pressure. The direction of the purge flow is also opposite to that of the feed flow. Steps 3 and 4 follow the same sequence but with the beds interchanged.

During the high-pressure adsorption step the gas phase behind the front has essentially the feed composition, while the composition beyond the front is enriched in the weak sorbate. Feeding continues until the product impurity level rises to the acceptable limit. In other words, the concentration front of the strong sorbate is allowed to break through to a preassigned limit. The idea behind the purge step is to flush the void spaces within the bed and to ensure that at least the end of the bed from which product will be withdrawn

PSA CYCLES: BASIC PRINCIPLES

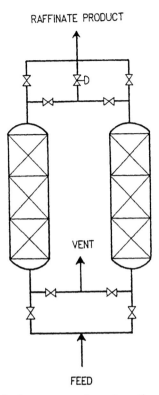

Figure 3.3 The basic two-bed pressure swing adsorption system.

in the next half-cycle is completely free of the strong sorbate. At steady state the concentration front is pushed back in the blowdown and desorption steps by a distance that is exactly equal to the distance it has advanced in the pressurization and high-pressure adsorption steps. Reverse-flow regeneration prevents retention of the more strongly adsorbed species at the product end, thereby reducing the purge requirement. For highly favorable systems, forward-flow regeneration would require an impractically large volume of purge to ensure complete cleaning of the bed through to the product end.

From the preceding discussion it is clear that increasing purge improves product purity but at the expense of a decrease in product recovery, and after a certain point the gain in product quality becomes marginal, relative to the loss of product quantity. The effects of incomplete purge have been studied in detail by Matz and Knaebel.[5]

The original Skarstrom cycle was used for air drying.[2] This process, with a silica gel desiccant, was shown to reduce water content from 3800 ppm to less than 1 ppm, with the recovery of dry air being about 73%.[4] The process details along with the product profile are shown in Figure 3.5. The Skarstrom

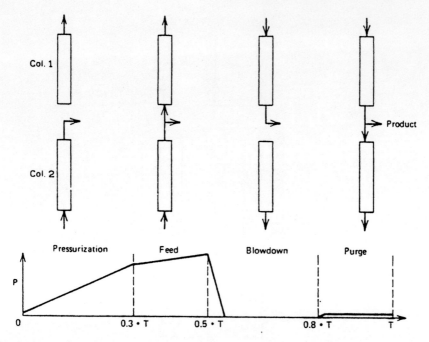

Figure 3.4 The sequence of steps in the basic Skarstrom PSA cycle.

cycle is still widely used for small-scale air drying, and this cycle has also proved successful for other similar separations where the impurities are present at low concentration and the selectivity of the adsorbent is high. Under these conditions the raffinate product behaves as a nonadsorbing inert.

Oxygen production from air using 5A or 13X zeolite as the adsorbent is an example of a bulk separation. The preferentially adsorbed species (nitrogen) is present at a relatively high concentration level, and there is significant coadsorption of the less strongly adsorbed species (oxygen). Such a separation can be achieved using the Skarstrom cycle, but a reasonably pure raffinate product can be achieved only at low fractional recovery, making the economics unattractive. In Skarstrom's original experiments using a 13X zeolite adsorbent, a 90% pure oxygen product was achieved only at a recovery of 10%.[6] The separation factor for this particular adsorbent appears to have been rather low (2.0), and a somewhat better performance can be expected with the higher separation factors (3–3.5) typically obtained with a well-dehydrated zeolite.[7] However, to improve the economics, further enhancement of the recovery–purity profile is obviously desirable.

In a Skarstrom cycle the column effluent during the blowdown and purge steps is normally waste gas (rich in the more strongly adsorbed species but containing a significant fraction of the less strongly held species). Skarstrom

PSA CYCLES: BASIC PRINCIPLES

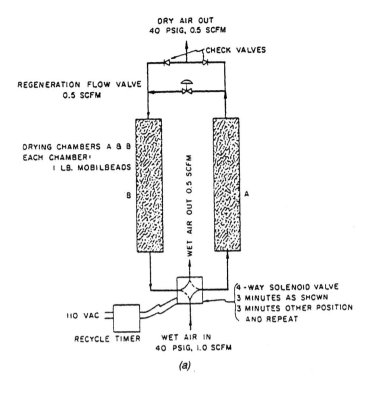

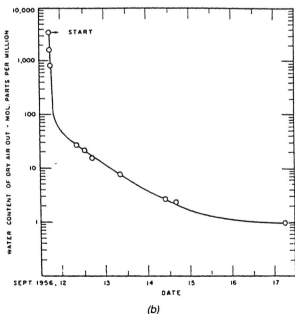

Figure 3.5 (a) Process details and (b) product profile for the PSA air drying system developed by Skarstrom. (From Ref. 4; reprinted with permission.)

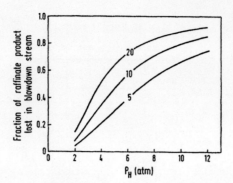

Figure 3.6 Variation of the blowdown stream contribution to the loss of raffinate product with feed pressure. A simple mass balance (assuming negligible adsorption of the raffinate product) yields for the fraction of the raffinate product lost in the blowdown stream

$$\left[1 + v_{\text{oH}} P_L G \overline{X}_{\text{pdt}} t_{\text{pu}} / L P_H (P_H - L) \overline{X}_{\text{bd}}\right]^{-1}$$

The profiles are shown for $G = 2.0$, $P_L = 1.0$ atm, $t_{\text{pu}} = 60$ s, $\overline{X}_{\text{pdt}} = 0.95$, and $\overline{X}_{\text{bd}} = 0.5$. The numbers on the curves indicate L/v_{OH} ratios.

suggested that the purge backwash volume (measured at the purge pressure) should exceed the feed volume (measured at the high operating pressure) at all points in the beds during each cycle in order to obtain a pure product.[4] In practice the purge-to-feed volume should generally be between one and two. The relative contributions from blowdown and purge streams to the total loss of raffinate product depend on the level of the higher operating pressure. Since the product emerges at the high pressure while purging takes place at atmospheric pressure, the actual fraction of the product stream lost as purge is quite small and becomes negligible when the pressure ratio is large. On the other hand, the contribution from the blowdown loss increases with increasing pressure, and becomes completely dominant at high operating pressure, as may be seen from Figure 3.6. The improved performance of most of the more complex cycles comes from reduction of the blowdown losses.

3.2.2 Pressure Equalization

The first improvement over Skarstrom's original cycle was the introduction of a pressure equalization step proposed by Berlin.[8] A schematic diagram of the improved process and the modified sequence of operations are shown in Figure 3.7. After the first bed has been purged and the second bed has completed the high-pressure adsorption step, instead of blowing down the second bed directly, the two beds are connected through their product ends to equalize the pressure. The first bed is thus partially pressurized with gas from the outlet region of the second bed. Following pressure equalization the beds are disconnected and the first bed is pressurized with feed gas while the

PSA CYCLES: BASIC PRINCIPLES 77

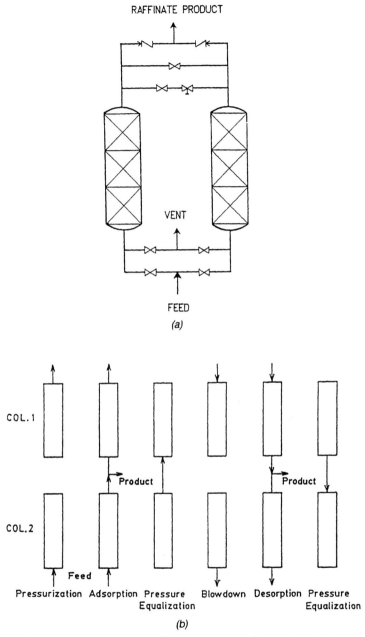

Figure 3.7 (a) Schematic diagram and (b) the sequence of operations in the modified Skarstrom cycle including pressure equalization.

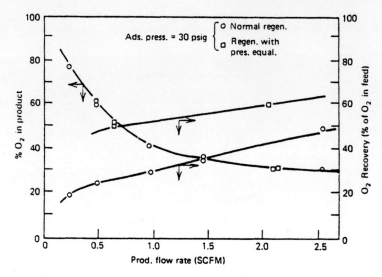

Figure 3.8 Purity and fractional recovery of O_2 in a two-bed PSA air separation unit showing improvement in recovery obtained by inclusion of pressure equalization step. (From Ref. 4; reprinted with permission.)

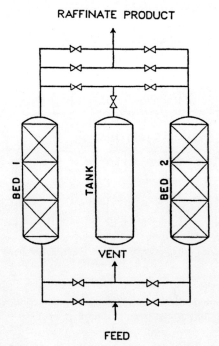

Figure 3.9 Schematic diagram of a PSA cycle showing the use of a third empty tank for reducing blowdown loss.

second bed is vented to complete the blowdown. The pressure equalization step conserves energy since the compressed gas from the high-pressure bed is used to partially pressurize the low-pressure bed and, since this gas is partially depleted of the strongly adsorbed species, separative work is also conserved. Blowdown losses are reduced to about half, with consequent improvement in the recovery of the raffinate product, as may be seen from Figure 3.8.

Prior to Berlin's modification of the Skarstrom cycle, another patent based on a different idea for reducing blowdown loss was assigned to Marsh et al.[10] The process scheme, shown in Figure 3.9, requires an empty tank in addition to the two adsorbent beds. At the end of the high-pressure adsorption step but well before breakthrough, the feed flow is stopped and the product end of the high-pressure bed is connected to the empty tank where a portion of the compressed gas, rich in the raffinate product, is stored. The blowdown of the high-pressure bed is completed by venting to the atmosphere in the reverse-flow direction. The stored gas is then used to purge the bed after which the bed is finally purged with product gas. The product purge requirement is reduced, thereby increasing the recovery, but the savings in the specific energy of separation is less with this arrangement than with a direct pressure equalization step.

3.2.3 Multiple-Bed Systems

Further improvements in efficiency are generally achieved by using multiple adsorbent beds with a sequence of pressure equalization steps incorporated into the cycle.[11,12] In fact, multiple-bed systems also use the blowdown gas for purging other beds. Since this is done at a pressure level where further pressure equalization is not worthwhile, the resulting gain in recovery yields an additional benefit. One such example is shown schematically in Figure 3.10, which shows a typical large-scale air separation process for oxygen production. In this system, which utilizes three or four columns, one column is in the adsorption step and the other two (or three) columns are in various stages of pressurization, depressurization, or purging. The process operates at two intermediate pressures between the feed pressure and the exhaust pressure (usually atmospheric). At the end of the adsorption step, column 1, which is at high pressure, is connected at the discharge end to column 2, and the pressures are equalized. Prior to pressure equalization column 2 has just completed the purge step and is essentially at atmospheric pressure. A fraction of the remaining gas from bed 1 is used for reverse-flow purging of bed 3. When the pressure in bed 1 has fallen to the required level, beds 1 and 3 are disconnected and the residual gas from bed 1 is vented to atmosphere from the bed inlet. Bed 1 is then purged in reverse flow with gas from the fourth bed and repressurized to the first intermediate pressure from the second bed, which has just completed the adsorption step. Final repressurization is accomplished using a part of the product gas and the feed is then

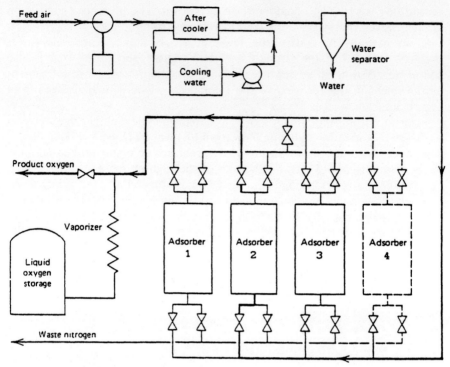

Figure 3.10 Schematic diagram of a three- or four-bed PSA system for air separation. (From Ref. 11; reprinted with permission.)

Vessel Number												
1	Adsorption			EQ1 ↑	CD ↑	EQ2 ↑	CD ↓	Purge ↓	EQ2 ↓	EQ1 ↓	R ↓	
2	CD ↓	Purge ↓	EQ2 ↓	EQ1 ↓	R ↓		Adsorption			EQ1 ↑	CD ↑	EQ2 ↑
3	EQ1 ↑	CD ↑	EQ2 ↑	CD ↓	Purge ↓	EQ2 ↓	EQ1 ↓	R ↓		Adsorption		
4	EQ1 ↓	R ↓		Adsorption			EQ1 ↑	CD ↑	EQ2 ↑	CD ↓	Purge ↓	EQ2 ↓

EQ—Equalization
CD —Cocurrent depressurization
CD —Countercurrent depressurization
R—Repressurization

↑—Cocurrent flow
↓—Countercurrent flow

Figure 3.11 Summary of the cycle for a four-bed PSA unit. (From Ref. 13; reprinted with permission.)

PSA CYCLES: BASIC PRINCIPLES

connected to the inlet of bed 1. The cycle configuration is summarized in Figure 3.11.

The idea of product repressurization was put forward for the first time in a very similar patent for hydrogen purification by Wagner.[14] Pressurization with product pushes the residual adsorbed components toward the feed end of the adsorber, thereby enhancing the product purity. The four-bed configuration allows continuous product withdrawal and eliminates the use of an empty tank for storing purge gas.

In multiple-bed systems greater conservation of energy and separative work are achieved at the cost of a more complex process scheme. In some large-scale hydrogen purification PSA systems up to twelve adsorbent beds are used.

3.2.4 Vacuum Swing Cycle

The simplest way to understand a vacuum swing cycle (VSC) is to consider it as a Skarstrom cycle in which the low-pressure countercurrent product purge step is replaced by a vacuum desorption. The product end of the column is kept closed and the vacuum is pulled through the feed end as shown in Figure 3.12. In a vacuum swing cycle, using the same high operating pressure as a Skarstrom cycle, for the same product purity, the loss of the less favorably adsorbed species in the evacuation step is normally less than the corresponding loss in the purge. The gain in raffinate recovery is achieved here at the expense of the additional mechanical energy required for the evacuation step. A significant energy saving is possible if the cycle is operated with the higher pressure slightly above atmospheric pressure and a very low desorption pressure. In the low-pressure (linear) range of the adsorption isotherm it is the pressure ratio and not the actual high- and low-pressure levels that determines the achievable purity and recovery. A vacuum swing

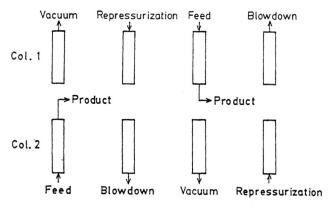

Figure 3.12 The sequence of steps in a vacuum swing cycle.

cycle will therefore be advantageous over a Skarstrom cycle if a low-pressure product is acceptable. However, this advantage in operating cost is to some extent offset by the increased capital cost arising from the increased size of the equipment.

The idea of vacuum regeneration was originally proposed by Guerin de Montgareuil and Domine in a patent assigned to Air Liquide.[3] There are, however, several differences between the pressure swing cycle proposed by Montgareuil and Domine and the simplified vacuum swing cycle shown in Figure 3.12. Depending on the nature of the gas mixture to be separated, the Air Liquide process can vary in the number of adsorbent beds, the type of bed association, and the scheme of cyclic operation, as well as in other operating conditions. The number of beds can vary from one to six or more. A two-bed illustration of this process is given in Figure 3.13. Bed 1 is pressurized to the high operating pressure by introducing a compressed feed gas from the inlet end. The inlet end is then closed and the gas is expanded cocurrently through bed 2 and the effluent from this bed is recovered as raffinate product. When the pressure in bed 1 reaches a predetermined intermediate pressure, the discharge end of bed 1 is closed and the vacuum line (located at the middle of the bed) is opened for regeneration. At the same time the inlet end of bed 2 is opened to high-pressure feed stream, with the discharge end closed, for repressurization. The major disadvantage of the Air Liquide cycle is that the product is delivered at a low (subatmospheric) pressure. (Air separation using this cycle produced 98% oxygen at 51% recovery. This result was markedly superior to the performance of the competing Skarstrom cycle for the same separation. Additionally, the Air

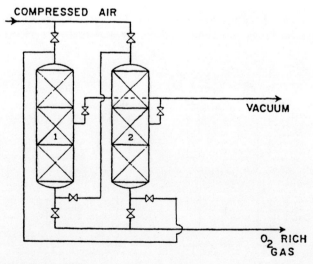

Figure 3.13 Schematic diagram of the two-bed Air Liquide PSA system. (After Montgareuil and Domine[3])

Liquide cycle also produces a nitrogen-rich stream (96.3% nitrogen at 58% recovery) from the evacuation step.

The gain in raffinate product recovery obtained here results from two improvements over the Skarstrom cycle. The recovery advantage of vacuum regeneration over purge has already been discussed. In addition, the cocurrent depressurization through an evacuated chamber actually conserves part of the raffinate product that would have otherwise been lost during evacuation. It is important to note that the gain in the raffinate recovery from cocurrent depressurization may not be achieved if this step is used in combination with (product) purge regeneration. Cocurrent depressurization contaminates the product end with the more strongly adsorbed species. Suh and Wankat[15] have shown that depending on the amount of the more strongly adsorbed species in the feed and the relative affinity of the components, the gain in the raffinate recovery from cocurrent depressurization may be outweighed by the increased purge requirement. Cocurrent depressurization is also beneficial in enhancing the purity of the strong adsorptive product (see Section 3.3).

3.3 Recovery of the More Strongly Adsorbed Species in Equilibrium-Controlled Separations

The vacuum swing cycle developed by Montgareuil and Domine and discussed in Section 3.2 was the first pressure swing process incorporating the provision for recovering the more strongly adsorbed species at high purity. The operations responsible for providing this additional benefit to the cycle are cocurrent depressurization and vacuum regeneration. The vacuum regeneration step produces the extract product. The region of the bed through which the feed penetrates during the high-pressure adsorption step is essentially at equilibrium with the feed gas. Since dispersive effects such as mass transfer resistance, axial dispersion, heat effects, and so on, are usually associated with equilibrium-controlled separation processes, there is always some spreading of the mass transfer zone. The region ahead of the mass transfer zone at the end of the high-pressure adsorption step is therefore available for further adsorption of the more strongly adsorbed species in the feed. By cocurrent depressurization the raffinate product remaining in the void space ahead of the mass transfer zone is pushed out of the bed and the more strongly adsorbed species is retained at the product end. Thus the mole fraction of the more strongly adsorbed species increases in both phases. The subsequent evacuation step therefore produces the extract product at high purity and recovery. If the situation is such that the length of unused bed prior to breakthrough is not sufficient to hold the strong adsorptive, then the high-pressure adsorption step is cut short well before breakthrough in order to provide with the additional capacity. Such an arrangement will reduce the adsorbent productivity and the optimum choice therefore depends primarily on the value of the extract product.

When the separation factor is low, the adsorbed phase concentrations of the light and heavy components are comparable. The reduction in the concentration of the lighter species in the bed as a result of cocurrent depressurization may not then be sufficient to meet the purity requirement for the extract product. A more effective method for improving extract purity is to purge (or rinse) the void spaces, after high-pressure adsorption, with the strongly adsorbed species in the direction of the feed.[16] The effluent gas during this step is produced at the feed pressure and has a feedlike composition so that the stream may easily be recycled.

The purity of the strongly adsorbed component depends critically on the use of cocurrent depressurization or purging by the more strongly adsorbed species. The use of vacuum desorption is not particularly crucial unless one is dealing with strongly adsorbed species with a type I isotherm. For bulk separations involving components with moderate isotherm curvature, it is possible to achieve high quality of both raffinate and extract products by cycling in a pressure range above atmospheric. Yang and Doong[17] separated a 50:50 hydrogen–methane mixture over activated carbon into 97.8% hydrogen (raffinate product) with 90% recovery and 90% methane (extract product) with 89.9% recovery. Cocurrent depressurization was employed and the pressure was cycled between 120 and 35 psig. The two-bed, five-step, cycle

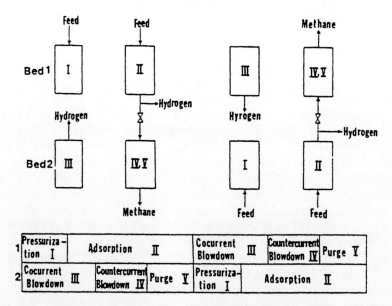

Figure 3.14 Schematic diagram of the two-bed, five-step PSA cycle used by Yang and Doong for recovery of both extract and raffinate products at high purity. (From Ref. 17; reprinted with permission.)

used by Yang and Doong is shown in Figure 3.14. Cen and Yang[18] in another study repeated the same separation and demonstrated that the purity of the extract product is improved further by replacing the cocurrent depressurization step with a cocurrent methane purge step conducted at the feed pressure.

3.4 Cycles for the Recovery of Pure Raffinate Product in Kinetically Controlled Separations

The PSA cycles discussed so far, for both purification and bulk separation, were developed for separations based on equilibrium selectivity. The cycles used for kinetic separations are somewhat different. In such systems the choice of contact time is critical. Since the idea is to exploit the difference in the diffusion rates of the adsorbing molecules, the contact time must be short enough to prevent the system from approaching equilibrium but not so short as to preclude significant uptake. The crucial element in any kinetic separation is therefore the duration of the adsorption and desorption steps.

The only widely used commercial PSA process based on kinetic selectivity is air separation for nitrogen production using a carbon molecular sieve or 4A zeolite adsorbent. Kinetic separation of air for nitrogen production using Union Carbide RS-10 molecular sieve (modified 4A zeolite) has been investigated by Shin and Knaebel.[19] In a recent study Kapoor and Yang[20] have shown that methane–carbon dioxide separation (from landfill gas or effluent gas from tertiary oil recovery) using carbon molecular sieve is another prospective candidate for kinetic separation.

Although with properly selected step times the Skarstrom cycle can be applied to a kinetic PSA separation, such a cycle is far from ideal. A major disadvantage is that the slowly diffusing raffinate product would be continuously adsorbed during the purge step. This difficulty can be avoided by use of vacuum desorption or by using a modified form of "self-purging" cycle.

3.4.1 Self-Purging Cycle

Equilibrium and kinetic data for the sorption of oxygen and nitrogen on the Bergbau–Forschung carbon molecular sieve[21] are shown in Figure 3.15 and summarized in Table 3.2. It is apparent that there is little difference in equilibrium but a large difference in diffusivity, with oxygen being the more rapidly adsorbed species. The high-pressure raffinate product in the carbon molecular sieve process is therefore nitrogen. In such a system purging with nitrogen to remove the faster diffusing oxygen from the bed (as in the Skarstrom cycle) is undesirable since, as well as wasting product, a certain fraction of the slowly diffusing nitrogen will be adsorbed, thus reducing the capacity for oxygen during the next adsorption step. The earlier kinetic

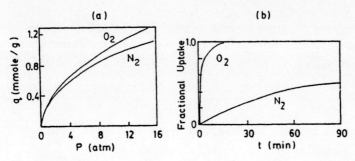

Figure 3.15 (a) Equilibrium isotherms and (b) experimental uptake curves for sorption of O_2 and N_2 on Bergbau–Forschung carbon molecular sieve. (From Ref. 21; reprinted with permission.)

nitrogen processes avoided this difficulty by using a vacuum to clean the bed rather than a purge, as illustrated in Figure 3.12. The general scheme for a vacuum swing carbon sieve process to produce nitrogen is shown in Figure 3.16. A better option is, however, available. At the end of the blowdown step the adsorbent contains both oxygen (fast diffusing) and nitrogen (slow diffusing). Thus, if the bed is simply closed at the product end and left for a period of time, the oxygen will diffuse out first, followed by nitrogen, so the system is, in effect, self-purging. The product purity is directly controlled by increasing feed pressure, and pressure equalization is incorporated to reduce the blowdown loss. A dual-ended pressure equalization is used in which the feed and product ends of the high-pressure bed are connected to the respective ends of the low-pressure bed. Most modern nitrogen PSA units therefore operate on the cycle shown in Figure 3.17, which incorporates both pressure

Table 3.2. Equilibrium and Kinetic Data of Oxygen, Nitrogen, Methane, and Carbon Dioxide on Bergbau–Forschung Carbon Molecular Sieve and 4A Zeolite at 25°C

	Diffusional time constant (s^{-1})	Henry's constant[a]	Saturation constant (g moles/cm^3)
O_2-CMS[b]	2.70×10^{-3}	9.25	2.64×10^{-3}
N_2-CMS[b]	5.90×10^{-5}	8.90	2.64×10^{-3}
O_2-4A[c]	8.51×10^{-3}	2.10	1.72×10^{-2}
N_2-4A[c]	8.99×10^{-5}	4.26	1.20×10^{-2}
CO_2-CMS[d]	9.00×10^{-4}	135.83	2.85×10^{-3}
CH_4-CMS[d]	5.00×10^{-6}	25.83	1.74×10^{-3}

[a] Dimensionless basis.
[b] *Source*: Kinetic data from Ref. 24 and equilibrium data from Ref. 22.
[c] *Source*: Ref. 23.
[d] *Source*: Ref. 20.

PSA CYCLES: BASIC PRINCIPLES 87

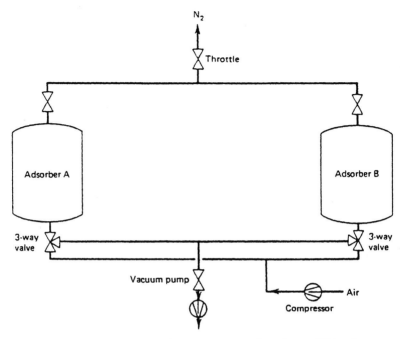

Figure 3.16 Schematic diagram of the Bergbau–Forschung PSA air separation process. (From Ref. 21; reprinted with permission.)

equalization and a self-purging desorption step. Connecting both ends of the beds allows rapid pressure equalization and improves product purity since the oxygen-rich gas remains at the feed end of the pressurized bed. The higher recovery advantage of dual-ended pressure equalization, however, decreases with increasing product purity.

Kinetic selectivity may be increased by increasing the diffusivity of the faster component, oxygen, or further decreasing the diffusivity of the slower component, nitrogen. An increase in oxygen diffusivity will increase the nitrogen product purity without seriously affecting the recovery (see Figure 5.9). Intuitively it may appear that nitrogen recovery will increase with decreasing nitrogen diffusivity, and the best situation would be reached when there is practically no penetration of nitrogen. What is overlooked in this intuitive argument is the role of the desorbing nitrogen in a self-purging cycle. The desorbing nitrogen cleans the void volume of the adsorption column, and, unless there is significant uptake of nitrogen during the high-pressure adsorption step, inadequate self-purging will result in increased oxygen contamination in the nitrogen product. There is therefore a lower limit of nitrogen diffusivity below which the self-purging cycle becomes ineffective.

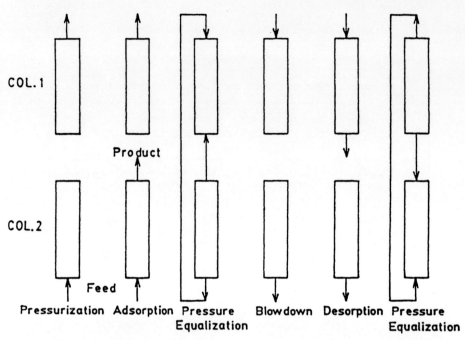

Figure 3.17 The sequence of operations in a two-bed PSA cycle including dual-ended pressure equalization and no purge.

3.4.2 Kinetic and Equilibrium Effects in Opposition

When equilibrium selectivity favors the slower-diffusing component, as in air separation for nitrogen production on modified 4A zeolite (see Table 3.2 for quantitative values), the system cannot be made self-purging since the amount of desorbing nitrogen is not sufficient to eliminate the residual oxygen concentration in the bed. For such a system conventional purge or vacuum regeneration is necessary. Shin and Knaebel[19] have reported an extensive experimental study of this system. They adopted the Skarstrom cycle in a single-bed apparatus and therefore enjoyed the additional freedom to optimize the cycle by varying independently the duration of the individual steps. Their experimental system is shown in Figure 3.18. In such an arrangement an additional tank was used for product storage (in order to supply the purge) and pressure equalization cannot be performed. However, the contents of the auxiliary tank may be used to partially repressurize the bed in a manner essentially equivalent to a pressure equalization step.

Blowdown and purge under vacuum were also investigated in this study for comparison. Some of the more important findings are summarized in Tables 3.3 and 3.4. As the pressure ratio increases, the purity increases, but the

PSA CYCLES: BASIC PRINCIPLES

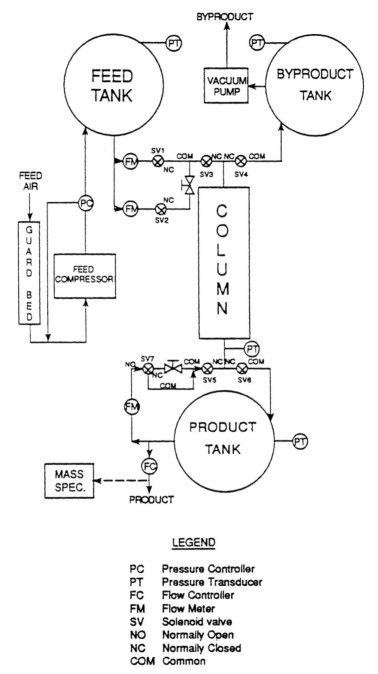

Figure 3.18 The single-bed experimental PSA system used by Shin and Knaebel to study air separation on a modified 4A zeolite. (From Ref. 19)

Table 3.3. Effect of Pressure Ratio on Separation of Air on Union Carbide RS-10 Molecular Sieve[a]

High pressure (kPa)	Low pressure (kPa)	Pressure ratio	Nitrogen recovery (%)	Product purity (mol % N_2)	Productivity
200	113	1.77	41.56	82.3	4.78
358	123	2.91	22.46	93.7	3.9
620	154	4.03	12.54	98.2	2.61
200	113	1.77	13.83	93.8	1.00
358	123	2.91	22.46	93.7	3.9
620	152	4.08	23.05	93.7	5.46

[a] High-pressure adsorption = 35 s, blowdown = 2 s, purge = 3 s, pressurization = 15 s. Purge volume = 0.0166×10^{-3} m^3 at STP.
Source: Ref. 19.

Table 3.4. Effect of Purge Pressure on Separation of Air on Union Carbide RS-10 Molecular Sieve[a]

High pressure (kPa)	200		358		620	
Low pressure (kPa)	113	60	123	78	152	119
Nitrogen recovery (%)	13.83	16.87	22.46	21.83	23.05	23.12
Product purity (mole % N_2)	93.8	93.8	93.7	93.7	93.7	93.7

[a] High-pressure adsorption = 35 s, blowdown = 2 s, purge = 3 s, pressurization = 15 s. Purge volume = 0.0166×10^{-3} m^3 STP.
Source: Ref. 19.

recovery is decreased. For the same purity both recovery and productivity increase with increasing feed pressure. The recovery gain is relatively less in the higher-pressure region, but the increase in productivity is consistent. Moreover, subatmospheric blowdown and purge are advantageous (in terms of recovery) at low feed pressures, but the advantage disappears as the feed pressure is increased. The latter finding further confirms that in a Skarstrom cycle blowdown loss becomes dominant at high feed pressure. Product purity may therefore be controlled by either the feed pressure or the purge rate or by a combination of both of these.

3.4.3 Equilibrium and Kinetic Effects Reinforce

Separation of methane from methane–carbon dioxide mixture on a carbon molecular sieve is an example where equilibrium is in favor of the faster-dif-

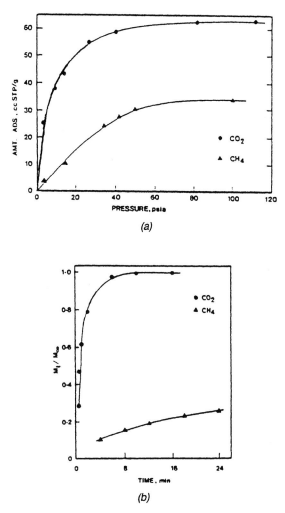

Figure 3.19 (a) Equilibrium isotherms and (b) experimental uptake curves for sorption of CO_2 and CH_4 on Bergbau–Forschung carbon molecular sieve. (From Ref. 20; reprinted with permission.)

fusing component, carbon dioxide. Relevant equilibrium and kinetic data are presented in Figure 3.19 and Table 3.2. A purity-versus-recovery plot for this separation (feed is 50:50 methane-to-carbon dioxide ratio) constructed from the data of Kapoor and Yang[20] in the region of their optimal operating point is shown in Figure 3.20. The effects of varying the high and low pressures and the product rates about their optimal values are also indicated. A cycle similar to that shown in Figure 3.14 was used, except that the countercurrent purge step was replaced by vacuum desorption through the feed end. It is clear from Figure 3.20 that there is an upper limit of the high

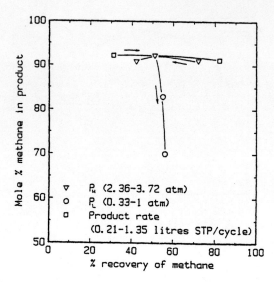

Figure 3.20 Effects of feed pressure, desorption pressure, and product rate on the experimental purity and recovery of CH_4 from CH_4/CO_2 separation by pressure swing adsorption on a carbon molecular sieve. The arrows indicate the direction of increasing parameter values. (Data taken from Ref. 20.)

operating pressure beyond which both purity and recovery decline. With atmospheric blowdown the maximum methane product purity that can be achieved by raising the adsorption pressure is therefore limited to about 70%. Further improvement in product purity is essentially controlled by the (subatmospheric) desorption pressure.

These observations provide an interesting contrast with similar performance profiles for the kinetic air separation process (Figure 3.21). In Figure 3.21 all profiles are monotonic with no evidence of an upper limit. While there must always be a theoretical limit for the high operating pressure, beyond which both recovery and purity decline, it is clear that this limit lies well beyond the normal range of operating pressures for air separation on a carbon molecular sieve and therefore does not limit the system performance. In contrast to the methane–carbon dioxide system, a high-purity nitrogen raffinate product can therefore be achieved simply by raising the adsorption pressure, without recourse to subatmospheric desorption.

The key difference between air separation and methane–carbon dioxide systems appears to lie in the shape of the equilibrium isotherm for the more strongly adsorbed species. For nitrogen–oxygen on CMS the isotherms are of linear or slightly favorable (type I) form, whereas, in the relevant pressure range, the isotherm for carbon dioxide on CMS is highly favorable, approaching the rectangular limit. If the isotherm for the faster-diffusing species is

PSA CYCLES: BASIC PRINCIPLES

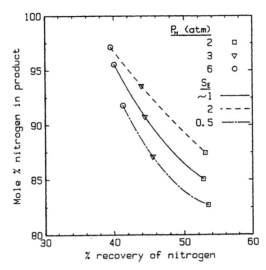

Figure 3.21 Performance of an air–carbon molecular sieve system operated on a modified Skarstrom cycle (no external purge) showing the effect of high operating pressure and equilibrium selectivity. Kinetic data and equilibrium data for $S_E \simeq 1$ are given in Table 3.2. S_E has been varied by changing the oxygen equilibrium. (From Ref. 25)

highly favorable, a very high feed pressure is not desirable, and the product purity is primarily controlled by the (subatmospheric) desorption pressure. The enhancement of performance to be expected in a kinetic PSA separation when equilibrium and kinetic effects reinforce will be observed only when the equilibrium relationship does not deviate too much from the linear form.

3.5 Cycle for Recovery of the Rapidly Diffusing Species

In the kinetic separation of methane and carbon dioxide mixture on a carbon molecular sieve, discussed in Section 3.4, the rapidly diffusing component, carbon dioxide, was also recovered at high purity (over 90% purity and recovery). Cocurrent depressurization and vacuum desorption, which are commonly employed in equilibrium-controlled separations to produce a high purity extract product, were, in this study, adapted to a kinetically controlled process by proper control of the contact time.

Some of the elementary steps discussed here are addressed in more detail in Chapter 4. The application of the basic principles in representative industrial PSA processes is discussed in Chapter 6.

References

1. G. F. Fernandez and C. N. Kenney, *Chem. Eng. Sci.* **38**, 834 (1983).
2. C. W. Skarstrom, U.S. Patent No. 2,944,627 (July 12, 1960), to Exxon Research and Engineering.
3. P. G. de Montgareuil and D. Domine, U.S. Patent No. 3,155,468 (1964), to Air Liquide.
4. C. W. Skarstrom, *Recent Developments in Separation Science*, Vol. 2, p. 95, N. N. Li, ed., CRC Press, Cleveland, Ohio, 1975.
5. M. J. Matz and K. S. Knaebel, *AIChE J.* **34**(9), 1486 (1988).
6. C. W. Skarstrom, U.S. Patent No. 3,237,377 (1966), to Exxon Research and Engineering.
7. S. Farooq, D. M. Ruthven, and H. A. Boniface, *Chem. Eng. Sci.* **44**(12), 2809 (1989).
8. N. H. Berlin, U.S. Patent No. 3,280,536 (Oct. 25, 1966), to Exxon Research and Engineering.
9. H. Lee and D. E. Stahl, *AIChE Symp. Ser.* **64**(134), 1 (1973).
10. W. D. Marsh, R. C. Hoke, F. S. Pramuk, and C. W. Skarstrom, U.S. Patent No. 3,142,547 (1964), to Exxon Research and Engineering.
11. J. C. Davis, *Chem. Eng.* Oct. 16 (1972), p. 88.
12. L. B. Batta, U.S. Patent No. 3,564,816 (1971), to Union Carbide Corporation.
13. R. T. Cassidy, in *A.C.S. Symp. Ser.* **135**, Adsorption and Ion Exchange with Synthetic Zeolites, W. H. Flank, ed., American Chemical Society, Washington, D.C., 1980, p. 275.
14. J. L. Wagner, U.S. Patent No. 3,430,418 (1969), to Union Carbide Corporation.
15. S.-S. Suh and P. C. Wankat, *AIChE J.* **35**(3), 523 (1989).
16. T. Tamura, U.S. Patent No. 3,797,201 (1974), to T. Tamura, Tokyo, Japan.
17. R. T. Yang and S. J. Doong, *AIChE J.* **31**(11), 1829 (1985).
18. P. L. Cen and R. T. Yang, *Separation Sci. Tech.* **21**(9), 845 (1986).
19. H.-S. Shin and K. S. Knaebel, *AIChE J.* **34**(9), 1409 (1988).
20. A. Kapoor and R. T. Yang, *Chem. Eng. Sci.* **44**(8), 1723 (1989).
21. K. Knoblauch, *Chem. Eng.* **85**(25), 87 (1978).
22. D. M. Ruthven, N. S. Raghavan, and M. M. Hassan, *Chem. Eng. Sci.* **41**, 1325 (1986).
23. O. W. Haas, A. Kapoor, and R. T. Yang, *AIChE J.* **34**(11), 1913 (1988).
24. S. Farooq and D. M. Ruthven, *Chem. Eng. Sci.* **46**(9), 2213 (1991).
25. S. Farooq and D. M. Ruthven, *Chem. Eng. Sci.* **47**(8), 2093 (1992).

CHAPTER

4

Equilibrium Theory of Pressure Swing Adsorption

4.1 Background

Since the introduction of PSA, a wide variety of mathematical models have been suggested, based on theories that extend from simple to complex. One of the first was a detailed model, presented by Turnock and Kadlec,[1] which accounts for nonlinear adsorption equilibrium, pressure drop, and temperature effects. That type of model is covered in the next chapter. Shortly thereafter, Shendalman and Mitchell[2] suggested a simpler type of PSA model, based on the assumption of local equilibrium. This type of model accounts mainly for mass conservation, and ignores transport phenomena. That might make the model seem trivial, but the basic equations of mass conservation account for time and axial position variations of flow, pressure, and composition, which are essential in pressure swing adsorbers.

The model of Shendalman and Mitchell was based on a four-step PSA cycle in which a trace contaminant was adsorbed from a nonadsorbing carrier. Thus, it applies only in very restrictive cases, such as cleanup of hydrogen or helium containing less than 1% of methane or nitrogen using activated carbon as the adsorbent. Over the past decade, there have been several extensions of the basic ideas proposed by Shendalman and Mitchell; many of these developments are explained in this chapter. For example, the model has now been extended to binary mixtures having arbitrary compositions, with both components adsorbing, and in cycles composed of diverse steps. The models are not perfect, but realistic PSA applications can be studied relatively easily.

Most PSA applications exploit equilibrium selectivity, and systems are usually designed to minimize the negative effects of mass transfer resistance. In such cases, the trends and, frequently, even precise measures of PSA performance can be predicted accurately from local equilibrium models. In particular, estimates of product recovery are often excellent, even when mass transfer resistances are large, because mass conservation predominates over diffusion and heat effects. That such models can be accurate, without requiring extensive experimental data, has made them valuable as a tool for PSA simulation and design. Other advantages of equilibrium theories are: they help to identify proper combinations of operating conditions; under certain conditions they reduce to very simple algebraic performance equations relating the operating and design parameters; they clarify the underlying links between steps, conditions, adsorbent properties, and performance; and as a result they may facilitate conception and optimization of novel cycles.

Another major advantage is that the model parameters may be obtained directly from equilibrium measurements; so it is not necessary to fit experimental PSA data. In the simplest case, that is, when isotherms are practically linear, the adsorbent–adsorbate interactions can be lumped together as a single parameter, roughly analogous to the inverse of selectivity. It is even possible to incorporate dispersion by accounting for dead zones at the entrance or exit of the adsorbent bed.

As explained in Chapter 3, the conventional four-step PSA cycle for separation of a binary mixture comprises feed, blowdown, purge, and pressurization, as illustrated in Figure 3.11. Each of these steps serves a vital function that contributes to successful operation of the PSA system. By accounting for the relations governing flow and transfer between the gas and adsorbent, equilibrium models are able to predict the phenomena occurring in each step. It is then easy to combine these relations to predict overall performance. Extending the basic equations, and modifying the conditions of the conventional steps allows complex properties, conditions, and cycles to be simulated.

Before proceeding with mathematical details, the reader may wish to scan Section 4.4 on Cycle Analysis. For example, Eqs. 4.27, 4.37, 4.44, and 4.45 are final equations that predict product recoveries for a variety of cycles. Those equations should convey the idea that, even though the PSA cycles and governing equations may seem complicated, they can be solved in closed forms that are simple and yet have broad applicability. A separate Section, 4.5, covers Experimental Validation, in which predictions of some of the models are compared with experimental data.

The restrictions of the equilibrium theory are evident in the following two groups of inherent assumptions. The first group is generally valid for equilibrium-based PSA separations, and they make the resulting equations amenable to solution by simple mathematical methods.

EQUILIBRIUM THEORY

1. Local equilibrium is achieved instantaneously between the adsorbent and adsorbates at each axial location.
2. The feed is a binary mixture of ideal gases.
3. Axial dispersion within the adsorbent bed is negligible.
4. Axial pressure gradients are negligible.
5. There are no radial velocity or composition gradients.
6. Temperature is constant.

The second group links the equations and conditions to a particular cycle and geometry. The assumptions of a simple four-step cycle and adsorption system are listed, though they may be modified to reflect other cycles or conditions without affecting the applicability of the equilibrium theory.

7. All of the adsorbent is utilized during the feed and purge steps.
8. Pressure is constant during the feed and purge steps.
9. The isotherms may be linear or nonlinear, but they are uncoupled.
10. Dead volume at the adsorber entrance and exit is negligible.

Actually, in the first group of assumptions, all but the first could be relaxed within the confines of an equilibrium model. To drop those assumptions, however, would complicate the mathematics and diminish the simplicity of the equilibrium approach. After all, if one resorts to orthogonal collocation or other potent mathematical methods to account for dispersion, pressure drop, etc., there is no point in restricting such a model to instantaneous mass transfer. Nevertheless, some aspects of detailed models are mentioned in this chapter, along with the effects of axial pressure drop (see Section 4.9). In addition, the assumption that PSA operation be isothermal is not necessarily rigid. In fact, heat effects appear in a variety of ways, some of which are covered in Section 4.8. The effects of relaxing many of the second set of assumptions are also discussed in this chapter (see Sections 4.4.3–4.4.6).

4.2 Mathematical Model

The simplest PSA cycle studied in this chapter is shown in Figure 4.1, which shows the basic steps and conventions of position and direction. The bottom of that figure will be discussed later, but it shows how compositions move through the adsorbent bed during each step, with the shaded region depicting the penetration of some of the more strongly adsorbed (or "heavy") component, while the plain region contains only the pure, less strongly adsorbed (or "light") component. In this chapter the heavy and light components are referred to as A and B, respectively.

The following individual component balance applies to binary mixtures in which both components adsorb, so they are coupled in the gas phase, but are

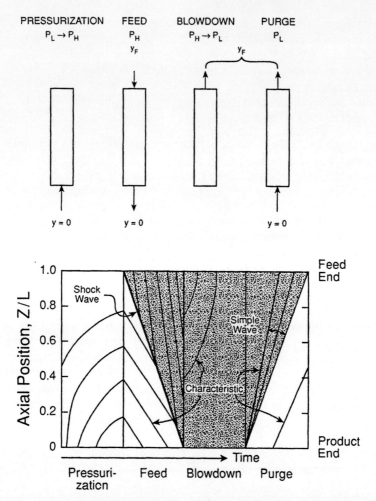

Figure 4.1 Steps in a conventional PSA cycle; orientation of column and streams. (Top) Flows and compositions associated with each step. (Bottom) Positions-versus-time representation of each step. Shaded regions depict penetration of heavy component.

assumed independent in the adsorbed phase:

$$\varepsilon\left(\frac{\partial Py_i}{\partial t} + \frac{\partial vPy_i}{\partial z}\right) + RT(1-\varepsilon)\frac{\partial q_i}{\partial t} = 0 \qquad (4.1)$$

The subscript i refers to component A or B. Shendalman and Mitchell[2] and Chan et al.[3] considered the case of a trace amount of component A being adsorbed from a carrier, B. Shendalman and Mitchell took the carrier to be nonadsorbing, while Chan et al. accounted for adsorption of the carrier. In both cases, the interstitial velocity, v, was taken to be independent of composition (and of both time and axial position at constant pressure). As it is written, however, Eq. 4.1 applies to bulk separations as well as removal of a dilute contaminant.

An important detail is the definition of the void fraction ε. If the adsorbed phase concentrations are expressed on a particle volume basis (i.e., including intraparticle porosity), then ε is simply the extra-particle or interstitial bed voidage. Such a definition is logical where mass transfer rates are finite and when the mass transfer resistance is associated with external film diffusion, macropore diffusion, or solid-side mass transfer resistance at the particle surface. With this definition, the bed density (ρ_B), particle density (ρ_p), and the solid or microparticle density (ρ_c), and the corresponding void fractions are related by:

$$\rho_B = (1-\varepsilon_B)\rho_p = (1-\varepsilon_B)(1-\epsilon_p)\rho_c, \qquad \varepsilon \cong \varepsilon_B$$

When micropore resistance is dominant, or indeed under the assumption of local equilibrium on which this chapter is based, it is more logical to regard the external void fraction as the sum of the particle macroporosity and the bed voidage (i.e., all gas space outside the micropores). The densities and void fraction are then related by:

$$\rho_B = (1-\varepsilon)\rho_c, \qquad \varepsilon \cong \varepsilon_B + (1-\varepsilon_B)\epsilon_p$$

ρ_c corresponds to the density determined with a fluid that penetrates the external voidage and the macropores but not the micropores. Since these definitions affect the magnitude of the isotherm parameters, it is essential to maintain consistency when equilibrium data are incorporated into a PSA model.

The balance equations are modified by substituting adsorption isotherm expressions for each component in the form,

$$q_i = f_i\left(\frac{Py_i}{RT}\right) = f_i(c_i) \qquad (4.2)$$

Examples are given in Chapter 2. Following assumption 6 above, the temperature is fixed, so the second term of Eq. 4.1 can be determined from Eq. 4.2, as:

$$\frac{\partial q_i}{\partial t} = \frac{1}{RT}f'(Py_i) \qquad (4.3)$$

Inserting this into Eq. 4.1 yields:

$$\frac{\partial P y_i}{\partial t} + \beta_i \frac{\partial v P y_i}{\partial z} = 0 \qquad (4.4)$$

where $\beta_i = \{1 + [(1 - \varepsilon)/\varepsilon] f_i'\}^{-1}$, which is a function of the component partial pressure. Note that when the isotherms of both components are linear, this parameter is constant, although here it is treated as if it depends on partial pressure. Simply adding the appropriate forms of Eq. 4.4 for the respective components yields the overall material balance, which governs the interstitial flow in the packed bed. Hence, this expression may be employed to determine the local velocity in terms of composition, pressure, and adsorbent–adsorbate interactions. The result is

$$\frac{v_2}{v_1} = \exp\left(\int_{y_{A_1}}^{y_{A_2}} \frac{1-\beta}{1+(\beta-1)y_A} dy_A\right) \cong \frac{1+(\beta-1)y_1}{1+(\beta-1)y_2} \qquad (4.5)$$

where the approximation is only exact for linear isotherms, and $\beta = \beta_A/\beta_B$, which for nonlinear isotherms may depend on pressure and composition, but for linear isotherms it is constant. During pressurization and blowdown, the composition, pressure, and interstitial velocity vary with time at each axial position. The details are beyond the current scope (see Section 4.9), but the result for linear isotherms is

$$v = \frac{-z}{\beta_B[1 + (\beta - 1)y_A]} \frac{1}{P} \frac{dP}{dt} \qquad (4.6)$$

The simplicity of equilibrium-based theories arises because the coupled first-order partial differential equations governing the material balances can be recast as two ordinary differential equations that must be simultaneously satisfied. The mathematical technique employed is called the *method of characteristics*. A brief explanation is given in Appendix A. The resulting equations are:

$$\frac{dz}{dt} = \frac{\beta_A v}{1 + (\beta - 1) y_A} \qquad (4.7)$$

and

$$\frac{dy_A}{dP} = \frac{(\beta - 1)(1 - y_A) y_A}{[1 + (\beta - 1) y_A] P} \qquad (4.8)$$

Equation 4.7 defines characteristic trajectories in the z, t plane. First of all, Eq. 4.8 shows that when pressure is constant, composition is also constant, and the characteristics given by Eq. 4.7 are straight lines. Conversely, when pressure varies with time, the composition varies according to Eq. 4.8, and the characteristics are curved. Examples of characteristics are shown in the bottom portion of Figure 4.1, for example, as lines during feed and purge and as curves during pressurization and blowdown. It is important to note that characteristics do not end at the end of a step; rather, they continue in an altered form. Since composition can be determined along any characteristic, knowing the initial composition, they provide an excellent means for performing stepwise material balances for PSA cycles.

The blowdown step is governed by Eqs. 4.7 and 4.8, while the subsequent purge step can be analyzed with Eq. 4.7 alone. These steps essentially regenerate the adsorbent. In many conventional adsorption systems, regeneration creates a wave that gradually moves through the bed, which is commonly referred to as a *proportionate pattern* or *dispersive front* (see Section 2.4). In the remainder of this chapter, it is referred to as a *simple wave*, partly because it contrasts with the term *shock wave* which is discussed below (see also Section 2.4.1), and to suggest an association with equilibrium behavior (as opposed to kinetic or dispersive effects).

To complete the analysis of even the simplest PSA cycle, it is necessary to account for the uptake step(s). In particular, the feed step involves uptake of the more strongly adsorbed component. As in conventional adsorption systems, a sharp concentration front is created by this uptake that is sometimes called a *constant pattern* or *self-sharpening* profile. At the extreme of local equilibrium behavior, the front is a step change, and is called a *shock wave*. Since equilibrium effects are emphasized in this chapter, the term *shock wave* will be used, even though in real systems dissipative effects may diminish the sharpness. A shock wave is shown in the bottom portion of Figure 4.1 in the feed step. It appears as a thick line at which characteristics intersect, and it separates the shaded region (depicting presence of the heavy component) from the plain region (depicting the pure light component). Examples of breakthrough data (from experiments in which methane was admitted to a bed of activated carbon previously pressurized with nitrogen) are shown in Figure 4.2, illustrating the sharpness attainable in many PSA applications.

It is the velocity of a shock wave through the packed bed that governs the duration of the feed and rinse steps. Similarly, that velocity is controlled by the rate of flow into and out of the system; hence, the material balance is also affected. The velocity of the shock wave depends on the interstitial fluid velocities at the leading and trailing edges of the wave. These are related by equating the shock wave velocities based on the conditions for both components A and B to get

$$v_{SH} = \theta_A \frac{v_2 y_{A_2} - v_1 y_{A_1}}{y_{A_2} - y_{A_1}} \tag{4.9}$$

where, in general,

$$\theta_i = \theta_i(P_H, y_1, y_2) = \left(1 + \frac{1-\varepsilon}{\varepsilon} \frac{f_{i_2} - f_{i_1}}{y_{i_2} - y_{i_1}} \frac{RT}{P_H}\right)^{-1}$$

1 and 2 refer to the leading and trailing edges of the wave, respectively, and f_{i_j} is given by Eq. 4.2 for component i at composition j.

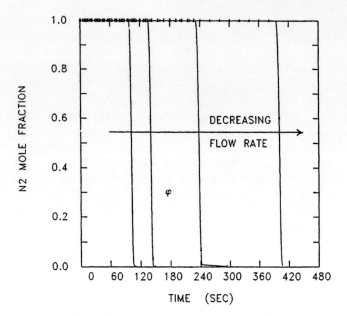

Figure 4.2 Breakthrough data of N_2 and CH_4 on activated carbon at 2.7 atm and 25°C.

A balance for component A around the shock front yields an expression for the velocities ahead of and behind it,

$$\frac{v_2}{v_1} = \frac{1 + (\theta - 1)y_{A_1}}{1 + (\theta - 1)y_{A_2}} \tag{4.10}$$

where $\theta = \theta_A/\theta_B$. Again, Appendix A treats this subject somewhat more generally and thoroughly.

4.3 Model Parameters

Two parameters provide the simplest means by which to express the impact of adsorbent–adsorbate interactions on PSA performance. The parameter β is evaluated at a specific composition using tangents of the respective isotherms. The parameter θ, however, is evaluated at a jump discontinuity using chords of the respective isotherms. For systems having linear isotherms, β and θ are identical.

The amount of adsorbate held in a unit volume of adsorbent is sometimes referred to as the *column isotherm*. In terms of the definitions given, the column isotherm is

$$\frac{c_i}{\theta_i} = \frac{y_i P}{RT}\left(1 + \frac{1-\varepsilon}{\varepsilon}\frac{f_i}{y_i}\frac{RT}{P}\right).$$

In terms of wave movements, this is the cumulative amount of adsorbate i admitted at P and y_i to an initially clean adsorbent (i.e., totally evacuated or prepressurized with a less strongly adsorbed component), stopping at breakthrough. The column isotherm is discussed later in other contexts.

The parameters β and θ are related to velocities via Eqs. 4.5, 4.9, and 4.10. Therefore, it is possible to measure their effective values in breakthrough experiments. In one type of experiment, a fixed bed of adsorbent is initially purged and pressurized with the light component. The feed is then admitted to the bed at the same pressure. By simply monitoring the influent and effluent flow rates, the value of θ (or β when both isotherms are practically linear) may be found from:

$$\theta = \theta(P, 0, y_F) = 1 - \frac{1 - Q_{\text{out}}/Q_{\text{in}}|_F}{y_{A_F}} \tag{4.11}$$

$(= \beta$ for linear isotherms$)$

where the product is pure (i.e., $y_{A_P} = 0$), the bed pressure is kept constant, $\Delta P = 0$, and the volumetric flow rates, Qs, are constant. In an alternative type of experiment the adsorbent bed is initially equilibrated with feed. Then the pure heavy component is admitted to the bed at the same pressure. By simply monitoring the influent and effluent flow rates, the value of θ (or β) may be found from:

$$\theta = \theta(P, y_F, y = 1) = \frac{1 - y_F}{Q_{\text{in}}/Q_{\text{out}} - y_F}$$

$(= \beta$ for linear isotherms$)$ \quad (4.12)

Note that when both isotherms are linear, the values obtained in the two types of experiments should be consistent, but if isotherm curvature is significant, the values of θ determined in the two types of experiments may be different. In that case, the first type of measurement would be more useful when the light component is desired, and the second type would be more useful when the heavy component is desired.

This approach has several advantages: a range of different operating conditions (feed composition, pressures, and cycle times) can be examined, the effects of minor variations in packing and/or adsorbent properties can be assessed directly, and even effects of dispersion and diffusion can be identified and easily resolved.

As mentioned earlier, examples of experimental breakthrough curves, for methane (A)-nitrogen (B) on activated carbon, are shown in Figure 4.2. The

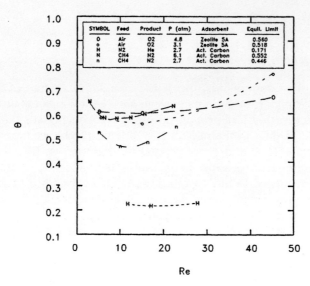

Figure 4.3 Effect of Re on adsorption selectivity parameter, θ, for air–O_2 on zeolite 5A (Ball[5]), and N_2 and He on activated carbon, and N_2 and CH_4 on activated carbon.

results are plotted as θ versus Reynolds number (Re $= \rho \varepsilon v d_P/\mu$, where ρ and μ are the gas density and viscosity, εv is the gas superficial velocity, and d_P is the particle diameter) in Figure 4.3. In that figure, absolute pressure is shown as a parameter, and results for air (A)–oxygen (B) on zeolite 5A, and nitrogen (A)–helium (B) on activated carbon are also included. Some of the pressures were sufficiently low for the isotherms to be nearly linear, while others were at pressures high enough for isotherm curvature to be significant. Furthermore, the dependence of θ (or β) on Re is similar among the different cases, reflecting the impact of dissipative effects. For each data set, the minimum value of θ corresponds to conditions in which the combined effects of diffusion and dispersion are minimal. At that point, θ is typically found to be larger (worse) than the value predicted from isotherm data alone by about 0.02 to 0.05. The optimal Reynolds number (Re) for all the cases, except nitrogen–helium, is in the range of 9 to 18. The elapsed time per breakthrough experiment is on the order of a few minutes, while batchwise isotherm measurements are much more time consuming.

For certain compositions, when the isotherms are quite nonlinear, there may be a selectivity reversal, indicated by $\theta > 1$. This can occur when the partial pressure of the heavy component is so large that θ_A becomes greater than θ_B. In that case, if a shock front existed, it would begin to disintegrate and a simple wave would form. The resulting breakthrough pattern would have a tail (at high partial pressures) that might be falsely diagnosed as an effect of mass transfer resistance (see Figure 2.23). Generally, other complications are possible due to multiple conceivable values of the velocities and the parameters θ_i and θ. The compositions bounding the shock front are constrained by the influent and effluent compositions, but due to entropic effects they may tend to lie somewhere between those. For multicomponent systems, the phenomenon of "rollup" can cause local maxima of the lighter components, resulting in a shock velocity slower or faster than expected. Subtleties arise because the choice is subject to a uniqueness condition. Applications to PSA have been discussed by Kayser and Knaebel.[4] When the curvature of the isotherms is not too severe (i.e., at low partial pressures), the uniqueness condition will be automatically satisfied, and the shock velocity predicted by Eq. 4.9 will be valid for $y_{A_2} = y_{A_F}$ (i.e., the feed composition).

4.4 Cycle Analysis

Certain preliminaries are essential for predicting PSA performance. First, one must determine basic properties, and among these, for the local equilibrium theory to be accurate, mass transfer resistances must be small. One must then decide on the steps comprising the PSA cycle, and choose operating conditions, such as feed composition, pressures, and step times. At that point, material balances and thermodynamic relationships can predict overall performance in terms of: flow rates, product recovery, byproduct composition, and power requirements.

The key concept involved in applying equilibrium models is that each step is intended to accomplish a specific change. For steps such as pressurization and countercurrent blowdown, the specific change is obvious. Other steps are more subtle because they may proceed until breakthrough is imminent, complete, or some fraction thereof. Such operating policies link the flow rates, step times, bed size, etc. of those steps, and depending on initial and final states, may impact other steps in the cycle. In that sense the goals of the steps are not at all open ended.

As an example of stepwise material balances, the number of moles contained in an influent or effluent stream during any step can be determined from appropriate velocities, as given by Eqs. 4.5, 4.7, 4.9, and 4.10. The choice depends on the nature of the step. The moles added to or removed from the column in each step can be expressed as the integral over time of the instantaneous molar flow rate(s), or as the difference between the final and initial contents of the column. General expressions are:

$$\Delta N_A = \left(\overline{Q_{A_{\text{in}}}} - \overline{Q_{A_{\text{out}}}}\right)t_{\text{step}} = N_A|_{\text{final}} - N_A|_{\text{initial}} \qquad (4.13)$$

$$\Delta N_{\text{TOTAL}} = \left(\overline{Q_{\text{in}}} - \overline{Q_{\text{out}}}\right)t_{\text{step}} = N_{\text{TOTAL}}|_{\text{final}} - N_{\text{TOTAL}}|_{\text{initial}} \qquad (4.14)$$

where the products of the average molar flow rates and time are:

$$\overline{Q_A}t_{\text{step}} = \int_0^{t_{\text{step}}} \frac{\varepsilon A_{\text{CS}} P v y_A}{RT} \, dt \qquad (4.15)$$

$$\overline{Q}t_{\text{step}} = \int_0^{t_{\text{step}}} \frac{\varepsilon A_{\text{CS}} P v}{RT} \, dt \qquad (4.16)$$

For steps in which pressure is constant, the influent flow rate or effluent flow rate can be set to complete the step within the allocated step time. No matter which is specified, the other can be easily determined if both compositions are known, via Eq. 4.5 or 4.10, depending on whether a shock front exists in the column.

For steps in which pressure varies, it is easier to specify the rate of pressure change, because the volumetric flow rate varies as pressure changes. Employing Eq. 4.5, Eq. 4.16 can be written as:

$$\overline{Q}t_{\text{step}} = \int_0^{t_{\text{step}}} Q \frac{dt}{dP} \, dP \qquad (4.17)$$

For some steps, it is convenient to determine the flows from the molar contents of the column, when the composition profiles are known. For either A or B, the contents are:

$$N_i = \int_0^L [\varepsilon c_i + (1 - \varepsilon) f_i(c_i)] A_{\text{CS}} \, dz \qquad (4.18)$$

where f_i is the isotherm function, given by Eq. 4.2. This relation is equivalent to the *column isotherm* mentioned in Section 4.3. In the same vein, the total column contents are obtained by summing the amounts of both components.

$$N_{\text{TOTAL}} = \int_0^L (\varepsilon c + (1 - \varepsilon)[f_A(c_A) + f_B(c_B)]) A_{\text{CS}} \, dz \qquad (4.19)$$

Now these balance equations can be combined to predict the overall performance of some PSA cycles. As discussed in Chapter 3, the simplest PSA cycles employ four steps, so they will be considered first. The steps comprise: pressurization either by feed or product, feed at constant pressure until breakthrough is imminent, countercurrent blowdown, and complete purge (so that all of the heavy component is exhausted). The version employing feed for pressurization is usually called the *Skarstrom cycle*,[6] and is discussed in Section 3.2. Several variations of the Skarstrom cycle have been analyzed via the local equilibrium theory, including steps with incomplete purge,[7-9] simultaneous pressurization and feed and simultaneous feed and cocurrent blowdown,[10] cocurrent blowdown,[11] rinse,[12] etc. As shown in Sections 4.4.1-6, the final results of those models are surprisingly simple. Futhermore, experiments conducted over a wide range of conditions have confirmed their validity, as shown in Section 4.5.

4.4.1 Four-Step PSA Cycle: Pressurization with Product

The cycle covered in this section is shown in Figure 4.1, and it is probably the simplest PSA cycle, at least from a mathematical viewpoint. One result of that simplicity is that it has been possible to extend the equilibrium theory, in closed form, to systems exhibiting nonlinear isotherms.[3] For the sake of clarity, although generality is sacrificed, the discussion is given here in terms of a specific type of mixture, viz., in which the light component has a linear isotherm while the isotherm for the heavy component is nonlinear (e.g., a Langmuir or quadratic isotherm). When both isotherms are practically linear, the equations presented here can be easily simplified, and when both are nonlinear, the preceding equations can be adapted, although the resulting expressions become somewhat more complicated.

As mentioned previously, step times, velocities, and molar flow rates are interrelated through material balances. Therefore, since the influent and effluent moles required for a certain step are fixed by Eqs. 4.13 through 4.19, the choice of step time really only affects the interstitial velocity. For some steps that choice is critical. For example, during the feed step the flow rate affects the apparent selectivity, as suggested in Figure 4.3, as well as pressure drop. In contrast, the time allotted for the purge step is usually chosen in order to synchronize steps occurring in parallel beds. Pressure drop and mass transfer rates are normally of little importance while purging. From a mathematical viewpoint, the velocities in the feed and purge steps are governed by the rates at which the shock and simple waves propagate through the bed, as given by Eqs. 4.5, 4.7, 4.9, and 4.10. These equations relate the interstitial velocity, the length of the bed, the step time, and the column isotherm, as follows:

$$v_{in}t|_F = L/\theta_A \tag{4.20}$$

$$v_{in}t|_{PU} = L/\beta_A \tag{4.21}$$

Note that the feed step is assumed to produce the pure light component at high pressure, so in Eq. 4.20 $\theta_A = \theta_A(P_H, y = 0, y_F)$. Furthermore, since in Eq. 4.21 the purge gas is also presumed to be pure, $\beta_A = \beta_{A_0}$. In this section, the light component is assumed to have a linear isotherm, so in the following treatment $\beta_B = \beta_{B_0} = \theta_B$.

Accordingly, the moles required for the feed and purge steps may be determined from Eqs. 4.13 through 4.21 as

$$\overline{Q_{in}}t|_F = \phi \mathscr{A} \wp \beta_{A_0}/\theta_A \tag{4.22}$$

$$\overline{Q_{in}}t|_{PU} = \phi\beta_{A_0}/\beta_A = \phi, \quad \text{if } y_{A_{in}} = 0 \tag{4.23}$$

where $\phi = \varepsilon A_{CS}LP_L/\beta_{A_0}RT$, $\wp = P_H/P_L$,

$$\beta_0 = \frac{\beta_{A_0}}{\beta_{B_0}} = \left(1 + \frac{1-\varepsilon}{\varepsilon}K_B\right) \bigg/ \left(1 + \frac{1-\varepsilon}{\varepsilon}K_A\right)$$

and K_i is the Henry's law coefficient of component i.

The moles of the pure, light product withdrawn during the feed step may be determined from Eqs. 4.10 and 4.22, as follows

$$\overline{Q_{out}t}|_F = \phi\wp\left[1 + (\theta - 1)y_{A_F}\right]\beta_{A_0}/\theta_A \tag{4.24}$$

where θ_A has the same value as in Eqs. 4.20 and 4.22.

The pressurization with product step in this cycle follows the purge step; so the bed is presumed to contain only the pure, light component. Thus, Eq. 4.17 yields

$$\overline{Q_{in}t}|_{PR} = \phi\beta_0(\wp - 1) \tag{4.25}$$

Consequently, the *rate* of pressurization is immaterial; only the initial and final pressures matter.

The definition of *recovery* of the light component for this four-step cycle, with pressurization by product is:

$$R_B = \frac{\overline{Q_{out}t}|_F - \overline{Q_{in}t}|_{PR} - \overline{Q_{in}t}|_{PU}}{\overline{Q_{in}t}|_F y_{B_F}} \tag{4.26}$$

Thus, by combining Eqs. 4.22 through 4.26 and rearranging, one obtains:

$$R_B = (1 - \theta)\left(1 - \frac{\zeta}{\wp y_{B_F}}\right) \tag{4.27}$$

where $\zeta = 1/[1 + (\xi P_H y_F \beta_{A_0})/(1 - \beta_0)]$ is a factor that deviates from unity only for nonlinear isotherms. For example, when component A follows a quadratic isotherm, $q_A = K_A c_A + M_A c_A^2$ and component B follows a linear isotherm, $q_B = K_B c_B$, one gets

$$\xi = \frac{1 - \varepsilon}{\varepsilon}\frac{M_A}{R}$$

Some specific results of Eq. 4.27 are shown in Figure 4.4 for the case of linear isotherms ($M_A = 0$ and $\zeta = 1$). This figure illustrates the effects of feed composition and pressure ratio on product recovery, for two adsorbent selectivities, $\beta = 0.1$ and 0.9. These selectivities span the range of very easy (e.g., hydrogen purification) to quite difficult (e.g., separation of argon from oxygen), respectively. The results are shown as three-dimensional surfaces that have quite similar shapes, despite the large difference in selectivities. Both surfaces approach an asymptote at high pressure ratios and, to a lesser extent, as the percentage of the heavy component in the feed approaches zero. Conversely, recovery of the light component always decreases as the amount of the heavy component in the feed increases. Later, in Section 4.6, additional comparisons are made that focus on the effect of isotherm curvature.

Another measure of overall PSA performance is the *enrichment* of the byproduct, $E_A = \bar{y}_{A_W}/y_{A_F}$. This may be of interest when the more strongly adsorbed component is valuable. If that component is very valuable, the

EQUILIBRIUM THEORY

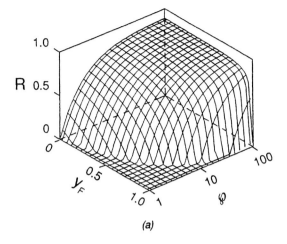

(a)

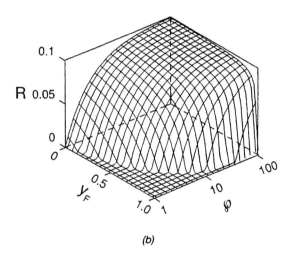

(b)

Figure 4.4 The effect on recovery of feed composition and pressure ratio, for β (a) = 0.1 and (b) 0.9, for pressurization with product.[13]

present four-step cycle (with complete purge) may not be the best choice. More will be said about alternatives later in this chapter. Nevertheless, a simple material balance can be used to determine this parameter, regardless of the operating conditions. The following relation applies for any cycle that splits a binary mixture in which the light component is obtained as a pure product, and the only other effluent stream is the byproduct:

$$E_A = \frac{1}{1 - y_{B_F} R_B} \qquad (4.28)$$

4.4.2 Four-Step PSA Cycle: Pressurization with Feed

What may seem to be the simplest modification of the four-step cycle outlined previously is to pressurize with feed rather than the light product. This arrangement was actually the cycle proposed by Skarstrom.[6] It seems intuitively possible, if not probable, that pressurizing with feed rather than product could produce more net product (i.e., have a higher recovery). That is, on physical grounds it is easy (but deceptive) to regard pressurization as a "parasitic" step, since no product evolves. At first glance, the basic mathematics would seem to confirm these expectations. For example, in the case of pressurization with feed, N_{PR}, the moles consumed for pressurization, appears in the denominator of the definition of recovery, as shown in Eq. 4.29. For the counterpart cycle (pressurization with product), N_{PR} appears as a negative term in the definition of recovery, Eq. 4.26. Thus, in both cases it would appear that pressurization is detrimental to performance. Following that notion, it might be deduced that the relatively less valuable feed should be employed for this purpose, as opposed to the pure product.

Following this reasoning, it is useful to examine pressurization with feed as an alternative to pressurization with product, primarily in order to better understand the PSA cycle, and secondarily to gain insight into the development of equilibrium models. As mentioned earlier, both Shendalman and Mitchell[2] and Chan et al.[3] ignored the effect of the heavy component on the molar flows and velocities. As a result, their models do not distinguish between the amount of gas required for pressurization with feed versus product. Their predicted recovery of the light component, restricted to the case of complete purge, is:

$$R_S \equiv \frac{N_{HP} - N_L}{y_{B_F}(N_H + N_{PR})} = \frac{\wp^{1-\beta} - 1}{y_{B_F}\left[\wp^{1-\beta} + \beta(\wp - 1)\right]} \qquad (4.29)$$

Note that $y_{B_F} \rightarrow 1$ was assumed by both Shendalman and Mitchell and Chan et al., so it would be superfluous in the denominator on the right-hand side in their versions. It is included here only for completeness. In addition, Shendalman and Mitchell assumed that $\beta = \beta_{A_0}$, and $\beta_B = 1$, while Chan et al. assumed $\beta = \beta_{A_0}/\beta_{B_0}$.

EQUILIBRIUM THEORY

Removing the restrictions on feed composition and including the impact of sorption on the interstitial gas velocity leads to a more widely applicable and accurate model for most PSA systems. That approach was followed by Knaebel and Hill[13] for a system having linear isotherms. Their relation to predict recovery of the light component, also restricted to the case of complete purge is:

$$R_B = \frac{\wp\Omega - 1}{y_{B_F}[\wp\Omega + \beta(A\wp - 1)]} \qquad (4.30)$$

In this equation the parameter, Ω ($= \Theta$ in the original paper), is determined by integration via Runge–Kutta or a similar approach, and is somewhat involved. In general, $\Omega \propto \wp^{-\beta} y_{B_F}$ when $\beta \to 0$ (e.g., 0.1), while for larger values of β, Ω is somewhat larger than that product. Since Ω is not a simple function of feed composition, adsorbent selectivity, or operating pressures, one might expect that the dependence of recovery would be equally complex. The discrepancies between the models, due to differences in their inherent assumptions, are discussed in Section 4.6. In addition, some of the subtleties of the pressurization step are discussed in Section 4.9.

As in the previous section, specific results calculated from Eq. 4.30 are shown in Figure 4.5, for the case of linear isotherms. That figure illustrates the effects of feed composition and pressure ratio on product recovery for two adsorbent selectivities, $\beta = 0.1$ and 0.9, which span the range of very easy to quite difficult PSA applications. The results, again are shown as three-dimensional surfaces that in this case have quite different shapes, due to the difference in selectivities. As for pressurization with product, recovery of the light component always decreases as the amount of the heavy component in the feed increases. In addition, both surfaces approach an asymptote at high pressure ratios, but, for the low-selectivity case ($\beta = 0.9$), there is a ridge representing maximum recovery at low pressure ratios.

Another relevant issue is the inherent differences between the pressurization methods discussed. Though there may be differences in mechanical complexity and other details, the most significant difference is, more than likely, between the recoveries of the light product. To expand on that point, Figure 4.6 shows the incremental improvement in recovery for pressurization by product versus pressurization by feed as affected by feed composition and pressure ratio, again for $\beta = 0.1$ and 0.9. The comparison is again limited to systems having linear isotherms. As can be seen, regardless of the conditions and parameters, the recovery of the light component that is attainable by pressurization with product is generally superior to that obtainable by pressurization with feed. The percentage difference is small for systems with high selectivities, but grows larger as selectivity drops. Perhaps surprisingly, the difference increases as the pressure ratio ($\wp$) increases.

This result underscores the fallacies of the previous intuitive arguments in favor of pressurization by feed, and shows that it is a misconception to view pressurization as a "parasitic" step. The primary underlying principle is that,

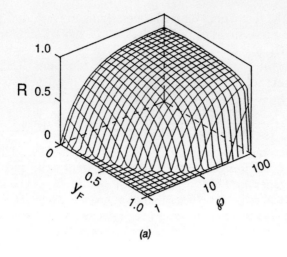

(a)

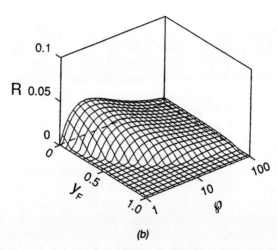

(b)

Figure 4.5 The effect on recovery of feed composition and pressure ratio, for β (a) = 0.1 and (b) 0.9, and for pressurization with feed.[13]

EQUILIBRIUM THEORY

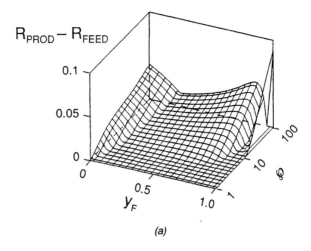

(a)

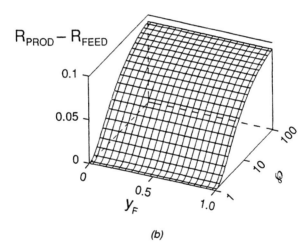

(b)

Figure 4.6 Difference in recoveries for pressurization with product versus pressurization with feed as a function of feed composition and pressure ratio, for β (a) = 0.1 and (b) 0.9.[13]

when the adsorbent is pressurized with feed, the adsorbent near the feed end contacts the feed at essentially the lowest pressure in the cycle. This allows the more strongly adsorbed component to penetrate farther into the bed, because less is adsorbed by that adsorbent than if the gas were fully pressurized. As a result, that adsorbent is less than fully utilized. Thus, despite the fact that pressurization does not contribute directly towards production, it can diminish the useful capacity of the adsorbent.

4.4.3 Four-Step PSA Cycle: Incomplete Purging

The two cycles considered so far in this section are simple, and relatively easy to analyze. Unfortunately, however, they are not particularly efficient in terms of their performance relative to power requirements. That is, they generally require high pressure ratios to attain high recoveries. For this section, we consider a simple modification of the four-step PSA cycle described in Section 4.4.1. It will be seen that this modification, which simply involves varying the extent of purging, can lead to remarkably higher recovery at relatively low pressure ratios.

Incomplete purging has been common in industrial practice, as mentioned by Wagner[14] and Wankat.[15] Quantitative studies of the extent of purge have focused mainly on flow rate ratios, especially the purge-to-feed ratio. For example, the effects of the purge-to-feed ratio on light-product purity were studied by Yang and Doong,[16] Doong and Yang,[17] and Yang.[18] Their results implied that it was not feasible to increase recovery by decreasing the purge-to-feed ratio and still maintain high product purity. Kirkby and Kenney[19] showed theoretically and experimentally that there is an optimum extent of purging, for which product purity and recovery are maximized. Their cell model suggested that the optimum corresponded to complete purge, but their experiments revealed that a lesser amount was optimal. An equilibrium-based model was developed by Matz and Knaebel[7] to assess the effect of purge on PSA performance, based on systems having linear isotherms, and that work is the basis of the following discussion. Recently, Rousar and Ditl[9] solved the same basic equilibrium-based equations analytically to determine the optimum purge amount, and they examined operation in a regime that yields impure light product.

By its nature, incomplete purging leaves a composition *tail*, or *heel*, at the feed end of the column, containing some of the more strongly adsorbed component. Subsequent pressurization with product (also countercurrent to the feed) pushes that residual material toward the feed end. More subtly, it also reduces the gas-phase mole fraction of the more strongly adsorbed component in that region, since the heavy component is preferentially taken up as pressurization proceeds. The presence of the compressed tail reduces the quantities of both the feed admitted and the gross product obtained during the feed step. The relative amounts consumed and/or produced during each step depend on the extent of purge, as well as on the conditions

EQUILIBRIUM THEORY

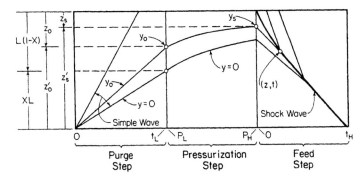

Figure 4.7 Paths of characteristics in the purge, pressurization, and feed steps, for a fractional extent of purge of X. Note: z' is measured from the bottom toward the top, and z is measured in the reverse direction.[7]

and parameter values. Generally, reducing the amount of purge always results in increased recovery, but beyond some limit (described below) there is no assurance that pure product can be obtained.

It may help to visualize the action of the steps in terms of wave movements, as in the discussion of Figure 4.1. A schematic diagram of the wave movements in the relevant steps is given in Figure 4.7, although in this figure the shading that represents the influx of the heavy component has been omitted. The purge step is still characterized by a simple wave. This wave spreads as it propagates over the length of the column, as shown on the left-hand side of the diagram.

The key feature of the extent of purging is the fraction of the column X that is completely purged. That is, X is the fraction of L over which $y = 0$ at the end of the purge step, t_L. Because they are linearly related, this is identical to the fraction of the amount of gas required to purge the bed completely, via Eqs. 4.21 and 4.23, as

$$X = \frac{\beta_A v_{in} t|_{PU}}{L} = \frac{\overline{Q}t|_{PU}}{\overline{Q}t|^*_{PU}} \tag{4.31}$$

where $\beta_A v_{in} t|_{PU}$ is the distance into the bed that is fully purged and $\overline{Q}t|^*_{PU}$ is the number of moles required for complete purge.

An arbitrary characteristic in the simple wave is denoted by its composition, y_0. The one that just reaches the effluent end of the column as the purge step ends is special. Its mole fraction is called $y_0|_{z=L}$. An operational constraint is that this should be less than the "expanded" feed mole fraction, or, if it is not, when it is repressurized it will simply revert to the feed composition. In other words, if this constraint is not met, regeneration will be incomplete, and the effective length of the column will be reduced, leading to premature breakthrough. Mathematically, this amounts to the following inequality $y_S|_{z=L} < y_F$. Generally, the value of y_S can be determined from its initial composition and the imposed pressure ratio,

$$y_S|_z = \left[y_0 \left(\frac{1-y_S}{1-y_0} \right)^\beta \right]_z \wp^{\beta-1} \tag{4.32}$$

The ultimate axial position, denoted z_S, is coupled to the original position z_0 by

$$z_S = z_0 \left(\frac{y_S}{y_0} \right)^{\beta/(1-\beta)} \left(\frac{1-y_0}{1-y_S} \right)^{1/(1-\beta)} \left(\frac{1+(\beta-1)y_S}{1+(\beta-1)y_0} \right) \tag{4.33}$$

Any greater extent of purge than the amount indicated by this inequality will drive off a sufficient portion of the more strongly adsorbed component so that net product is possible. At any rate, the composition at the end of the purge step can be determined from the fractional extent of purge as follows,

$$y_0|_{z=L} = \frac{1-X^{1/2}}{1-\beta} \tag{4.34}$$

Thus, beginning with values of X and y and inserting them into Eqs. 4.7 and 4.34, one can determine the composition profile in the column at the end of the purge step. One can subsequently employ Eqs. 4.32 and 4.33 to predict, by tracing characteristics, the profile after pressurization.

The feed step is affected by the profile in the column at the end of pressurization because characteristics having composition y_S encounter characteristics at the feed composition, forming a shock wave. Since the composition at the leading edge varies nonlinearly along the shock path, it may be necessary to determine the path by integration using a Runge–Kutta routine.

Since the composition profile at the beginning of the feed step is complicated, it is conceivable that variations of composition and velocity, along with the differing adsorption selectivities of the components, could lead to unusual waveforms (e.g., the formation of double shock fronts), which are possible for a single adsorbate that has a Type IV isotherm. If that were the case, column behavior would be difficult to understand and analyze. Applying the *entropy condition* and the method of characteristics, however, leads to the conclusion that multiple shock waves cannot occur at conditions typically encountered in PSA cycles.[7]

Having summarized the necessary modifications to the basic model, and discussed some of the subtleties, it is appropriate to look at some results. The recovery of pure product can be predicted for this cycle by combining the foregoing analysis with Eq. 4.26. From an engineering standpoint, the most interesting cases to consider are those were dramatic improvements are possible, for example, in relation to the simpler cycles discussed earlier. Perhaps the most interesting type of application at the present state of PSA technology is the situation in which both the feed composition and adsorbent selectivity are moderate. Separation of oxygen from air using zeolite 5A is a realistic example of such a system. Air is composed mainly of nitrogen

(78.03%), oxygen (20.99%), and argon (0.94%) (for simplicity the minor constituents are omitted here), and the temperature is taken to be 45°C, which assures that the isotherms are nearly linear (up to about 6 atm). Furthermore, the adsorption isotherms of argon and oxygen on 5A zeolite practically coincide, so argon and oxygen are not separated. The adsorbent–adsorbate interactions are characterized by $\beta = \theta = 0.593$ (Kayser and Knaebel[20]).

Two types of comparisons are possible: fixing the extent of purge and varying the pressure ratio, or vice versa. The results are shown in Figures 4.8 and 4.9, respectively. The former shows extents of purge of 100% and 50%, and pressure ratios from 1.45 to 100. The recovery based on complete purge passes the break-even point at a pressure ratio of 4.6, reaches 22% at a pressure ratio of 10, and approaches about 39% as the pressure becomes very large. Conversely, at 50% purge the recovery at a pressure ratio of 1.45 is 23%, rises to nearly 39% at a pressure ratio of 10, and attains the maximum value of about 40% at a high pressure ratio. Figure 4.9 shows pressure ratios of 2.0 and 4.0, with extents of purge from 45% to 100%. Both cases show nil recovery for high extents of purge, and over 30% recovery as the extent of purge reaches the minimum value for which pure product is attainable.

To summarize these results: reducing the extent of purge to about 50% of completion allows recovery to pure oxygen at low pressure ratios. As the pressure ratio increases, the improvement is still significant, though the

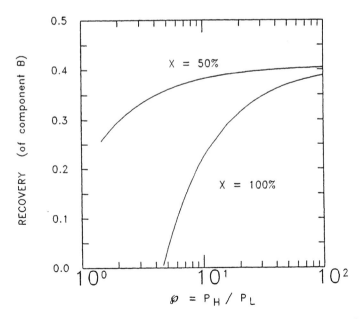

Figure 4.8 Effect on light product recovery of pressure ratio, for extents of purge of 50% and 100%, for $\beta = 0.593$ and $y_F = 0.78$.

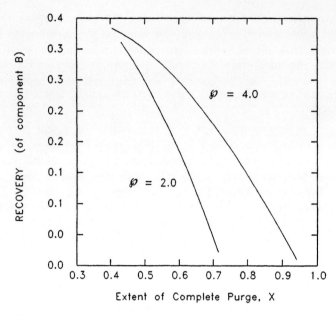

Figure 4.9 Effect of extent of purge on light product recovery for pressure ratios of 2.0 and 4.0, for $\beta = 0.593$ and $y_F = 0.78$.

returns diminish. The only potential advantage foreseen for purging more than the minimum amount is to compensate for any transport resistances or dispersive effects that could cause contamination of the product. Such effects would be greater for faster cycling, so there is bound to be an optimum at which product purity and recovery are balanced against adsorbent productivity and the power requirement.

4.4.4 Four-Step PSA Cycle: Pressure Variation During Feed

Modifications of certain PSA steps could lead to simplified equipment or superior performance. For example, the simplest PSA cycle is a two-step cycle that combines the pressurization and feed steps, and the blowdown and purge steps. This cycle requires the minimum number of valves and very simple control logic. Conversely, a number of studies have shown that cocurrent blowdown can significantly increase the recovery of the light product. Therefore, it seems promising to combine the feed and blowdown (cocurrent) steps, even though doing so would involve some mechanical complications. The local equilibrium theory is a natural choice to study such cycles because it can focus on the impact of major parameters and operating conditions, without the intrusion of extraneous effects which would involve adjustable parameters.

The interstitial flow rate can be found from the total material balance (i.e., the sum of Eq. 4.1 or 4.4 for both components). These equations can be integrated with boundary conditions (y_F, v_F, specified at the feed end of the bed) in order to evaluate the velocity and composition at other points in the adsorbent bed. The individual component balance for component A, can then be solved, yielding results that are slightly more complicated characteristics than those described by Eqs. 4.7 and 4.8:

$$\frac{dz}{dt} = \frac{\beta(\psi_O - P'z)}{P[1 + (\beta - 1)y_A]^2} \tag{4.35}$$

$$\frac{dy_A}{dz} = \frac{(\beta - 1)[1 + (\beta - 1)y_A](1 - y_A)y_A}{\beta[(\psi_O/P') - z]} \tag{4.36}$$

where $P' = dP/dt = (\psi_O - \psi_F)/L$, and $\psi_O = \beta_B v_O P[1 + (\beta - 1)y_{AO}]$, in which the subscript O refers to the outlet, and the subscript F refers to the feed end of the packed bed. When pressure varies linearly with time, P' is constant, as are the molar flow rates. A hypothetical problem could arise if imposing a pressure shift caused the shock wave to degrade into a diffuse front. The step time would then have to be shortened to maintain high product purity, which would reduce recovery. On that point, Kayser and Knaebel[10] concluded via the *entropy condition* that unless the pressure shift causes a dramatic increase in pressure drop or mass transfer resistance, the equilibrium tendency should preserve the shock front.

When the pure, less strongly adsorbed component, B, is used to pressurize the column from P_L to P_F, and when the feed step involves a pressure shift (e.g., partial pressurization by feed or partial cocurrent blowdown) to P_H, the recovery of the pure, less strongly adsorbed component can be expressed as:

$$R = \frac{1 + \left(\beta \wp_I^{\beta-1} - 1\right)y_{A_F}}{\wp_I^{\beta}(1 - y_{A_F})} \tag{4.37}$$

$$\times \left[1 - \frac{1/\beta + (\wp_F - 1)}{\wp_H - \wp_F}\left(\frac{\wp_I^{\beta}[1 + (\beta - 1)y_{A_F}]}{1 + \left(\beta \wp_I^{\beta-1} - 1\right)y_{A_F}} - 1\right)\right]$$

where $\wp_I = P_H/P_F$, $\wp_F = P_F/P_L$, and $\wp_H = P_H/P_L$. Only two pressure ratios of the three mentioned are independent; the latter quantities are preferred because they are constrained to be greater than unity. Suh and Wankat[21] studied separate feed and cocurrent blowdown steps, and found that the distinct steps can yield better recovery than when combined.

Figure 4.10 shows predictions of product recovery as affected by the latter pressure ratios. Three cases are shown involving separations that are "difficult," either because the feed is predominately the heavy component, or because the adsorption selectivity is poor, or both. These are represented by: (a) $\beta = 0.1$ and $y_F = 0.9$; (b) $\beta = 0.9$ and $y_F = 0.1$; and (c) $\beta = 0.9$ and

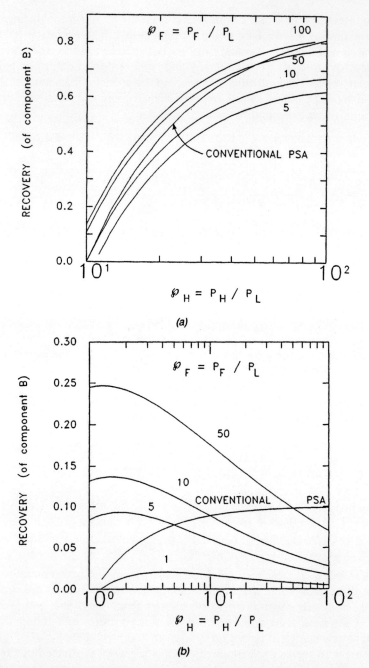

Figure 4.10 Recovery as affected by pressure ratios, for a four-step cycle in which the feed step pressure rises or falls. (a) $\beta = 0.1$, $y_F = 0.9$; (b) $\beta = 0.9, y_F = 0.1$; (c) $\beta = 0.9, y_F = 0.9$.

EQUILIBRIUM THEORY

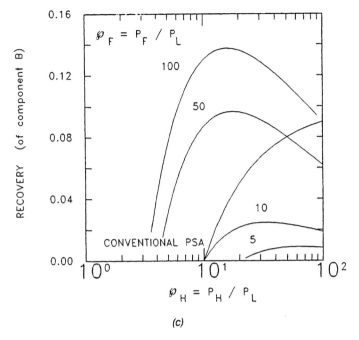

Figure 4.10 (*Continued*).

$y_F = 0.9$, respectively. It can be seen that recovery always increases by combining feed and partial cocurrent blowdown: the increase can be dramatic for a large value of $\wp_F$ and a low or moderate value of $\wp_H$. Conversely, recovery always declines when pressurization and feed are combined (i.e., $\wp_F < \wp_H$). The first plot shows the case of excellent selectivity, but the feed is very contaminated with the heavy component; the results suggest that, for this system, combined feed and cocurrent blowdown will yield only a small improvement over conventional PSA. The second plot, in which the selectivity is poor and the feed is predominately the light component, shows that very large improvements in recovery are possible for combined feed and cocurrent blowdown. For example, the maximum recovery via conventional PSA is 10% (at a very high pressure ratio). Combining feed and cocurrent blowdown can match that recovery at a pressure ratio of only 6, or vastly exceed it (e.g., reaching 25% recovery at a pressure ratio of 50). The third example shows a system having poor adsorbent selectivity and heavily contaminated feed. Recovery is improved for this case, too, by, for example, initially pressurizing to a pressure ratio of 50 then feeding while blowing down cocurrently to a pressure ratio of 15. The result exceeds the maximum recovery of the conventional cycle (at very high pressure ratios). Increasing the initial pressure ratio to 100 yields a 50% increase in recovery. The advantages of increased recovery must, of course, be weighed against the

4.4.5 Five-Step PSA Cycle: Incorporating Rinse and Incomplete Purge

In this section a rinse step is added to the four-step PSA cycle discussed in Section 4.4.3. The added rinse step follows the feed step, and it begins by admitting the pure heavy component to the bed. This displaces the residual feed, which is recycled. In so doing, the adsorbent bed becomes saturated with the heavy component. Therefore, during the blowdown step, the heavy component is recovered as the pressure drops from P_H to P_L. At least part of the heavy component must be recompressed (to P_H) for use in the subsequent rinse step. The cycle is shown schematically in Figure 4.11.

Though called rinse, the action of this step could also be thought of as a high-pressure purge. A major formal distinction between rinse and purge is that rinsing involves a composition shock wave, while purging involves a simple wave. Arguments can be made for directing the flow during the rinse step either cocurrent or countercurrent to the feed. Factors such as mechanical complexity and product purity affect the choice. For now, since the mathematical model to be discussed assumes local equilibrium, implying that ideal shock fronts exist during the feed and rinse steps, the direction does not affect performance. To be definite (and to favor the mechanically simpler version), the rinse flow is taken to be counter to that of the feed.

The present PSA cycle also includes incomplete purging. The equations that govern the purge, pressurization, and feed steps in this cycle are identical to those that apply in the four-step cycle covered in Sections 4.4.1 and 4.4.3. Similarly, the equations that govern the rinse step in this cycle are analogous to those for the feed step. That is, the relation between the interstitial velocity, the length of the bed, the step time, and the column isotherm is obtained from Eqs. 4.5, 4.7, 4.9, 4.10, and follows along the same lines as Eq. 4.20:

$$v_{\text{out}} t|_R = L/\theta_B \tag{4.38}$$

Similarly, the corresponding expression involving the rinse step molar effluent rate, step time, pressure ratio, and coefficients that represent geometry and adsorbent–adsorbate interactions is similar to Eq. 4.22. As a result, the molar quantities in the effluent and influent are:

$$\overline{Q_{\text{out}}} t|_R = \phi \wp \beta_{A_0}/\theta_B \tag{4.39}$$

$$\overline{Q_{\text{in}}} t|_R = \phi \wp \left[1 + (\theta - 1) y_{A_F}\right] \beta_{A_0}/\theta_A \tag{4.40}$$

where $\theta = \theta(P_H, y_F, y = 1)$. Note that the form of Eq. 4.40 is identical to that of Eq. 4.24. For nonlinear isotherms, since the parameters are evaluated under different conditions, the results may be different.

EQUILIBRIUM THEORY

5-STEP PSA CYCLE

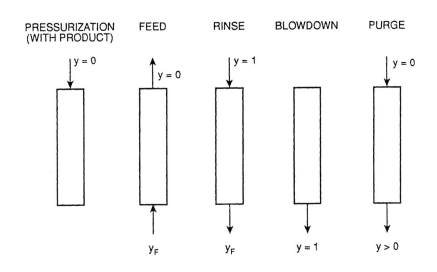

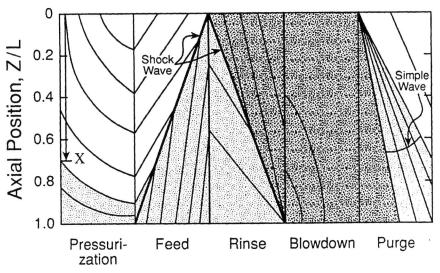

Figure 4.11 Five step cycle-including a rinse step. (Top) Flows and composition associated with each step. (Bottom) Position-verus-time representation of each step. Shaded region depicts penetration heavy component. X denotes fraction of completion of purge step.

The blowdown step is critical to this cycle since it is the source of the heavy component product. It follows the rinse step, so the bed is presumed to contain only the pure, heavy component. The most direct approach to determine the net effluent is simply to determine the difference the initial and final states, according to Eq. 4.18 (since the contents are assumed to be pure A). The result is:

$$\overline{Q_{out}t}|_{BD} = \phi \beta_{A_0} \left(\frac{\wp}{\theta_A(P_H, y = 0, y = 1)} - \frac{1}{\theta_A(P_L, y = 0, y = 1)} \right) \tag{4.41}$$

Thus, as for pressurization, the *rate* of blowdown is immaterial; only the initial and final pressures matter.

The light and heavy product recoveries from the current five-step cycle, based on complete purge and nonlinear isotherms are:

$$R_B = 1 - \frac{\theta_A/\beta_{A_0} - \theta}{\wp y_{B_F}(1 - \theta)} \tag{4.42A}$$

where $\theta = \theta(P_H, y_F, y = 0)$ and $\theta_A = \theta_A(P_H, y_F, y = 0)$.

$$R_A = \frac{\wp[\varphi_1 + (1 - \theta)y_F - 1] - \varphi_2}{\wp y_F(\varphi_3 - \theta)} \tag{4.43}$$

where $\theta = \theta(P_H, y = 1, y_F)$, $\varphi_1 = \theta_A(P_H, y = 1, y_F)/\theta_A(P_H, y = 1, y = 0)$, $\varphi_2 = \theta_A(P_H, y = 1, y_F)/\theta_A(P_L, y = 1, y = 0)$, and $\varphi_3 = \theta_A(P_H, y = 1, y_F)/\theta_A(P_H, y_F, y = 0)$.

For linear isotherms, these equations simplify to:

$$R_B = 1 - \frac{1}{\wp y_{B_F}} \tag{4.44}$$

$$R_A = 1 - \frac{1}{(1 - \beta)\wp y_{A_F}} \tag{4.45}$$

In this cycle, just as for the four-step cycle mentioned in Section 4.4.3, the recovery of the light component can be improved by reducing the amount of pure light product consumed in the purge step. The analysis of incomplete purge here is identical to that presented earlier, since the steps are basically the same. A minor detail is that, at the outset of the purge step, more of the heavy component remains in the column than in the four-step cycle, when the blowdown step follows the feed step.

The present cycle differs somewhat from the cycle suggested by Sircar,[22] which also includes a rinse step and is described in Section 6.4. Differences between the cycles are mainly in the details, such as the flow direction in certain steps. Both cycles were reduced to practice for the purpose of splitting air, and the ranges of pressure ratios are nearly identical, implying that the power required would be equivalent. Experimentally, however, these cycles performed differently, and the resulting recoveries and product purities are compared in Section 4.5. That section also compares the experimental results with the theoretical predictions presented here [see Figures 4.14(a) and 4.14(b)].

4.4.6 Four-Step PSA Cycle: Columns with Dead Volume

This section covers a method for estimating the magnitude of effects caused by dead volume, via an equilibrium model. The effects of dead volume are diverse—they vary depending on which end of the adsorbent bed is affected, and they depend on which step of the cycle is being considered. As shown here, the effects are also typically more severe when adsorbent selectivity is poor.

Notwithstanding the last assumption stated in Section 4.1, an inescapable feature of adsorption columns is dead volume at both the feed end and product end of the fixed bed. The primary reason is that, if plug flow is to be established in the adsorbent bed, the gas must be allowed uniform access in the direction of flow. For an ordinary cylindrical column, with axial feed and discharge nozzles and bad retention plates, this could amount to a dead volume (at each end) of 5 to 10% or more of the volume of the adsorbent bed.

Breakthrough curves obtained from most commercial columns exhibit "rounding" arising from axial dispersion, mass transfer resistance, or back mixing before or after the adsorbent bed. If dead volume is largely responsible for the rounding, it should be easy to diagnose. At a given pressure, temperature, velocity, etc., one can simply compare the breakthrough curve of a conventional column with one in which dead volume has been minimized. Examples of the latter, shown in Figure 4.2, were obtained with commercial adsorbents at velocities and pressures typical of commercial systems, but in a column in which dead volume was minimized. Of course, if rounding exists, it could also be an artifact of the slow response of the sampling instrument.

Dead volume at the product end of a PSA column affects the steps differently, as follows:

Blowdown. If, during the previous feed step, breakthrough had not begun, retained pure product partially purges the bed during blowdown; if breakthrough had begun, the gas that expands from this volume is merely additional (unnecessary) waste.

Purge. If breakthrough had not begun during the feed step, less gas is consumed during purge because of the contribution during blowdown; if breakthrough had begun, excess gas is necessary to cleanse this volume before the adsorbent can begin to be purged.

Pressurization. Extra pure light product is required to pressurize this space, though it is recovered during the feed step.
Feed. No effect except to retain part of the product gas, if breakthrough was prevented; if breakthrough had begun, mixing in this region dampens the contamination of the product stream leaving the column.

In comparison, dead volume at the feed end of a PSA column has the following effects:

Blowdown. No effect except to retain part of the waste gas.
Purge. Same as blowdown.
Pressurization. Excess pure light component is necessary to pressurize this space.
Feed. Feed gas mixes with the pure light component, resulting in a "diffuse" front entering the adsorbent bed.

Thus, there are several disadvantages to dead volume, and few advantages. In fact, the only positive aspect is the possibility that the dead volume at the product end of the column may hold sufficient pure product to partially purge the bed during blowdown. That benefit is balanced by the fact that the gas will be less effective at purging the column than gas admitted during the purge step (at low pressure), because it needlessly expels gas while desorption is in progress.

The following development is based on linear isotherms and the four-step cycle described in Section 4.4.1. The cycle is composed of pressurization with product, constant pressure feed, countercurrent blowdown, and complete purge. To the extent possible, both feed-end and product-end dead volumes are considered. Again in this section, compositions are expressed in terms of the heavy component, A. Some of the concepts to be presented have been examined by Kolliopoulos.[23]

For simplicity, the analysis begins with the blowdown step. This is because the gas retained in the product-end dead volume and exhausted in that step contributes to purging. It is presumed that the feed step stops at the point of imminent breakthrough, so that the gas retained in that dead volume, having volume $= V_{DV_P}$, is not contaminated (i.e., $y_{DV_P} = 0$). The approach taken is to determine the extent of purging that occurs during blowdown, and then to determine the additional amount needed during the purge step. The amount of pure light component required for pressurization is increased by that needed to fill the dead volumes. Under these assumptions, the feed step is not directly affected by product-end dead volume, but only by the feed-end dead volume.

To predict the composition profile at the conclusion of blowdown, one must follow characteristics representing different initial compositions and axial positions. The characteristics velocity is given by Eq. 4.35. Perhaps the most important characteristic is the one that identifies the extent of purging due to expansion of gas from the dead space:

$$\left.\frac{dz}{dt}\right|_{y=0} = \frac{\beta}{P}(\psi_0 - P'z) \tag{4.46}$$

where $\psi_0 = \beta_B v_0 P$, which is nil if $V_{DV_P} = 0$, $P' = (P_L - P_H)/t_{BD}$, and t_{BD} is the time allotted for blowdown. This can be rearranged to get the distance into the bed (measured from the product end) that is completely purged,

$$z_0 = z|_{y=0} = \frac{\psi_0}{P'}(1 - P^{-\beta}) \tag{4.47}$$

A material balance for the dead volume demands that

$$\psi_0 = \beta_B V_{DV_P} P'/A_{CS}\varepsilon = \beta_B \lambda_P L P'/\varepsilon \tag{4.48}$$

where $\lambda_P = V_{DV_P}/V_C$ and $V_C = \pi d_C^2 L/4$ is the volume of the adsorbent bed. When this is combined with the preceding relations, the fraction of complete purge that is attained during blowdown is found.

$$X = \frac{z_0}{L} = \frac{\lambda_P \beta_B}{\varepsilon}(1 - \wp^{-\beta}) \tag{4.49}$$

The bed is assumed to have been saturated with feed at high pressure during the preceding feed step. If the feed-end dead volume is very large, however, the actual concentration of A may be less. That possible discrepancy is neglected here. Thus, as blowdown proceeds, the residual gas becomes enriched in component A and is pushed towards the outlet end. The composition shifts to y_{BD}, as in Eq. 4.32, which yields:

$$y_{BD} = 1 - (1 - y_F)\left(\frac{y_{BD}}{y_F}\wp^{\beta-1}\right)^{1/\beta} \tag{4.50}$$

The characteristic having this composition propagates according to Eq. 4.33, which may be combined with a material balance to obtain:

$$X^* = \left.\frac{z}{L}\right|_{y_{BD}} \tag{4.51}$$

$$= \frac{\lambda_P \beta_B}{\varepsilon}\left[1 - \frac{1 + (\beta - 1)y_{BD}}{1 + (\beta - 1)y_F}\left(\frac{1 - y_F}{1 - y_{BD}}\right)^{1/(1-\beta)}\left(\frac{y_{BD}}{y_F}\right)^{\beta/(1-\beta)}\right]$$

The dimensionless distance, X^*, reached by the expanding feed gas is measured from the product end. Thus, Eqs. 4.49 and 4.51 define the partially purged region between completely purged and expanded feed.

Before proceeding, it may be enlightening to consider the potential impact of dead volume at both ends of the column, and the variety of possibilities that arise. For example, the adsorbent selectivity, feed composition, pressure ratio, and size of the product-end dead volume all affect the ultimate position of the partially purged region. So, assuming that there is dead volume at the feed end, the contents of this space may be: unaffected, partly affected, or completely purged as a result of expansion of the pure light component from the product-end dead volume. Similar additional possibilities arise in other steps of the cycle, leading to many permutations and contingencies. Hence, the overall problem is difficult to generalize.

To determine the effluent composition during the countercurrent blowdown step, the material balance for the feed-end dead volume is combined with the foregoing balances of the adsorbent bed and product-end dead volume. The former yields $y_{BD}(P(t))$ at $z = L$ via Eq. 4.50. The overall balance equation is:

$$\frac{V_{DV_F}}{RT}\frac{d(Py_{out})}{dt} = \frac{P}{RT}\left[\dot{V}_L y_{BD}(P(t)) - \dot{V}_{out} y_{out}\right] \quad (4.52)$$

By applying the chain rule for differentiation and simplifying, this reduces to:

$$\frac{dy_{out}}{dt} = \frac{\varepsilon/\beta_B - \lambda_P}{\lambda_F}\left(\frac{y_{BD} - y_{out}}{(1-\beta)(1-y_{BD})y_{BD}}\right) \quad (4.53)$$

where $\lambda_F = V_{DV_F}/V_C$. This expression may be integrated either analytically or via a Runge–Kutta routine. The result of the former approach is given by Eq. 4.54, although in terms of time and effort, the numerical approach may well be easier.

$$y_{BD_{out}} = y_F\left(\frac{1-y_{BD}}{1-y_F}\frac{y_F}{y_{BD}}\right)^\eta \left(1 - \eta \sum_{i=0}^{\infty}\frac{y_F - 1}{\prod_{j=0}^{i}(\eta - j)}\right) \quad (4.54)$$

$$+ \eta y_{BD}\sum_{i=0}^{\infty}\frac{y_{BD} - 1}{\prod_{j=0}^{i}(\eta - j)}$$

where $\eta = (\varepsilon/\beta_B - \lambda_P)/\lambda_F$.

The dead volume at the feed end is assumed to be well mixed, and thus cannot be *completely* purged. For practical purposes, it is sufficient to purge the adsorbent completely. In fact, in view of the earlier discussion of incomplete purge, even this is often unnecessary. The quantity of light component essential for complete purging of the adsorbent bed is:

$$\overline{Q_{in}}t|_{PU} = \phi\left(1 - \frac{\beta_B\lambda_P}{\varepsilon}(1 - \wp^{-\beta})\right) \quad (4.55)$$

This is less than the amount required for a column having $V_{DV_P} = 0$, by the fraction of the bed that had been purged by residual product during the blowdown step.

To continue the analysis, it is necessary to find the final composition in the feed-end dead volume during the purge step. Two phases are relevant: during the first, the heavy component concentration increases (reflecting residual material in the feed-end dead volume), and during the second there is diminishing concentration of the heavy component (which continues until the adsorbent is purged). The column outlet concentration during the first phase is given by:

$$y_{PU_{out_I}} = y_{BD} - (y_{BD} - y_{BD_{out}}) \\ \times \exp\left(-\frac{\varepsilon}{\beta_A \lambda_F}[1 + (\beta - 1)y_{BD}](1 - X^*)\right) \quad (4.56)$$

The second phase is characterized by gradual cleansing of the feed-end dead volume, which depends on the relative volume at the product end, as follows:

$$y_{PU_{out_{II}}} = y_{PU_{out_I}} e^{\alpha(\theta_0 - 1)} + \frac{1}{1 - \beta}\left[\alpha e^{-\alpha \theta_0} - \left(1 - \frac{1}{\alpha}\right)(1 - e^{\alpha(\theta_0 - 1)})\right] \quad (4.57)$$

where $\alpha = 2\varepsilon/\lambda_F \beta_A$, $\theta_0 = (t^*/t_L)^{1/2} = [1 + (\beta - 1)y^*](1 - X^*)^{1/2}$, and X^* is given by Eq. 4.51. When the product-end dead volume is small, the ultimate composition leaving the column during purge is:

$$y_{PU_{out_{II}}} = \left(y_{PU_{out_I}} - y^* - \frac{\beta_A \lambda_F}{2\varepsilon(1 - \beta)}\right)\exp\left(-\frac{2\varepsilon}{\beta_A \lambda_F}(1 - \beta)y^*\right) \\ + \frac{\beta_A \lambda_F}{2\varepsilon(1 - \beta)} \quad (4.58)$$

The pressurization step demands excess material to pressurize the dead volume. The precise amount, according to the ideal gas law, is independent of composition. Nevertheless, at the feed end, since material flows into the dead volume as pressure increases, the composition changes. The moles required for pressurization become:

$$\overline{Q}t|_{PR} = \phi(\wp - 1)\left(\beta + \frac{\beta_A(\lambda_F + \lambda_P)}{\varepsilon}\right) \quad (4.59)$$

The composition in the feed-end dead volume at the end of pressurization becomes:

$$y_{PR_{final}} = \frac{y_{PU_{out_{II}}}}{\wp} \quad (4.60)$$

Finally, the feed step is potentially somewhat involved due to the presence of the heavy component in the feed-end dead volume. The feed immediately begins to shift that composition, but it displaces material into the bed simultaneously. The composition in the feed-end dead volume varies with time according to:

$$y_{in} = y(z = 0, t') = y_b(t) = y_F + (y_{PR_{final}} - y_F)e^{-(\varepsilon v_F A_{CS}/V_{DV_F})t'}$$
(4.61)

Material at this composition enters the bed at t', and it reaches the shock wave at t. The shock wave velocity varies as the composition at its trailing edge varies, that is,

$$\left.\frac{dz}{dt}\right|_{SH} = \beta_A v_F \frac{1 + (\beta - 1)y_F}{1 + (\beta - 1)y_b}$$
(4.62)

Likewise, the velocity of the characteristic that intersects the shock wave at the trailing edge varies depending on its composition, as follows:

$$\left.\frac{dz}{dt}\right|_{y_b} = \beta_A v_F \frac{1 + (\beta - 1)y_F}{[1 + (\beta - 1)y]^2} = \frac{z}{t - t'}$$
(4.63)

Solving Eqs. 4.61 and 4.63 simultaneously for y_b yields an expression for t' in terms of z and t. The path of the shock wave can be determined by integrating Eq. 4.62, for example

$$\left.\frac{dz}{dt}\right|_{SH} = K[1 + (\beta - 1)y_b] = f(y_b(z, t, t'(z, t)))$$
(4.64)

To simulate a complete PSA cycle when both dead volumes are significant calls for only two more parameters to be specified than for a system without dead volumes. Despite that, the pressure ratio, feed composition, and adsorbent selectivity all affect the impact of both dead volumes. Hence, to present a general perspective would require more space than is available here. As an alternative, it is possible to keep details to a minimum, yet get a sense of the important factors, by restricting attention to a single dead volume. Since it is conceivable that the product-end dead volume may improve PSA performance by its passive purging action, it is more interesting than feed-end dead volume alone, for which all the foreseen effects are negative. For that reason, the discussion that follows is focused mainly on product-end dead volume.

Figures 4.12(a)–(d) show the combined effects of product-end dead volume and pressure ratio on recovery of the light component in a four-step PSA cycle. The cycle is the same as that shown in Figure 4.1, except for dead volume in the column. Each figure applies to a different feed composition and adsorbent selectivity. While examining the details, it may be revealing to keep in mind some historical facts. Early PSA systems for hydrogen purification used modest pressure ratios (e.g., $\wp \leq 10$), had high adsorbent selectivity (e.g., $\beta \leq 0.1$), and the feed was predominately the light component ($y_F \leq 0.1$). Figure 4.12(a) is based on those conditions, and it shows that the effect of dead volume on recovery is small. Thus, hindsight affirms these theoretical results; that is, hydrogen purification systems could contain significant dead

EQUILIBRIUM THEORY

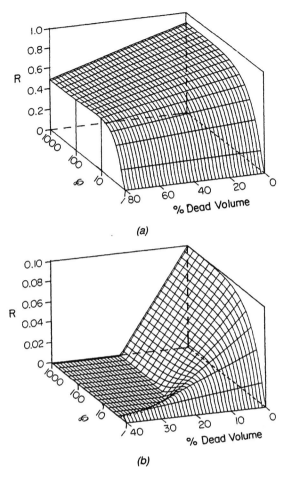

Figure 4.12 Predicted recovery versus pressure ratio and percent dead volume: (a) $\beta = 0.1$, $y_F = 0.1$, (b) $\beta = 0.9$, $y_F = 0.1$.[23]

volume, as in conventional adsorbers, without suffering much loss in recovery. At the present time, increasingly more difficult PSA applications are being considered, and an appropriate question is whether column designs need to be modified to accommodate them. For example, the predicted recovery for a system having low adsorbent selectivity is shown in Figure 4.12(b). To be specific, the conditions are the same as for Figure 4.12(a), except that $\beta = 0.9$ in that figure instead of $\beta = 0.1$. The effect of dead volume is severe, except at low pressure ratios (e.g., $\wp \leq 3$). Conversely, at moderate to large pressure ratios (e.g., $\wp \geq 10$), there is a 50% drop in recovery for only 10% dead volume, and nil recovery for dead volumes of 20% or greater.

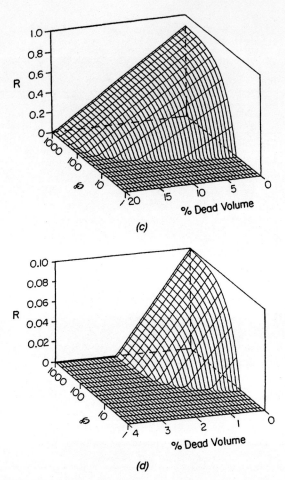

Figure 4.12 Predicted recovery versus pressure ratio and percent dead volume: (c) $\beta = 0.1$, $y_F = 0.9$, (d) $\beta = 0.9$, $y_F = 0.9$.[23]

In the same vein, another type of difficult PSA separation exists when the feed is mostly the more strongly adsorbed component (e.g., $y_F = 0.9$). In that case, the loss of recovery as shown in Figure 4.12(c), on a fractional basis, follows the same trends as in Figure 4.12(b), even though in this case the adsorbent selectivity is large. Applications that are difficult in both regards, that is, they have both low adsorbent selectivity and a high level of the more strongly adsorbed component in the feed, are extremely sensitive to dead volume, as shown in Figure 4.12(d). In that case, merely 2% dead volume is sufficient to destroy the potential recovery. From these results a rule-of-thumb is apparent: the fractional dead volume that will lead to nil recovery is: $\Lambda_p|_{R=0} \cong 0.02/y_F\beta$.

4.5 Experimental Validation

The beauty of local equilibrium theories lies in their simplicity and their ability to draw attention to the most important operating conditions, geometric parameters, and physical properties. Unless these predictions agree with reality, however, that beauty is superfluous. This section reviews some of the experimental work that has been deliberately aimed at verifying local equilibrium theories.

The early experimental studies appear to have been conducted as "black-box" studies, in which cycling frequency, feed-to-purge flow rate ratios, pressures, etc. were varied systematically. As these parameters were manipulated, the performance (flow rates, product and byproduct purities) and other variables were monitored. Comparisons with theory were made retrospectively. Perhaps the first comparison of this sort was that of Mitchell and Shendalman.[24] Their application was the removal of 1% CO_2 from a helium carrier using silica gel. They did not find close correspondence with their equilibrium theory; so they introduced a mass transfer resistance to account for the discrepancy. This modification allowed them to bracket the observed behavior, but neither model was accurate over the entire range of conditions. Flores-Fernandez and Kenney[25] developed a more broadly applicable equilibrium theory, and solved it via finite differences and a commercial package known as CSMP. They tested their model by experimentally separating oxygen from air using 5A zeolite, and obtained fair agreement: within 15% for the prediction of feed flows, and within 12% for the prediction of recovery.

More recently a different approach has been taken, which is to build an experimental system in such a way that the inherent assumptions of the equilibrium theory are closely approached, then to operate it in such a way that the best possible performance is expected. This approach has the advantage of examining conditions that are of most practical interest, as well as using the theory as a tool to guide the experiments. Several different cycles have been evaluated this way, but to conserve space only three are discussed here. They are: the four-step cycle employing pressurization with product, a four-step cycle with combined feed and cocurrent blowdown, and a five-step cycle incorporating a rinse step in order to obtain two pure products. The underlying theory of all these cycles was discussed earlier in this chapter.

The first test determined the validity of the theory that was described in Section 4.4.1, which applies for the four-step pressurization-with-product cycle, shown in Figure 4.1. The experimental system was a two-bed apparatus containing zeolite 5A, designed to separate oxygen from dry air (Kayser and Knaebel[20]). The temperature and pressures were such that nearly linear isotherms were expected, and the equipment was designed so that the assumptions cited in Section 4.1 were valid (including minimal dead volume). Six sets of experimental conditions were tested, and for each, the system was operated until cyclic steady state was achieved. The pressure ratio range was

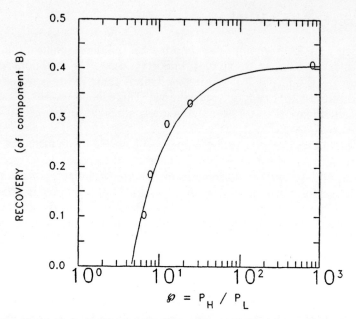

Figure 4.13 Experimental recovery versus predicted recovery (according to Eq. 4.27) for oxygen (and argon) separation from air using 5A zeolite at 45° C.[20]

from 6.5 to 840, and the average temperatures were 45 and 60° C. It should be noted that argon, which appears as about 1% in dry air, adsorbs nearly identically to oxygen, so the targeted "pure" light product is actually about 95% oxygen and 5% argon. Product recovery was of primary interest, since it is predicted quantitatively by the theory. Results of the experiments and the predictions of Eq. 4.27 are shown in Figure 4.13. Recall that there are no adjustable parameters in the model, so the extent of agreement is not due to empirical fitting. The average absolute deviation between the experimental and predicted recoveries was 1.5%, and the product purity averaged 99.6% (nitrogen-free oxygen and argon). These experiments provide strong evidence that, for this system, the equilibrium theory is essentially correct.

Although the results described are reassuring, they focus on the high-pressure feed step; the pressurization, purge, and especially blowdown are ancillary steps. So a major question still remains as to the validity of equilibrium theories when *pressure changes* are vital to the cycle, rather than practically immaterial. This issue was examined in two separate types of experiments. The first type of experiment looked at the acceleration or deceleration of the shock wave during the feed step, in conjunction with decreasing or increasing pressure, respectively.[26] Comparing experimental results with predictions of the theory showed nearly perfect agreement for both increasing and decreasing pressure. Second, PSA experiments involving simultaneous feed and cocurrent blowdown have been conducted in a two-bed

Table 4.1. Conditions and Results of PSA Experiments with Combined Feed and Cocurrent Blowdown[27]

Run	$\wp_F$	$\wp_H$	R_{expt}	R_{theory}	Δ (%)
1	12.77	10.43	25.4	26.5	−1.1
2	13.18	10.95	28.4	27.1	1.3
3	17.41	14.35	29.6	31.4	−1.8
4	22.91	19.03	34.1	34.6	−0.5
5	33.73	25.12	35.9	38.5	−2.6
6	33.29	21.06	37.4	39.1	−1.7

apparatus.[27] The theory for this type of cycle was discussed in Section 4.4.4. The application was to split oxygen from air with zeolite 13X. Six experiments were conducted in which two pressure ratios were varied independently. A summary of the conditions and results is given in Table 4.1.

The average absolute deviation of the light product recovery between theory and these experiments was only 1.5%, just as it was for the experiments having constant pressure during the feed step. This close agreement provides additional evidence that the local equilibrium theory is indeed valid, and is relatively insensitive to the cycle and operating conditions.

Finally, a set of experiments has been conducted in which a rinse step was introduced, in order to extract the heavy component as a pure product. In addition, the purge step was left incomplete, in order to achieve high recovery of the light component. The relevant theory is described in Section 4.4.5. The specific application was to split dry air to get oxygen (with residual argon) and nitrogen using zeolite 5A as the adsorbent. In those experiments, a single bed was used, and pressure ratios were varied between 6 and 20. Generally, it was possible to reduce the level of impurities in the products to about 1% (i.e., O_2 in N_2, and N_2 in O_2) and to achieve corresponding product recoveries between 27% and 90%. The predicted and experimental recoveries of both products agreed well, even though they varied with the applied pressure ratio, as shown in Figures 4.14(a) and (b).

In addition, Figure 4.15 shows a cross-plot of the experimental results from those experiments, along with results obtained by Sircar[22] for a very similar cycle. The axes depicted are product purity and recovery. Sircar's results show a commonly observed trend: as recovery increases, product purity decreases. The data from the experiments described, however, show that purity can be maintained at high levels as recovery increases, without a significant increase in power consumption.

To conclude this section, it appears that the equilibrium theory is accurate and reliable for different PSA cycles, even for relatively difficult separations. Parenthetically, it should be mentioned that other experimental evidence, shown in Figure 4.3, indicates that the equilibrium theory should be valid for a wide range of applications, although the degree of agreement depends on

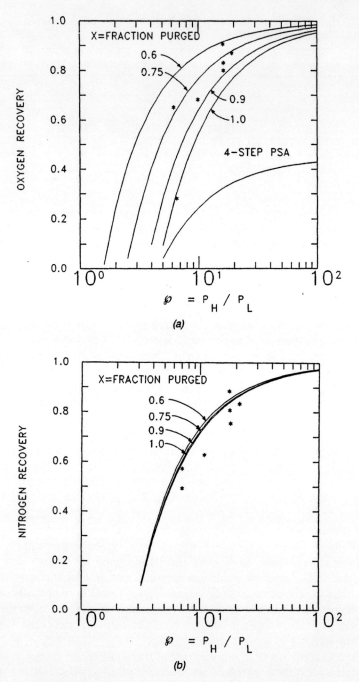

Figure 4.14 Recoveries of (a) oxygen and (b) nitrogen from air, for a five-step cycle, with an incomplete purge step.[12]

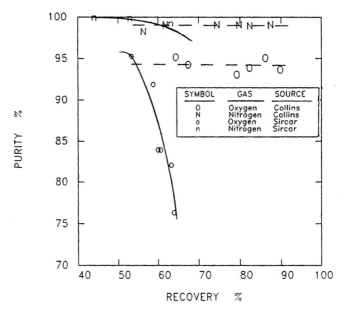

Figure 4.15 Recoveries and purities of oxygen and nitrogen from air, comparing the present five-step cycle[12] with a similar cycle proposed by Sircar.[22]

the Reynolds number in the bed. Finally, across the spectrum of potential PSA applications the intrusion of heat effects can be expected to vary widely. Such effects may be either beneficial or detrimental, as discussed in Section 4.8.

4.6 Model Comparison

This section examines the salient features of four distinct local equilibrium-based models that have been developed over the past several years. In general, it is fair to say that the obvious differences between these models are, remarkably, not in the final equations. Rather the differences lie in the *allowable* values of certain variables, which affect the parameters of the models. The original derivations contained subtle assumptions that imposed tight restrictions on the parameters. Thus, it was not merely an oversight that larger ranges of the parameters were not examined.

4.6.1 Four-Step PSA Cycle: Pressurization with Product

The first and simplest model, that of Shendalman and Mitchell,[2] can be applied to the PSA cycle shown in Figure 4.1, and the resulting equation for recovery can be expressed in the same form as Eq. 4.27. This model assumes that the more strongly adsorbed component is a trace contaminant and that it follows Henry's law, while the carrier is not adsorbed. In terms of the parameters used above, these are: $y_{B_F} \cong 1$ (i.e., $y_{A_F} \to 0$), $\zeta = 1$, $\beta = \theta = \beta_{A_0} = \beta_A$, and $\beta_B = \theta_B = 1$. The next equilibrium model, developed by Chan, Hill, and Wong,[3] is less restrictive, though it also assumes that the more strongly adsorbed component is very dilute. It allows for adsorption of the carrier gas (following a linear isotherm). These restrictions amount to the following: $y_{B_F} \cong 1$, $\zeta = 1$, and $\beta = \theta = \beta_{A_0}/\beta_{B_0}$. Both of these theories ignore the effects of uptake (and release) on the interstitial gas velocity. That assumption allows the cycle to be analyzed easily, but it leads to potentially serious errors, because it implies that the molar flows of feed and product are identical, and similarly that the amount of gas required for pressurization is equal to the amount exhausted during blowdown.

A similar model, suggested by Knaebel and Hill,[13] incorporated adsorption of both components of an arbitrary binary mixture [i.e., $y_{B_F} \in (0, 1)$]. Thus, the variation of velocity arising from composition variations was taken into account. Adsorption equilibrium, however, was still restricted to linear isotherms. Hence, the parameters of that model are: $\zeta = 1$, $\beta = \theta = \beta_{A_0}/\beta_{B_0}$, and $\beta_{i_0} = \theta_i$. Finally, the model of Kayser and Knaebel,[4] which is the basis of Eqs. 4.1–4.27, allows for nonlinear isotherms, as well as arbitrary composition. Thus, in that model the parameters are distinct, and the following combinations are allowed: $\zeta \neq 1$, $\beta \neq \theta \neq \beta_{A_0} \neq \beta_A$, and $\beta_B \neq \theta_B$ (although the last inequality is dropped in the examples to follow).

If we use the result of the most complete derivation to predict recovery, viz., Eq. 4.27, the differences between the models are reflected in the allowable values of the parameters. To emphasize the impact of the different parameters on the four models, two different adsorbent–adsorbate systems, which are both simulated at two different sets of conditions, will be discussed in the following paragraphs.

In the first PSA system, a very light gas, helium, is to be removed from nitrogen (cf. Figures 4.2 and 4.3). Nitrogen is much more adsorbable than helium, but not to the point that the isotherm is very nonlinear (at or below 2 atm). The first set of conditions represents what might be thought of as an *ideal* PSA application, since the light gas is taken as the major component. It is perhaps not surprising that there is excellent agreement (i.e., within about 3%) among the Shendalman–Mitchell, Chan et al., Knaebel–Hill, and Kayser–Knaebel models for that situation, and that the predicted recovery is high. The second set of conditions involves a significant shift: now the heavy component is the major component of the feed, and the pressure is high enough so that curvature of its isotherm is important. As a result, the two simplest models agree, but they must be incorrect since the heavy component is *not* merely a trace contaminant. The Knaebel–Hill model accounts correctly for composition, yielding a 15% reduction of recovery. Correcting for curvature of the nitrogen isotherm, via the Kayser–Knaebel model, reduces the recovery again by 23%.

EQUILIBRIUM THEORY

Table 4.2. Comparisons of Four Local Equilibrium Models for a Four-Step PSA Cycle Employing Pressurization with Product[a]

System and model	Low $\wp$ and low concentration				High $\wp$ and high concentration			
	θ_A	θ_B	θ	R	θ_A	θ_B	θ	R
N_2–He	$y_{A_F} = 0.10, \wp = 5$				$y_{A_F} = 0.90, \wp = 50$			
1	0.171	1	0.171	0.663	0.171	1	0.171	0.812
2	0.171	0.956	0.179	0.657	0.171	0.956	0.179	0.805
3	0.171	0.956	0.179	0.638	0.171	0.956	0.179	0.657
4	0.172	0.956	0.180	0.636	0.286	0.956	0.299	0.426
N_2–O_2	$y_{A_F} = 0.10, \wp = 5$				$y_{A_F} = 0.79, \wp = 20$			
1	0.0579	1	0.0579	0.754	0.0579	1	0.0579	0.895
2	0.0579	0.163	0.356	0.515	0.0579	0.163	0.356	0.612
3	0.0579	0.163	0.356	0.501	0.0579	0.163	0.356	0.490
4	0.0583	0.163	0.359	0.497	0.0795	0.163	0.489	0.301

[a] Model 1 is the Shendalman–Mitchell model, $y_{A_F} \to 0$. Model 2 is the Chan–Hill–Wong model, $y_{A_F} \to 0$. Model 3 is the Knaebel–Hill model. Model 4 is the Kayser–Knaebel model. For N_2–He and N_2–O_2, the value of P_L used for the "low-$\wp$" comparison was 1 atm. For N_2–He, the value of P_L used for the "high-$\wp$" comparison was 0.5 atm. For N_2–O_2, the value of P_L used for the "high-$\wp$" comparison was 0.25 atm. These values, and the pressure ratios, allowed P_H to stay within the range of equilibrium data. He–N_2: Adsorbent = activated carbon, temperature = 20° C, $q_A = 10.63c_A - 12,826c_A^2$, $q_B = 0.1c_B$, $\varepsilon = 0.6874$. O_2–N_2: Adsorbent = Zeolite 13X, temperature = 0° C, $q_A = 15.013c_A - 24,547c_A^2$, $q_B = 4.7542c_B$, $\varepsilon = 0.480$.

The second PSA system is intended to separate oxygen (the light gas) from nitrogen using zeolite 13X. The first set of conditions in Table 4.2, are *less ideal* than for helium and nitrogen, because oxygen is much more adsorbable than helium. This shows the inadequacy of the Shendalman–Mitchell model. The heavy component is dilute, however, so there is little difference between the Chan–Hill–Wong and Knaebel–Hill models. Since the total pressure is relatively low, the curvature of the nitrogen isotherm is not significant, so the Kayser–Knaebel model is in good agreement with the previous two models. The second set of conditions, again, differs significantly from the first set: the feed is taken to be air, so the heavy component, nitrogen, is the majority of the feed. In addition, the pressure is high enough so that the curvature of the nitrogen isotherm is important. This leads to serious discrepancies among all the models. The Kayser–Knaebel model accounts for all the effects, and yields a significantly lower, but more realistic prediction of the recovery.

Figures 4.16 and 4.17 expand the scope of the oxygen–nitrogen example cited by showing a larger range of operating pressures. The basis is the same as described in Table 4.2. For the hypothetical feed composition of 10% nitrogen and 90% oxygen, shown in Figure 4.16, all the model predictions, except those of Shendalman and Mitchell (which does not account for

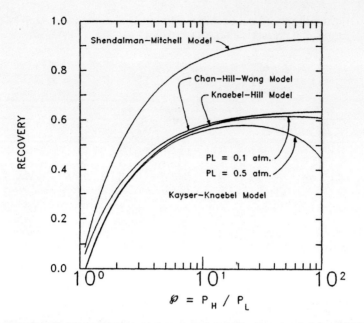

Figure 4.16 Predicted recoveries versus pressure ratio for separation of 10% nitrogen from oxygen with zeolite 13X.

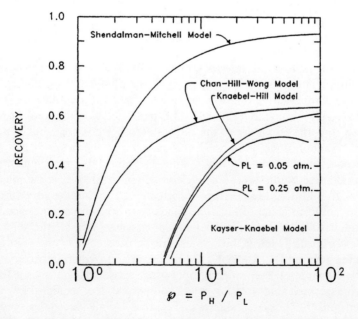

Figure 4.17 Predicted recoveries versus pressure ratio for separation of 79% nitrogen from oxygen with zeolite 13X.

EQUILIBRIUM THEORY

sorption of oxygen), are in good agreement at low to moderate pressure ratios. Even when curvature of the nitrogen isotherm is taken into account, the differences are minor, as long as P_L is small. When it is raised to just 0.25 atm, to correspond to the comparison in Table 4.2, the effect of curvature becomes pronounced at high pressure ratios.

For the more typical feed composition of 79% nitrogen and 21% oxygen, shown in Figure 4.17, there is practically no agreement among the models. That figure clearly shows the magnitude of deviations caused by ignoring: (1) sorption of the light component (the difference between the Shendalman-Mitchell model and that of Chan et al.), (2) composition dependence of interstitial velocity (the difference between the model of Chan et al. and the Knaebel-Hill model), (3) the effect of isotherm curvature (the difference between the model of Knaebel-Hill and the Kayser-Knaebel model), and finally (4) the impact of absolute pressure when the isotherms are not linear (the difference between the values of P_L within the Kayser-Knaebel model). In all of these comparisons, recovery (at a given pressure ratio) is always diminished by taking into account more of the effects mentioned.

4.6.2 Four-Step PSA Cycle: Pressurization with Feed

The differences between the assumptions of the models of Chan et al.[3] and Knaebel and Hill[13] are also evident in PSA cycles that employ pressurization

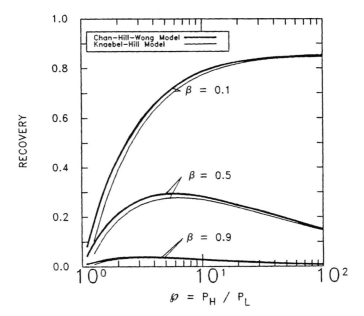

Figure 4.18 Recovery versus $\wp$ for various adsorbent selectivities, $y_{B_F} = 0.9$, comparing the models of Chan et al. and Knaebel and Hill.

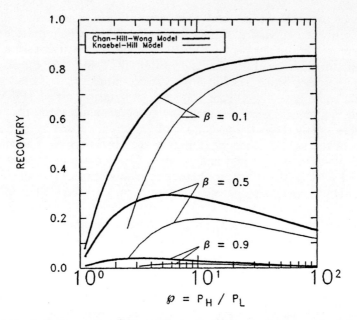

Figure 4.19 Recovery versus $\wp$ for various adsorbent selectivities, $y_{B_F} = 0.1$ according to the model of Knaebel and Hill.

with feed. This section briefly discusses the qualitative and quantitative differences between Eqs. 4.29 and 4.30.

The approach follows that of the previous section in which conditions are chosen first to be reasonably valid for both models; the conditions are then altered to violate the simpler model. In both cases the model predictions are compared to show the magnitudes of the potential errors due to oversimplification. Generally, Figures 4.18 and 4.19 compare the models of Chan et al. and Knaebel and Hill by showing the dependence of recovery on $\wp$ for various adsorbent selectivities. Specifically, Figure 4.18 shows relatively close predictions for $y_{A_F} = 0.1$ ($y_{B_F} = 0.9$). Conversely, Figure 4.19 shows larger discrepancies between the predicted recoveries, especially at lower pressure ratios, for $y_{A_F} = 0.5$ ($= y_{B_F}$). It should be emphasized that the lower values are considered to be more accurate, because the assumptions are less restrictive. In both figures, the differences diminish as the pressure ratio increases, implying that the impact of composition dependence on gas velocity becomes relatively smaller. Both of those figures also show that maximum recovery occurs at an intermediate pressure ratio for systems with moderate to low selectivities (i.e., $\beta > 0.4$). Such behavior is not observed for the case of pressurization with product, except for nonlinear isotherms at relatively high values of P_L.

EQUILIBRIUM THEORY

4.7 Design Example

Ordinarily, *open-ended* design of a PSA system might mean an involved optimization of the production rate, pressures (and other operating conditions), adsorbent, steps within the cycle, and equipment. Conversely, to design a PSA system for a *specific* application is more straightforward, because one would be given specifications for the feed and purified product streams and perhaps the adsorbent. Important steps for that case, which is really a subset of an open-ended design, include: obtaining relevant isotherms and other properties, selecting a PSA cycle, estimating recovery of the desired product at various operating pressures (assuming that sufficient product purity could be attained), determining power requirements and adsorbent bed sizes associated with each operating pressure range, and finally estimating the costs of the adsorbent, power, vessels, valves, etc. to arrive at the total cost. Detailed design considerations would address the product purity question, and ancillary details that affect the optimum conditions, and therefore the cost. Such details, though, are beyond the scope of this example. The purpose here is to illustrate rough sizing of a specific PSA system, generally following the steps listed, and employing the equilibrium theory.

The application considered here is production of 100 Nm^3 per hour of oxygen (at 4.0 bar) from air, which is at ambient pressure and temperature. To avoid possible confusion in dealing with too many degrees of freedom, a number of arbitrary choices are made so that the focus can be kept on PSA rather than auxiliary issues. For example, one would ordinarily consider pretreatment of the feed to remove contaminants that could reduce adsorption capacity (e.g., in this case integrating desiccant with this system). For simplicity, however, the air fed to this system is assumed to be dry and free of contaminants. Thus, only the main constituents [nitrogen (78.03%), oxygen (20.99%), and argon (0.94%), i.e., 99.96% of "standard" air] are considered. In addition, the adsorbent is chosen to be zeolite 5A. Coincidentally, the adsorption isotherms of argon and oxygen on 5A zeolite are practically identical, so argon is treated as oxygen in the PSA analysis. Selecting the operating temperature to be 45° C ensures that the isotherms are essentially linear up to about 6 atm (so there is no effect of absolute pressure on PSA performance). In addition, at that temperature the adsorbent–adsorbate interactions are characterized by $\beta_A = 0.10003$, $\varepsilon = 0.478$, $\rho_B = 810$ kg/m^3, and $\beta = \theta = 0.593$.[20]

Even within these constraints, there remain many options: for operating conditions, one could optimize the pressure ratio, extent of purge, and possibly the pressure manipulation scheme (cf. Sections 4.4.1–4.4.4). One could also consider producing purified nitrogen as a byproduct via a five-step cycle, since it offers very high potential oxygen recovery (cf. Section 4.4.5). To be concise, however, only two four-step PSA cycles are evaluated here, both operating at pressures sufficient to meet the product pressure specification. Both cycles employ pressurization with product, the first with complete purge, and the second with minimal purge.

In reviewing the steps mentioned pertaining to design of a specific PSA system, one might get the impression that, as far as PSA calculations are concerned, simply estimating PSA recovery is sufficient. It is not. An overall balance is required to determine the amount of adsorbent needed, then to size the bed and to predict the necessary power requires stream flow rates and pressures for individual steps. To refine the performance estimates further (e.g., to see the relationship between bed dimensions, flow rates, recovery, and pressure drop), one can use the Reynolds number in the fixed bed. Considering briefly Figure 4.3, one can see that there is a broad optimum of apparent adsorbent selectivity versus Reynolds number. In this example, the Reynolds number is stipulated to be 15, to be close to the equilibrium limit. One can see from Figure 4.3, however, that *doubling* it would increase the apparent value of $\theta = \beta$ only from about 0.59 to about 0.64, which means a loss of recovery of only about 15%. Hence, considering that less adsorbent can be used if velocities are raised, there is an economically optimum Reynolds number that is higher than at the apparent equilibrium limit.

Once the cycle and adsorbent are set, the variables that affect the optimum pressure ratio are the recovery and the power requirement. For the sake of simplicity, we will restrict consideration to adiabatic compression of an ideal gas,

$$\wp = \frac{\gamma}{\gamma - 1} \frac{QRT}{\eta} \left(\wp_C^{(\gamma-1)/\gamma} - 1 \right) \qquad (4.65)$$

where γ is the ratio of specific heats at constant pressure and constant volume, Q is the molar flow rate, η is the mechanical efficiency, and $\wp_C$ is the compression ratio, P_{C_H}/P_{C_L}. For a four-step cycle, it is common to operate above atmospheric pressure so that only a feed compressor is needed, so $P_L = P_{C_L} \cong 1$ atm, and $\wp_C \cong \wp$. It is possible, however, to extend to subatmospheric pressure for blowdown and purge. This operating range is often referred to as *vacuum swing adsorption* (VSA), as if to imply that there is something inherently different about it from PSA. In that case, $P_L < 1$ atm, and there are $\wp_C$s for both a gas compressor and vacuum pump to be considered, along with equipment costs and power requirements, though $\wp$ (and R_B) for the PSA system would be unaffected by the absolute pressure. While on the subject of costs, those of the major components are taken to be: $5 per kg of 5A zeolite, $0.05 per kWh, and $0.20 per N m³ of 95% oxygen at 4 bar.

For the case of complete purge, recovery can be estimated via Eq. 4.27. It is easy to see that if $P_L = 1$ atm and $P_H = 4$ atm so that $\wp = 4$, the predicted recovery is negative. Therefore, rather than considering the gamut of possibilities, we can set $P_L = 0.25$ atm, and $\wp = 16$, which implies a recovery of 29.1%. The required net product, 100 N m^3/h is equivalent to 1.24 mol/s. According to the definition of recovery, given by Eq. 4.26, the molar feed rate, Q_{in_F}, for the case of complete purge, is 19.36 mol/s. This is continuously fed, we will use two parallel columns to allow one to go through blowdown, purge, and pressurization while the other column receives feed air.

From Eq. 4.22 we find that for a given bed of adsorbent: $\phi/t_F = 1.21$ mol/s, where $\phi = \varepsilon A_{CS} L P_L / \beta_A RT = 4.58 \times 10^{-5} V_{ads}$ and V_{ads} is the volume of the adsorbent in cm^3. Accordingly, $V_{ads}/t_F = 26,442.0$ cm^3/s. The Reynolds number is defined as Re $= \varepsilon \rho v_{in_F} d_P/\mu$, where $\varepsilon \rho v_{in_F} = Q_{in_F} M_F/A_{CS}$; thus Re $= 148\,000./A_{CS}$. Setting Re $= 15$, we find that $A_{CS} = 9850$ cm^2; hence the column diameter is 112 cm. Returning to the ratio V_{ads}/t_F, we can determine $L/t_F = 2.685$ cm/s. Now, we are free to choose the length of the bed, say, $L = 161$ cm. The result is that the feed-step duration, $t_F = 60$ s, and the volume of adsorbent (for a two-bed system) is $V_{ads} = 3.17$ m^3, which, based on its bulk density of 810 kg/m^3, is equivalent to 2.57 metric tons of adsorbent.

Based on these values and a mechanical efficiency of 80%, the required power can be estimated. First of all the power for compression of the feed from 1 to 4 atm is 109 kW, and that to compress the subatmospheric blowdown and purge effluent from 0.25 atm to atmospheric pressure is 28.8 kW. Note that the amount of gas evolved during blowdown at any intermediate pressure can be determined from Eqs. 4.19 and 4.50, assuming that the bed is saturated with feed prior to blowdown. Hence the costs associated with the complete purge case would be: $55,079 per year for power and $12,851 for the adsorbent. Additional costs (for vessels, compressor, vacuum pump, valves, piping, instrumentation, site development, maintenance, and fees) should be proportional to these. These are balanced against a projected value of $160,000 per year for the product.

For the case of incomplete purge, the calculations are somewhat more involved. The minimum extent of purge for the adsorbent properties and conditions cited and for $P_L = 1$ atm so that $\wp = 4$, is $X = 41\%$. This corresponds to a recovery of 36.55% (cf. Figure 4.9), and a required feed flow rate of $Q_{in_F} = 15.41$ mol/s. Again, setting the Reynolds number at 15 leads to a column diameter of 100 cm. Using the same feed duration as before, $t_F = 60$ s leads to a column length of $L = 176$ cm. Thus, the volume of adsorbent needed to fill two columns is $V_{ads} = 2.76$ m^3. Based on the bulk density of 810 kg/m^3, this is equivalent to 2.24 metric tons of adsorbent. The necessary power, in this case, is only for compression of the feed from 1 to 4 atm. The result is 88.7 kW (about two-thirds of that required in the first case).

Therefore, the costs associated with the incomplete purge case would be: $34,668 per year for power and $11,197 for the adsorbent. As before, additional costs (for vessels, etc.) should be proportional to these. These are also balanced against a projected value of $160,000 per year for the product.

In this case the power cost has been reduced by 37%, and the adsorbent cost has been reduced by 13%. In addition, this system does not require a vacuum pump or as many valves (e.g., for blowdown in successive stages), so the entire system is simpler and less expensive. A final, subtle point that has not been taken into account is that the cycle time could possibly be reduced for this case (since less uptake and release occur), which would lead to less required adsorbent, and a further reduction in cost.

4.8 Heat Effects

The term *heat effect* is frequently applied to a conspicuous change in performance that coincides with a temperature fluctuation. Although the term is introduced here, it is covered more quantitatively in Section 5.4. Viewing a packed bed pressure swing adsorption system, there are two main heat effects: heat is released as a heavy component displaces a light component due to the preferential uptake, and compression raises the gas temperature. Of course, heating due to adsorption and compression is at least partially reversible, since desorption and depressurization both cause the temperature to drop. The relative magnitudes of such temperature swings are affected by heats of adsorption, heat capacities, and rates of heat and mass transfer. Hence, the potential effects on performance are many, and they depend on operating conditions, properties, and geometry, sometimes in complicated ways. For example, at a constant pressure, a cycle of adiabatic adsorption followed by adiabatic desorption involves less uptake and release than the isothermal counterpart. As a result, one might expect poorer performance from an adiabatic system as opposed to an isothermal system, but that is not necessarily true for PSA, as shown later.

The term *heat effects* has acquired a connotation of mystery and confusion. This is especially true in the field of PSA since many different effects occur simultaneously. Thus, determining cause-and-effect relationships is not trivial. In fact, some unusual thermal behavior was revealed in a patent disclosure by Collins[28] that continues to perplex some industrial practitioners. Collins stated that, "Contrary to the prior art teachings of uniform adsorbent bed temperature during pressure swing air separation, it has been unexpectedly discovered that these thermally isolated beds experience a sharply depressed temperature zone in the adsorption bed inlet end. ... The temperature depression hereinbefore described does not occur in adsorbent beds of less than 12 inches effective diameter." Without attempting to unravel those observations, it may be instructive to consider what happens in a small column, to see detailed effects directly and to infer their causes, and to become aware of the range of potential effects.

For most systems, the principal heat effect arises from simultaneous axial bulk flow, adsorption, and heat release due to adsorption. These phenomena

EQUILIBRIUM THEORY

lead to composition and thermal waves that propagate toward the product end of the column. For many PSA applications, these fronts coincide, and the induced temperature shifts just due to adsorption are often greater than 10°C and may exceed 50°C (Garg and Yon[29]). It is shown later that temperature shifts during a typical PSA cycle need not significantly affect the adsorption selectivity, even though relatively large changes in absolute capacity may occur. For systems with small amounts of a strongly adsorbed contaminant or a very weakly adsorbed carrier, however, the thermal wave may lead the composition wave (see Section 2.4) and in such cases the adsorption selectivity can be dramatically affected.

Temperature profiles for a typical bulk separation application are shown in Figure 4.20. That figure shows the internal column temperatures during a PSA cycle in which oxygen is being separated from air using zeolite 5A with a pressure swing of 1 to 5 atm. Five thermocouples were placed at the centerline of the column at equal spacings. They were small and had exposed junctions for fast response. The pressurization step (up to 39 s) appears as a linear increase of temperature measured at all the thermocouples. During the feed step (continuing to 149 s), the temperature is stable until the composition front passes, and then a sharp rise occurs and a new plateau is reached. Simultaneous measurements indicate that for this system the tem-

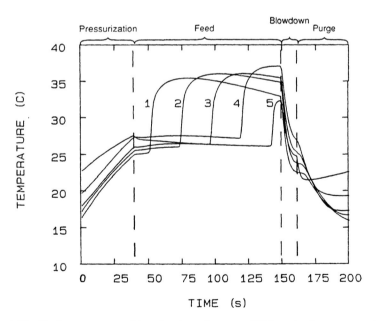

Figure 4.20 Bed temperature histories for a four-step PSA cycle, from pressurization through purge. The numbers 1 through 5 indicate thermocouple locations in the bed, from near the feed end to near the product end. The application is production of oxygen from air with zeolite 5A. The pressure range is from 1.0 to 5.0 atm.

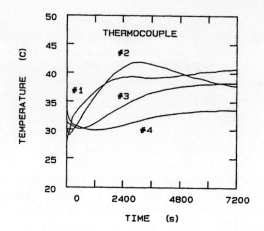

Figure 4.21 Bed temperature histories during a four-step PSA cycle in which water vapor is removed from air with silica gel at 25°C and a pressure ratio of 4.

perature front coincides with the composition front. Blowdown causes a nearly instantaneous temperature drop due to simultaneous depressurization and desorption (until 161 s). Finally the purge step exhibits a small drop in temperature as desorption of the heavy component is completed, followed by a gradual rise back towards the ambient (influent) temperature (completed at 205 s).

For comparison, temperature profiles for the feed step in a PSA air dryer, which is a typical contaminant removal application, are shown in Figure 4.21. That figure shows internal column temperatures for air drying by silica gel with a pressure range of 1 to 4 atm. The four thermocouples were identical to those in the previous case. In this figure the front that propagates through the bed is barely discernible, and is certainly not sharp.

Knowing some of the details of temperature fluctuations during PSA cycles, it is appropriate to explore the effect of temperature on overall PSA performance for a bulk separation. The clearest and simplest indication of temperature dependence comes from the limiting case of linear isotherms in the four-step cycle discussed in Section 4.4.1. Each stream depends differently on individual component β values which are themselves dependent on temperature. The overall recovery, however, depends only on β_0 (which is written simply as β here). First of all, consider the overall dependence of recovery on temperature from Eq. 4.27,

$$\frac{\partial R_B}{\partial T} = \left(\frac{1}{\wp y_{B_F}} - 1\right)\frac{\partial \beta}{\partial T} \tag{4.66}$$

EQUILIBRIUM THEORY 149

where

$$\beta = \frac{\beta_A}{\beta_B} = \left(\frac{\varepsilon}{1-\varepsilon} + K_B\right) \bigg/ \left(\frac{\varepsilon}{1-\varepsilon} + K_A\right)$$

and

$$K_i = K_{i_0} \exp\left[\frac{\Delta H}{R}\left(\frac{1}{T_0} - \frac{1}{T}\right)\right]$$

is the temperature-dependent Henry's law coefficient of component i. Combining these yields

$$\frac{\partial \beta}{\partial T} = \frac{K_A \Delta H_A}{RT^2[\varepsilon/(1-\varepsilon) + K_A]^2} \left[\frac{\varepsilon}{1-\varepsilon}\left(\frac{K_B \Delta H_B}{K_A \Delta H_A} - 1\right) \right. \quad (4.67)$$
$$\left. + K_B\left(\frac{\Delta H_B}{\Delta H_A} - 1\right)\right]$$

To cite a specific example, the parameters for separating oxygen from air using zeolite 5A at 45°C are: $K_A = 8.24$, $K_B = 4.51$, $\Delta H_A \approx -6.0$ kcal/mol, $\Delta H_B \approx -3.0$ kcal/mol, and $\epsilon = 0.478$, so $\partial \beta_A/\partial T = 0.00269$ K^{-1}, $\partial \beta_B/\partial T = 0.00209$ K^{-1}, and $\partial \beta/\partial T = 0.00856$ K^{-1}.[20,30] At very high pressure ratios, only the second term in Eq. 4.66 is important, and the limit is: $\partial R_B/\partial T = -0.00856$ K^{-1}; that is, recovery would decrease by slightly less than 1% if the average temperature increased 1°C. At a more reasonable pressure ratio of 5, however, the first term in Eq. 4.66 is about -0.05, which would require an average temperature increase of 20°C for recovery to decrease 1%. At lower pressure ratios, Eq. 4.66 predicts that the recovery would increase, rather than decrease, if the average temperature increased.

Next, the flows involved in each step can be examined to see how temperature fluctuations from step to step may affect the overall recovery. That is, the temperature dependence of each stream can be found from the isotherm parameters. For the sake of discussion, quantities are identified only by orders of magnitude, and the temperature of the high-pressure product is taken as the base temperature. Relative to that, the temperature reached at the end of pressurization is practically the same. The fact that pressurization begins at a lower temperature is less important, since equilibration at the final temperature and pressure determines the quantity of gas admitted. The temperature encountered by the feed is higher (due to the heat released by uptake). Finally, the temperature during blowdown (which does not affect recovery) and purge is lower due to depressurization and desorption. Hence, looking at each term in Eq. 4.26 reveals the effects of temperature shifts for individual steps:

$$\Delta R_B(\Delta T_{\text{STEP}}) \approx \frac{0 - 0 - [\Delta(1/\beta_A)/\Delta T][\Delta T|_{\text{PU}}]}{[\Delta(1/\beta_A)/\Delta T][\Delta T|_{\text{F}}]} = -\frac{[-][-]}{[-][+]} \quad (4.68)$$

Thus, the net effect of temperature shifts during this four-step cycle is positive. That is, the recovery of the light component should be somewhat greater when the natural temperature shifts occur than if the system were forced to remain isothermal. The net effect is expected to be minor since, according to the basis chosen, only the purge and feed streams are affected, and the increase of the purge stream should be small relative to the decrease of the feed stream.

Before leaving this subject, it is worthwhile to point out that the conclusion just reached is not general: different cycles will respond to temperature shifts differently. For example, for cycles in which the heavy component is produced during blowdown (cf. Section 4.4.5), the most important stream in determining recovery is the blowdown step. This step involves a large temperature drop (relative to the pressurization, high-pressure product, and rinse steps) due to desorption and depressurization. As can be seen from Eq. 4.41, the net effect is that the adsorbent retains more of the heavy component than it would under isothermal conditions (i.e., the magnitude of the second term on the right-hand side is larger), so recovery is diminished.

Aside from gaining a better understanding of PSA systems via their inherent thermal response, there is an even greater incentive to understand this behavior. To elaborate, in many PSA systems it is important to prevent complete breakthrough, (e.g., during feed and cocurrent blowdown steps), which would reduce the purity of the product. Conversely, if breakthrough is not imminent at the end of these steps, the product recovery cannot be as high as possible, since any purified gas left in the column is exhausted with the byproduct. Similarly, a rinse step should be allowed to proceed until breakthrough is just complete. To go further would reduce recovery, and to stop prematurely would reduce purity. Accordingly, both high product purity and high-recovery PSA performance can be achieved by terminating such steps very precisely. A minor problem is that many composition sensing instruments have long response times or large sample volumes, so that on-line measurements are often impractical. That is where the thermal response comes in.

The fact that the shock wave of temperature usually coincides with the composition front can be exploited to control the timing. Evidence for that is shown both in Figure 4.20, which was described previously, and in Figure 4.22, in which, again, oxygen is being separated from air. In the latter figure, the pressure increases from 2.0 to 5.2 atm. as feed is being admitted to and product is being slowly released from the column. As can be seen, the average temperature in the bed rises, but the sharpness and magnitude of the temperature front are essentially the same as when pressure was constant. Equivalent, but reverse effects occur when the bed pressure decreases.[26] The possibility of controlling the step times in this manner can prevent reduced recovery [e.g., due to diminished adsorbent capacity or when operating conditions (or ambient conditions) vary significantly]. In fact, this concept was

EQUILIBRIUM THEORY

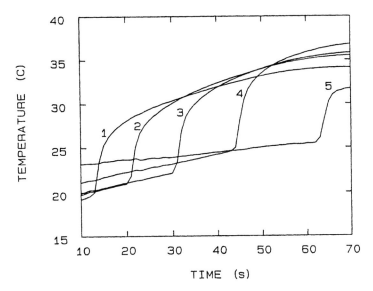

Figure 4.22 Bed temperature histories for a combined pressurization and feed step. The numbers 1 through 5 indicate thermocouple locations in the bed, from near the feed end to near the product end. The application is production of oxygen from air with zeolite 5A. The pressure increases linearly with time from 2.0 to 5.2 atm.[26]

used to control feed and rinse step times for the five-step cycle experiments described in Section 4.5 [cf. Figures 14(a) and (b)].

4.9 Pressurization and Blowdown Steps

Until now, attention has been focused on complete PSA cycles and overall effects. There are, however, cases in which the individual steps are important. For example, when the heavy component of a mixture is valuable, it may be desired as the sole product or as a co-product. In that case, the blowdown step, in particular, is vital to the performance of the PSA system, and it is important to know the composition of the effluent as a function of pressure, or to predict the composition profile in the bed at the end of blowdown. In other situations (e.g., involving pressurization by feed or by an intermediate product from a parallel bed) it may be of interest to predict the composition profile in the bed during pressurization.

In that vein, perhaps the first treatment of composition profiles at various extents of pressurization was given by Flores Fernandez and Kenney.[25] They assumed local equilibrium and neglected pressure drop, and they solved the equations by finite differences. A subsequent model that examined pressurization and included effects of axial dispersion, but not of mass transfer resistance or pressure drop, was developed by Rousar and Ditl.[31] Another local equilibrium model was proposed by Kumar,[32] and it was one of very few to incorporate an energy balance. That model was used to analyze adiabatic blowdown behavior. More detailed models are discussed later in this section.

This section first examines the simplest cases of pressurization and blowdown, and suggests that the key features predicted by more sophisticated models can be obtained analytically. In such cases, much less effort is required, and reasonably accurate estimates of the expected behavior can be obtained. That approach neglects axial pressure drop and mass transfer resistance, as is the case throughout this chapter. Later in this section, however, the impact of pressure drop on pressurization and blowdown is considered.

It turns out that the coupling of velocity, composition, and pressure is graspable for systems governed by linear isotherms, but when nonlinear isotherms are involved the additional complexity makes the set of equations unwieldy and to get detailed simulations via the method of characteristics is not practical. In addition, as noted in the previous section, the pressurization and blowdown steps also may give rise to significant temperature shifts that affect the validity of predictions based on isothermal models.

For pressurization a number of possibilities exist, and the two simplest extremes have already been covered, viz., pressurization with product (see Section 4.4.1), and with feed (see Section 4.4.2). These were assumed to begin with an initial condition in which the bed was purged with the pure light component. When the bed has not been completely purged, simulating pressurization is slightly more complicated (see Section 4.4.3). All three cases represent cyclic steady-state outcomes of operating a PSA cycle at local equilibrium, without dispersive effects. A slightly more complex situation occurs during startup, when the gas used for purging (and possibly pressurization) is not pure. Other possible complications arise from minor variations in operating procedures. For example, a pressure equalization step employs the gas evolved from one column (as it depressurizes) for pressurizing a parallel column. Such gas may have a slowly or suddenly varying composition, due to uneven rates of desorption, poor synchronization of the valves, or contamination by residual material in the connecting fittings.

Pressurization at startup or with gas having variable composition can be regarded as fitting the following possible scenarios: (1) Pressurizing with gas that gradually becomes leaner in the heavy component, during which a shock wave cannot form, or (2) pressurizing with gas that gradually becomes richer in the heavy component, during which a shock wave *may* form if the composition change is sufficiently large, or (3) pressurizing with gas that is significantly richer in the heavy component than the initial interstitial gas, for which formation of a shock wave is unavoidable. For simplicity, let us restrict

EQUILIBRIUM THEORY

consideration to a cycle having complete purge. To determine whether a shock wave may form, one must simply examine whether the characteristics intersect (in the space–time domain being considered). The paths of the characteristics are given by Eq. 4.33, and the composition as a function of pressure can be determined from Eq. 4.32. When the characteristics do not intersect, regardless of the initial and boundary conditions, that pair of equations is sufficient to predict the composition profile during pressurization. On the other hand, situations that result in the formation of shock waves require a few additional steps to predict the ultimate composition profile. When pressure varies, the shock wave trajectory can be determined by employing Eq. A.7 from Appendix A, as follows

$$v_{SH} = \theta_A \frac{\Delta v P y_A}{\Delta P y_A} \tag{4.69}$$

The interstitial velocity is obtained by summing Eq. 4.4 for components A and B.

$$\frac{\partial P}{\partial t} + \beta_B \frac{\partial vP}{\partial z} + (\beta_A - \beta_B)\frac{\partial v P y_i}{\partial z} = 0 \tag{4.70}$$

In this equation, the axial dependence of pressure can be neglected, which, given the dependence of pressure on time, leaves a separable ordinary differential equation,

$$\frac{d \ln P}{dt} + \beta_B \frac{dv}{dz} + (\beta_A - \beta_B)\frac{dvy_i}{dz} = 0 \tag{4.71}$$

Integration requires boundary conditions, and a convenient set is: $v = v_F$ at $z = L$, and $v = 0$ at $z = 0$. The result is the expression that was simply stated earlier in this chapter:

$$v(P, y, z, t) = \frac{-z}{\beta_B[1 + (\beta - 1)y]} \frac{d \ln P}{dt} \tag{4.6}$$

When Eqs. 4.6 and 4.69 are combined, the shock wave velocity may be found from

$$v_{SH} = \frac{-\beta z}{[1 + (\beta - 1)y_1][1 + (\beta - 1)y_2]} \frac{d \ln P}{dt} \tag{4.72}$$

The dependence of y_1 on P, and that of y_2 on z, can be expressed via Eqs. 4.7 and 4.8, respectively. By combining those with Eq. 4.72, the coupled material balances can be solved to get:

$$y_2 = \frac{\alpha y_1}{1 + (\alpha - 1)y_1} \tag{4.73}$$

where $\alpha = y_{20}(1 - y_{10})/[y_{10}(1 - y_{20})]$, and y_{10} and y_{20} are the initial compositions at the leading and trailing edges of the shock wave, respectively. Thus, α is a sort of selectivity, analogous to relative volatility for vapor–liquid equilibrium.

Thus, to find the ultimate axial position of the shock front at any pressure requires a sequence of steps: the initial conditions give α; then the composition ahead the shock front can be found from

$$y = 1 - (1 - y_0)\left(\frac{y}{y_0}\wp^{\beta-1}\right)^{1/\beta} \qquad (4.74)$$

(which is essentially the same as Eqs. 4.32 and 4.50); next the composition behind the shock can be determined from Eq. 4.73; and finally the axial position can be computed from

$$z = z_0\left(\frac{y}{y_0}\right)^{\beta/(1-\beta)}\left(\frac{1-y_0}{1-y}\right)^{1/(1-\beta)}\left(\frac{1+(\beta-1)y}{1+(\beta-1)y_0}\right) \qquad (4.75)$$

(which is essentially the same as Eq. 4.33) using the composition shift at the trailing edge of the shock front.

To illustrate the point, it is appropriate to compare such results with those obtained by finite difference techniques. Figure 4.23 shows predictions of the previous equations and those presented by Rousar and Ditl[31] for oxygen enrichment. It can be seen that the endpoints coincide, as does much of all three profiles. The principal distinction is due to rounding, which is inherently due to dispersion being included in the numerically derived results. All the results shown for the equilibrium theory were obtained with a calculator in several minutes time. Similar results have been obtained by Flores Fernandez and Kenney,[25] and are shown in Figure 3.2.

If we turn our attention now to blowdown, and still restrict conditions to local equilibrium, it is clear that blowdown is simpler than pressurization. That is because there is no interaction between the initial and boundary conditions. Furthermore, in previous sections of this chapter, the initial condition prior to blowdown was usually taken to be uniform, which led to relatively simple material balance calculations. Some subtleties arise for the case of pressurization with feed (cf. Section 4.4.2), when dead volume at the product end was considered (cf. Section 4.4.6), and when the heat effects accompanying blowdown were found to be deleterious (cf. Section 4.8). The most important feature of blowdown is the effluent composition, which in most instances continuously changes as pressure falls, except when a rinse step precedes blowdown, yielding the pure heavy component throughout blowdown. Another topic of practical interest is the ultimate composition profile in the bed following blowdown.

If the initial composition profile is known, applying the local equilibrium model for systems with linear isotherms is fairly simple. For example, in order to predict the effluent composition during blowdown or the residual interstitial gas composition following blowdown, one need only apply characteristic equations such as Eqs. 4.74 and 4.75, where y_0 and z_0 represent the

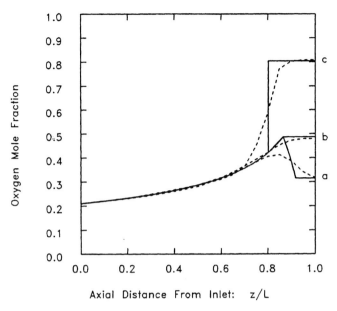

Figure 4.23 Composition profiles during pressurization of a bed of zeolite 5A, to which various mixtures of oxygen and nitrogen are admitted. $P = 6$ atm. Initial conditions: (a) $y_{O_2} = 0.1$, (b) $y_{O_2} = 0.21$, (c) $y_{O_2} = 0.60$. $\beta = 0.517$. Numerical results from Rousar and Ditl.[31]

initial composition and position of a particular characteristic. For a given set of initial conditions, it is easiest to choose a final composition y, then to determine the necessary pressure ratio $\wp$, and finally the ultimate axial position z. Otherwise (i.e., given the pressure ratio), a root-finding procedure is needed to determine the final composition. When the initial composition profile is uniform, Eq. 4.75 indicates that there will be no axial composition gradient as the pressure falls. Regardless, Eqs. 4.18 and 4.19 can be used with the composition–position–pressure information to determine the average composition and quantity of the effluent during blowdown. To relate these to flow rates, it would be necessary to select a depressurization rate: simplistically $dP/dt = $ constant, or somewhat more realistically, $d \ln P/dt = $ constant, although any operating policy can be accommodated.

In the models presented earlier in this chapter, the pressure gradient through the column is assumed to be negligible. In that situation the basic equations governing pressurization and blowdown steps are not much more complicated than those for steps at constant pressure. Neglecting pressure drop is reasonable for most conventional PSA units, but it is clearly inappropriate for single column, rapid pressure swing processes, which are discussed in Section 7.3. The detailed modeling of pressurization and blowdown steps, taking into account pressure gradients through the column, has attracted much attention recently. For example, a detailed model was suggested by Lu et al.,[33-35] who studied both pressurization and blowdown. Their model included mass transfer resistances, axial dispersion, intraparticle convection, and axial pressure drop, but not heat effects, and was solved by finite differences. Other similar models have been suggested,[36-41] and a brief summary of the major conclusions from that work is given here.

To a first approximation, the pressure drop through a packed adsorbent bed can be represented by Darcy's Law:

$$v = -\frac{B}{\mu}\frac{\partial P}{\partial z} \tag{4.76}$$

Coupling this with the differential fluid phase mass balance for a plug flow system (cf. Eq. 5.2) with rapid equilibration yields

$$\left.\frac{\partial P}{\partial t}\right|_z = \frac{B}{\mu[\varepsilon + (1-\varepsilon)(dq^*/dc)]} \frac{\partial}{\partial z}\left[P\left(\frac{\partial P}{\partial z}\right)_t\right]_t \tag{4.77}$$

where, for an isothermal system, dq^*/dc represents simply the local slope of the equilibrium isotherm. The appropriate initial and boundary conditions are, for pressurization:

$$P = P_i, \quad t = 0, \quad \text{for all } z \tag{4.78}$$

$$P = P_H, \quad t > 0, \quad \text{for } z = 0$$

$$\left.\frac{\partial P}{\partial z}\right|_{z=L} = 0$$

with P_H and P_L interchanged for blowdown.

When pressure drop through the column is negligible, Eq. 4.77 is equivalent to Eq. 4.6, which was discussed previously. For a pure gas, A, it reduces to:

$$\frac{dP}{dt} = \pm\beta_A Pv/z \tag{4.79}$$

where the sign reflects the orientation of the column and the direction of flow. This may be integrated directly to obtain the dimensionless time required to pressurize or depressurize the bed:

$$\tau = \frac{\text{volume of gas fed to the column}}{\text{holdup in the column}} = \ln \wp \tag{4.80}$$

which was obtained originally by Cheng and Hill.[37] In this situation, pressurization and blowdown are symmetric processes. This symmetry is lost, however, when the pressure gradients are significant since the pressure response is then governed by Eq. 4.77, which is nonlinear. This is illustrated in Figure 4.24, which shows pressure profiles for pressurization [Figure 4.24(a)]

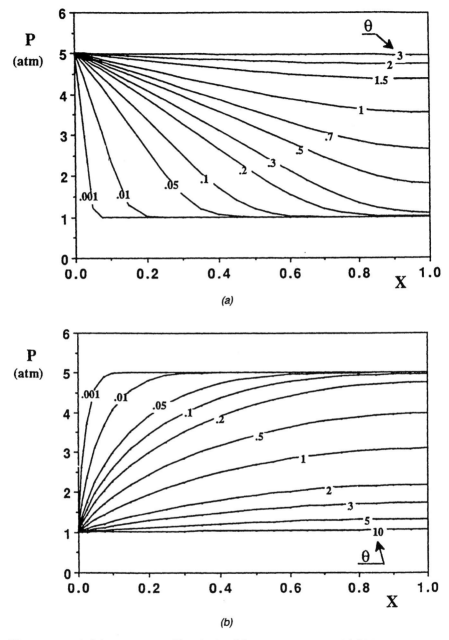

Figure 4.24 Axial pressure profiles during (a) pressurization and (b) blowdown of an adsorbent bed with a nonadsorbing gas. $P_H = 5$ atm, $P_L = 1$ atm, $L = 1$ m, reference velocity -0.2 ms^{-1}, θ = dimensionless time. (From Rodrigues et al.,[36] with permission.)

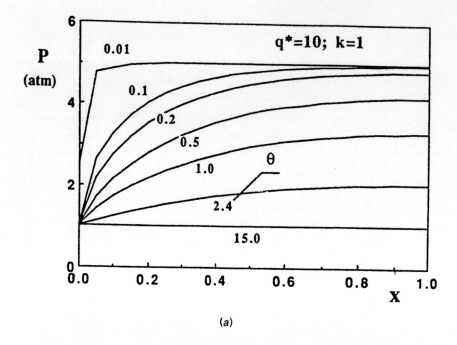

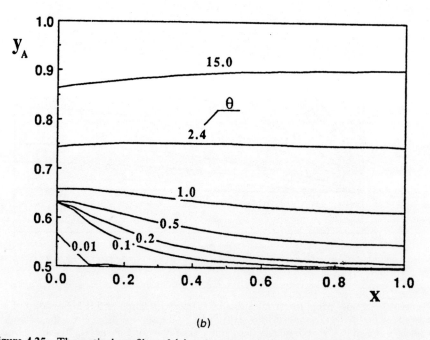

Figure 4.25 Theoretical profiles of (a) reduced total pressure and (b) mole fraction of the more strongly adsorbed component (y_A) during blowdown of an adsorbent bed. $P_H = 5$ atm, $P_L = 1$ atm, $y_0 = 0.5$, $K = 10$ (linear isotherm), θ = dimensionless time. (From Lu et al.,[35] with permission.)

EQUILIBRIUM THEORY

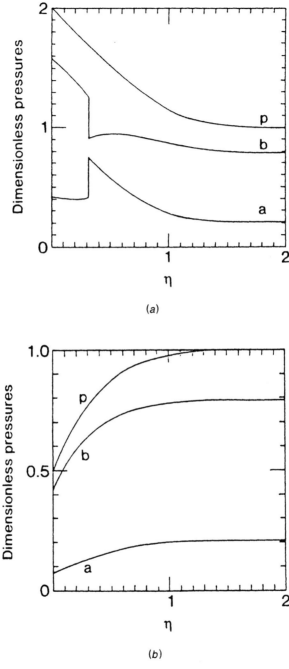

Figure 4.26 Theoretical profiles of total pressure (p), mole fraction of O_2 (a) and mole fraction of N_2 (b) during pressurization and blowdown of a 5A zeolite bed equilibrated with air, calculated from the analytic solution. (From Scott,[38] with permission.)

and blowdown [Figure 4.24(b)], over identical pressure ratios, calculated by numerical integration of Eq. 4.77.[36] Blowdown is clearly slower than pressurization. The dimensionless times for pressurization and blowdown are, respectively, about three and ten times the characteristic time defined by:

$$\tau_c = \frac{\mu \varepsilon L^2}{B(P_H - P_L)} \qquad (4.81)$$

In this example, the pressure profiles during pressurization and blowdown both assume the form of simple propagating waves. With mixtures of differently adsorbed components (or with mixtures of inert and adsorbing species) the profiles assume more complex forms. Many such ramifications, including

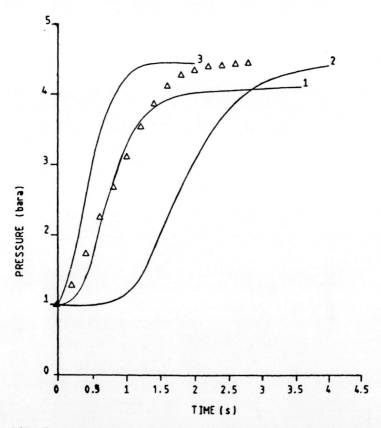

Figure 4.27 Comparison of experimental pressure profile for pressurization of an activated carbon bed with CO_2 and theoretical profiles predicted from (1) finite mass transfer model (LDF), (2) equilibrium adsorption model, and (3) model for no adsorption. (From Hart et al.,[39] with permission.)

the effect of a nonlinear isotherm and the effect of using the Ergun equation for pressure drop (in place of Darcy's Law) have been investigated by Rodrigues et al.[33-36]

One situation of special interest is to use a PSA process to concentrate strongly adsorbed component(s) from a feed of low concentration. An example of this was studied by Rodrigues et al.[36] The bed is initially at equilibrium with a linearly adsorbed light component, at mole fraction 0.5 and a total pressure of 5.0 atm. As the bed is depressurized to atmospheric pressure, the pressure along the axis of the bed responds as shown in Figure 4.25(a). Simultaneously, as shown in Figure 4.25(b), the mole fraction of the more strongly adsorbed component rises rapidly at the open end of the bed, to about 0.65, and the profile then pivots about this point until it reaches a more or less uniform profile through the bed. Thereafter, the profile remains almost uniform through the bed, rising asymptotically towards $y_A \approx 0.90$. At this point, essentially all the less strongly adsorbed species have been removed from the bed, and it would be possible, in principle, to recover the strongly adsorbed species in highly concentrated form by deep evacuation of the bed (see Section 6.10).

An alternative approach has been followed by Scott,[38] who has shown that, if the column can be regarded as infinitely long, a relatively simple analytical solution may be obtained. Profiles of pressure and composition for pressurization and blowdown, calculated with parameters representative of air–zeolite 5A, are shown in Figure 4.26. The composition profiles for pressurization [Figure 4.26(a)] show a complex wave form that includes a partial shock, which appears as an inflection at small values of θ. The profiles for blowdown [Figure 4.26(b)] contain only simple waves; that is, they are everywhere concave downwards.

The effect of finite mass transfer rate has been investigated by Hart et al.[39] Their experimental data for pressurization of an activated carbon bed with CO_2 are shown in Figure 4.27 together with theoretical curves calculated according to three different assumptions: negligible adsorption, instantaneous adsorption, and adsorption at a finite rate according to the linear driving force (LDF) model ($k = 0.2$ s^{-1}). The experimental data lie closest to the finite rate model, although there is a significant deviation in the long-time region. It seems likely that this deviation may be attributed to heat effects, which will reduce adsorption in the long-time region.

4.10 Conclusions

The local equilibrium theory approach is the simplest available for simulating or designing PSA systems. Furthermore, when data are sparse (e.g., no PSA pilot plant data), it is the most reliable method because it does not depend on semiempirical parameters that can only be determined from data. Such methods can lead to excellent designs when kinetics are fast. Even when kinetics are slow (though not controlling), such methods can predict overall performance (e.g., in terms of recovery and byproduct enrichment) very well. The main and perhaps only drawback is that, when kinetic constraints are important, it becomes impossible to estimate product purity reliably.

Nevertheless, a variety of aspects of PSA operation can be taken into account by equilibrium-based theories. Some that are illustrated in this chapter are: a variety of cycle and step options, wide ranges of operating conditions, isotherm nonlinearity, heat effects, and deadzones in PSA columns. In several instances, the simple theories have been verified experimentally, so there is little doubt as to their reliability when the assumptions are reasonably valid.

Chilton once said, "The simpler things become in a piece of research or development, the closer one has come to the truth." Pigford added to that, "The simpler an explanation is, the more widely it will be understood, appreciated, and used."[42] In pressure swing adsorption systems, it is impossible to achieve greater simplicity than local equilibrium models provide and still retain fundamental understanding of the process. Whether the model predictions are close to the truth or not depends on the extent to which the assumptions are valid. For all that, it is seldom possible to improve performance beyond the capability predicted by an equilibrium model because dissipative effects nearly always diminish performance. Hence, striving to conform to those assumptions can be worthwhile, not only because it will be possible to predict performance accurately and simply, but, more important, because performance will be superior. Whether modeling via the local equilibrium approach can be "understood, appreciated, and used" depends mainly on whether the implied superior PSA performance can be achieved in real applications.

Future efforts should be directed towards a unified treatment of equilibrium based separations accounting for the several diverse factors that until now have been accounted for separately. An example would be to account for the effects of nonlinear isotherms on sequential pressurization by feed and feed steps. Another facet to examine is the coupling of isotherms, since it is widely observed that the light component adsorbs proportionately less in a mixture than the amount due to uptake of the heavy component. This effect would tend to improve PSA performance, and might partially compensate for the effects of being nearly adiabatic (which is physically realistic), as opposed to the assumed isothermal condition of the adsorbent bed. Furthermore, experimental work could be done to validate the combined effects, to analyze cycles of more complex steps, and to account for a wider variety of properties and conditions. One example would be to generalize the dependence of the observed "effective" separation capacity of an adsorbent, especially with respect to flow conditions, as alluded to in Figure 4.3.

References

1. P. H. Turnock and R. H. Kadlec, "Separation of Nitrogen and Methane via Periodic Adsorption," *AIChE J.* **17**, 335 (1971).

2. L. H. Shendalman and J. E. Mitchell, "A Study of Heatless Adsorption in the Model System CO He, I.," *Chem. Eng. Sci.* **27**, 1449–58 (1972).

3. Y. N. I. Chan, F. B. Hill, and Y. W. Wong, "Equilibrium Theory of a Pressure Swing Adsorption Process," *Chem. Eng. Sci.* **36**, 243–51 (1981).

4. J. C. Kayser and K. S. Knaebel, "Pressure Swing Adsorption: Development of an Equilibrium Theory for Binary Gas Mixtures with Nonlinear Isotherms," *Chem. Eng. Sci.* **44**, 1–8 (1989).

5. D. J. Ball, M.S.Ch.E. Thesis, Ohio State University, Columbus, OH, 1986.

6. C. W. Skarstrom, "Use of Phenomena in Automatic Plant Type Gas Analyzers," *Ann. N.Y. Acad. Sci.* **72**, 751–63 (1959).

7. M. J. Matz and K. S. Knaebel, "Pressure Swing Adsorption: Effects of Incomplete Purge," *AIChE J.* **34**(9), 1486–92 (1988).

8. P. C. Wankat, "Feed–Purge Cycles in Pressure Swing Adsorption," *Separ. Sci. and Tech.* (submitted 1992).

9. I. Rousar and P. Ditl, "Pressure Swing Adsorption: Analytical Solution for Optimum Purge," *Chem. Eng. Sci.* (submitted 1992).

10. J. C. Kayser and K. S. Knaebel, "Integrated Steps in Pressure Swing Adsorption Cycles," *Chem. Eng. Sci.* **44**, 3015–22 (1988).

11. S.-S. Suh and P. C. Wankat, "Combined Cocurrent-Countercurrent Blowdown Cycle in Pressure Swing Adsorption," *AIChE J.* **35**, 523–26 (1989).

12. J. E. Collins and K. S. Knaebel, Paper presented at the AIChE Annual Meeting, San Francisco, CA (1989).

13. K. S. Knaebel and F. B. Hill, "Pressure Swing Adsorption: Development of an Equilibrium Theory for Gas Separations," *Chem. Eng. Sci.* **40**, 2351–60 (1985).

14. J. L. Wagner, "Selective Adsorption Process," U.S. Patent No. 3,430,418 (1969).

15. P. C. Wankat, *Large Scale Adsorption and Chromatography*, I, CRC Press, Boca Raton, FL., 95 (1986).

16. R. T. Yang and S. J. Doong, "Gas Separation by Pressure Swing Adsorption: A Pore Diffusion Model for Bulk Separation," *AIChE J.* **31**, 1829–42 (1985).

17. S. J. Doong and R. T. Yang, "Bulk Separation of Multicomponent Gas Mixtures by Pressure Swing Adsorption: Pore/Surface Diffusion and Equilibrium Models," *AIChE J.* **32**, 397–410 (1986).

18. R. T. Yang, *Gas Separation by Adsorption Processes*, Butterworths, Boston, MA, 328 (1987).

19. N. F. Kirkby and C. N. Kenney, "The Role of Process Steps in Pressure Swing Adsorption," *Fundam. of Adsorption*, Engng. Foundation, New York, 325 (1987).

20. J. C. Kayser and K. S. Knaebel, "Pressure Swing Adsorption: Experimental Study of an Equilibrium Theory," *Chem. Eng. Sci.* **41**, 2931–38 (1986).

21. S.-S. Suh and P. C. Wankat, "A New Pressure Swing Adsorption Process for High Enrichment and Recovery," *Chem. Eng. Sci.* **44**, 567–74 (1989).
22. S. Sircar, "Air Fractionation by Adsorption," *Separ. Sci. and Tech.* **24**(14 & 15), 2379–96 (1988).
23. K. P. Kolliopoulos, M.S.Ch.E. Thesis, Ohio State University, Columbus, OH, 1987.
24. J. E. Mitchell and L. H. Shendalman, "A Study of Heatless Adsorption in the Model System CO in H. II," *AIChE Symp. Ser.* **69**, 25 (1973).
25. G. Flores Fernandez and C. N. Kenney, "Modelling the Pressure Swing Air Separation Process," *Chem. Eng. Sci.* **38**, 827–34 (1983).
26. M. J. Matz and K. S. Knaebel, "Temperature Front Sensing for Feed Step Control in Pressure Swing Adsorption," *Ind. Eng. Chem. Res.* **26**(8), 1638 (1987).
27. R. R. Hill and K. S. Knaebel, "Effects of Combined Steps in Pressure Swing Adsorption Cycles: An Experimental and Theoretical Study," *Adsorption: Fundam. and Applic.*, Proc. China-Jap.–USA Symp. on Adv. Ads. Separ. Sci. and Tech., Zhejiang Univ. Press (1988).
28. J. J. Collins, "Air Separation by Adsorption," U.S. Patent No. 4,026,680 (1977).
29. D. R. Garg and C. M. Yon, *Chem. Eng. Prog.* **82**(2), 54–60 (1986).
30. G. W. Miller, K. S. Knaebel, and K. G. Ikels, "Equilibria of Nitrogen, Oxygen, Argon, and Air in Molecular Sieve 5A," *AIChE J.* **33**, 194–201 (1987).
31. I. Rousar and P. Ditl, "Optimization of Pressure Swing Adsorption Equipment: Part I.," *Chem. Eng. Commun.* **70**, 67–91 (1988).
32. R. Kumar, "Adsorption Column Blowdown: Adiabatic Equilibrium Model for Bulk Binary Gas Mixtures," *Ind. Eng. Chem. Research* **28**, 1677–83 (1989).
33. Z. P. Lu, J. M. Loureiro, M. D. LeVan, and A. E. Rodrigues, "Intraparticle Convection Effect on Pressurization and Blowdown of Adsorbers," *AIChE J.* **38**, 857–67 (1992).
34. Z. P. Lu, J. M. Loureiro, M. D. LeVan, and A. E. Rodrigues, "Intraparticle Diffusion/Convection Models for Pressurization and Blowdown of Adsorption Beds with Langmuir Isotherm," *Separ. Sci. Tech.* **27**, 1857–74 (1992).
35. Z. P. Lu, J. M. Loureiro, A. E. Rodrigues, and M. D. LeVan, "Pressurization and Blowdown of Adsorption Beds," *Chem. Eng. Sci.* **48**, 1699 (1993).
36. A. E. Rodrigues, J. M. Loureiro, and M. D. LeVan, *Gas Sep. and Purification*, **5**, 115 (1991).
37. H. Cheng and F. B. Hill, *AIChE J.* **31**, 95 (1985).
38. D. M. Scott, *Chem. Eng. Sci.* **46**, 2977 (1991).
39. J. Hart, M. J. Baltrum, and W. J. Thomas, *Gas Sep. and Purification*, **4**, 97 (1990).
40. N. Sundaram and P. C. Wankat, *Chem. Eng. Sci.* **43**, 123 (1988).
41. S. J. Doong and R. T. Yang, *AIChE Symp. Ser.* **84**(284), 145 (1989).
42. R. L. Pigford, Private communication. July 23, 1986.

CHAPTER

5

Dynamic Modeling of a PSA System

The simplest approach to the modeling of a PSA separation process involves the use of equilibrium theory, which has been discussed in the previous chapter. The advantage of this approach is that it allows analytic solution of the governing material balance equations by the method of characteristics. The closed-form equilibrium theory solutions provide preliminary design guidance and useful insight into the system behavior. The quantitative value of this approach is, however, restricted to idealized systems in which the adsorption selectivity is based on differences in equilibrium and there are no significant dispersive effects such as axial mixing or finite resistance to mass transfer. Under these conditions a perfectly pure raffinate product is obtained. Equilibrium theory does not allow easy extension to the more realistic situation where dispersive effects are significant and product purity is limited. Moreover, in real PSA systems (equilibrium controlled) there are two problems with this approach. In bulk separations the velocity varies through the bed, and, although an analytic solution for the concentration front may still be obtained, except in the case of a linear isotherm, the solution is in the form of a cumbersome integral which generally requires numerical evaluation.[1] A more serious difficulty arises in tracking the concentration waves for adsorption and desorption in partially loaded beds since, depending on the initial profile and the form of the equilibrium relationship, one may observe the formation of combined wave fronts (e.g., partial shock plus simple waves). Under these conditions the simple model is no longer adequate and it is necessary to track both waves and the transition point simultaneously.[2] During pressure changes the characteristic lines are curved and the task of tracking the shock and simple waves becomes even more difficult. Furthermore, the equilibrium theory approach is clearly not applicable to systems in which separation is based on kinetic selectivity.

The alternative route, which is discussed in this chapter, is to develop a dynamic simulation model, including the effects of axial mixing and mass transfer resistance. Such dispersive effects are always likely to be present in real systems, even when equilibrium controlled. The dynamic simulation model is therefore more realistic and sufficiently general to be applied for detailed optimization studies of both classes of process. However, unlike the equilibrium theory approach, dynamic simulation involves tracking the transient by repeated numerical integration of the governing equations. This approach, therefore, provides the advantages of flexibility and greater accuracy at the expense of increased computation. Both the simpler linear driving force (LDF) approximation and more detailed Fickian diffusion equations have been used to model the effect of mass transfer resistance. In an equilibrium-controlled process the detailed form of the kinetic model is of only secondary importance, and it is found that very little advantage is gained from using the more realistic pore diffusion model. Therefore, for equilibrium-controlled separations the LDF model proves adequate for all operating conditions, whereas a more detailed mass transfer model is sometimes necessary for separations based on kinetic selectivity.

5.1 Summary of the Dynamic Models

The theoretical modeling of a PSA system has been widely studied in order to gain a clearer understanding of this rather complex process. A summary of the published dynamic models for PSA systems in chronological order is compiled in Table 5.1. These models are based on a one- or two-bed process operated on a Skarstrom cycle or on a modified cycle depending on the requirements of the particular system. Because of the transient nature of the process and the complexity of the equations describing the system dynamics, the growth of PSA modeling has followed the route of gradual development by progressive elimination of the simplifying restrictions. Starting from very simple models, which are valid for only a few real PSA systems, it is now possible to include an adequate representation of all the more important factors that may affect performance, and thus to obtain an adequate quantitative model which can be extended to almost any PSA process.

Detailed numerical simulations have in general been developed only for single-bed or two-bed systems, but since the simulation gives the effluent composition as a function of time, the extension to a multiple-bed process is, in principle, straightforward. However, although multiple-bed systems are widely used in industry, a detailed report of a multibed process simulation has been published only for hydrogen purification.[27]

Table 5.1. Summary of Dynamic Models for Pressure Swing Adsorption Systems

Authors	Experimental system	Selectivity	Operating cycle[a]	Fluid flow Model[b]	Approximations (pressurization and depressurization)	Equilibrium isotherm	Mass transfer model	Heat effects	Numerical method
Mitchell and Shendalman[3]	Purification of He by removing trace CO_2 on silica gel	Equilibrium controlled; CO_2 is strongly absorbed	Skarstrom cycle	Plug flow, constant velocity along the column	Frozen solid approximation and square wave change in column pressure	Linear	LDF, $\Omega = 15$; Knudsen control	Isothermal	Method of characteristics
Chihara and Suzuki[4]	Drying of air on activated alumina (theoretical study only)	Equilibrium controlled; moisture is strongly adsorbed	Skarstrom cycle	Plug flow, constant velocity along the column	Frozen solid approximation and square wave change in column pressure	Linear	LDF, $\Omega = 15$; Molecular diffusion control	Heat balance included	Finite difference
Chihara and Suzuki[5]	Drying of air on silica gel	Equilibrium controlled; moisture is strongly adsorbed	Skarstrom cycle	Plug flow, constant velocity along the column	Frozen solid approximation and square wave change in column pressure	Freundlich isotherm	LDF, cycle-time-dependent Ω; surface diffusion control	Heat balance included	Finite difference
Carter and Wyszynski[6]	Drying of air on silica gel (Sorbead R)	Equilibrium controlled; moisture is strongly adsorbed	Skarstrom cycle	Plug flow, constant velocity along the column	Frozen solid approximation and square wave change in column pressure	Approximated by two straight lines	LDF; both external film and surface diffusion taken into account	Isothermal	Finite difference
Raghavan, Hassan, and Ruthven[7]	Purification of He by removing trace CO_2 on silica gel (experimental) data from Ref. 3	Equilibrium controlled; CO_2 is strongly adsorbed	Skarstrom cycle	Axial dispersed plug flow, constant velocity along the column	Frozen solid approximation and square wave change in column pressure	Linear	LDF, $\Omega = 15$, molecular diffusion control	Isothermal	Orthogonal collocation
Cen, Chen and Yang[8]	Separation of $H_2/CH_4/H_2S$ mixture into three useful products on activated carbon	Equilibrium controlled; equilibrium affinity $H_2S > CH_4 > H_2$	One-bed, modified cycle with cocurrent blowdown, countercurrent vacuum desorption; H_2 pressurization from product end	Plug flow, velocity varies along the column; column pressure constant in H.P. feed step only	Full solution allowing for mass transfer between fluid and solid	Loading ratio correlation	LDF (best fit); compared with (numerically solved) equilibrium model to show that the mass transfer resistance is negligible	Heat balance included	Finite difference

(*Continued*)

Table 5.1. (*Continued*)

Authors	Experimental system	Selectivity	Operating cycle[a]	Fluid flow Model[b]	Approximations (pressurization and depressurization)	Equilibrium isotherm	Mass transfer model	Heat effects	Numerical method
Yang and Doong[9]	Separation of H_2/CH_4 mixture into two useful products on activated carbon	Equilibrium controlled; CH_4 is preferentially adsorbed	One-bed, modified cycle with cocurrent depressurization, countercurrent blowdown (followed by H_2 purge)	Plug flow, velocity varies along the column; column pressure changes in all the steps	Full solution allowing for mass transfer between fluid and solid	Loading ratio correlation	Pore model (macropore diffusion); pore diffusion unimportant (comparsion with equilibruim model)	Heat balance included	Finite difference; parabolic concentration profile assumed within a particle
Hassan, Raghavan, Ruthven, and Boniface[10]	Purification of He by removing trace C_2H_4 on 4A and 5A zeolites	Equilibrium controlled; C_2H_4 is strongly adsorbed	Skarstrom cycle	Axial dispersed plug flow, constant velocity along the column	Frozen solid approximation and square wave change in column pressure	Langmuir isotherm	LDF, $\Omega = 15$; micropore control in 4A; molecular diffusion control in 5A	Isothermal	Orthogonal collocation
Raghavan and Ruthven[11]	Air separation for N_2 production on carbon molecular sieve	Kinetically controlled; oxygen is the faster component	Skarstrom cycle	Axial dispersed plug flow, velocity varies along the column	Frozen solid approximation and square wave change in column pressure	Linear	LDF, $\Omega = 15$; micropore control	Isothermal	Double collocation
Doong and Yang[12]	Separation of $H_2/CH_4/CO_2$ mixture into three useful products on activated carbon	Equilibrium controlled with equilibrium affinity $CO_2 > CH_4 > H_2$	One-bed, modified cycle with cocurrent depressurization, countercurrent blowdown (followed by H_2 purge); H_2 used for product end pressurization	Plug flow, velocity varies along the column	Full solution allowing for mass transfer between fluid and solid	Loading ratio correlation and IAS theory; PSA results using these different models were close	Pore model accounting for surface and Knudsen diffusion	Heat balance included	Finite difference; parabolic concentration profile assumed within a particle

(*Continued*)

Table 5.1. (*Continued*)

Authors	Experimental system	Selectivity	Operating cycle[a]	Fluid flow Model[b]	Approximations (pressurization and depressurization)	Equilibrium isotherm	Mass transfer model	Heat effects	Numerical method
Hassan, Ruthven, and Raghavan[13]	Air separation for N_2 production on a carbon molecular sieve	Kinetically controlled; oxygen is the faster component	Skarstrom cycle	Axial dispersed plug flow, velocity varies along the column	Frozen solid approximation and square wave change in column pressure	Binary Langmuir isotherm	LDF, cycle time dependent Ω; micropore control	Isothermal	Orthogonal collocation
Raghavan, Hassan, and Ruthven[14]	Drying of air on activated alumina (theoretical study with physical parameters from Ref. 4)	Equilibrium controlled; moisture is strongly adsorbed	Skarstrom cycle	Axial dispersed plug flow, constant velocity along the column	Frozen solid approximation and square wave change in column pressure	Langmuir isotherm	Pore model (macropore diffusion); compared with LDF model for correlation of Ω vs cycle time	Isothermal	Orthogonal collocation
Cen and Yang[15]	Separating H_2/CO mixture into two useful products on activated carbon	Equilibrium controlled, CO is preferentially adsorbed	One-bed modified cycle with cocurrent depressurization, countercurrent blowdown (follwed by H_2 purge); H_2 used for product end pressurization	Plug flow, velocity varies along the column; column pressure remains constant during high pressure feed step only	Full solution allowing for mass transfer between fluid and solid	Loading ratio correlation	LDF model, $\Omega = 15$; diffusional time constant obtained by fitting experimental data	Heat balance included	Finite difference
Shin and Knaebel[16]	Air separation for N_2 production on 4A zeolite (theoretical study)	Kinetically controlled; oxygen is the faster component	Skarstrom cycle	Axial dispersed plug flow, velocity varies along the column	Full solution allowing for mass transfer under linearly changing pressure	Linear	Pore model (micropore diffusion); constant diffusivity	Isothermal	Orthogonal collocation

(*Continued*)

Table 5.1. (*Continued*)

Authors	Experimental system	Selectivity	Operating cycle[a]	Fluid flow Model[b]	Approximations (pressurization and depressurization)	Equilibrium isotherm	Mass transfer model	Heat effects	Numerical method
Doong and Yang[17]	Separation of H_2/CH_4 mixture for high purity H_2 production on 5A zeolite	Equilibrium controlled; CH_4 is more strongly adsorbed	One-bed, modified cycle with cocurrent depressurization, countercurrent blowdown (followed by H_2 purge); H_2 used for product end pressurization	Plug flow, velocity varies along the column	Full solution allowing for mass transfer between fluid and solid	Loading ratio correlation	Bidisperse pore diffusion model	Heat balance included	Finite difference; parabolic concentration profile assumed in both crystals and pellets
Hassan, Raghavan, and Ruthven[18]	Air separation for N_2 production on a carbon molecular sieve	Kinetically controlled; oxygen in the faster component	Modified cycle with pressure equalization and no external purge	Axial dispersed plug flow, velocity varies along the column	Frozen solid approximation and square wave change in column pressure	Binary Langmuir isotherm	LDF, cycle time dependent Ω; micropore control	Isothermal	Orthogonal collocation
Farooq, Hassan, and Ruthven[19]	Purification of He by removing trace C_2H_4 on 5A zeolite	Equilibrium controlled; C_2H_4 is strongly adsorbed	Skarstrom cycle	Axial dispersed plug flow, constant velocity along the column	Frozen solid approximation and square wave change in column pressure	Langmuir isotherm	LDF, $\Omega = 15$; molecular diffusion control	Heat balance included	Orthogonal collocation
Shin and Knaebel[20]	Air separation for N_2 production on a RS-10 molecular sieve	Kinetically controlled; oxygen is the faster component	Single bed, Skarstrom cycle with purge from product tank	Axial dispersed plug flow, velocity varies along the column	Full solution allowing for mass transfer between fluid and solid	Linear	Pore model (micropore diffusion); constant diffusivity	Isothermal	Orthogonal collocation
Kapoor and Yang[21]	Separation of CH_4/CO_2 mixture into two useful products on a carbon molecular sieve	Kinetically controlled; CO_2 is the faster component	One-bed, modified cycle with cocurrent depressurization, countercurrent blowdown and countercurrent evacuation	Plug flow, velocity varies along the column; column pressure changes in all the steps	Full solution allowing for mass transfer between fluid and solid	Binary Langmuir isotherm	LDF, cycle time dependent Ω; micropore control	Isothermal	Finite difference

(*Continued*)

Table 5.1. (*Continued*)

Authors	Experimental system	Selectivity	Operating cycle[a]	Fluid flow Model[b]	Approximations (pressurization and depressurization)	Equilibrium isotherm	Mass transfer model	Heat effects	Numerical method
Farooq, Ruthven, and Boniface[22]	Air separation for O_2 production on 5A zeolite	Equilibrium controlled; N_2 is more strongly adsorbed	Skarstrom cycle	Axial dispersed plug flow, velocity varies along the column	Full solution allowing for mass transfer under linearly changing pressure	Binary Langmuir isotherm	LDF; $\Omega = 15$; molecular diffusion control	Isothermal	Orthogonal collocation
Liow and Kenney[23]	Air separation for O_2 production on 5A zeolite	Equilibrium controlled; N_2 is more strongly adsorbed	Backfill cycle	Axial dispersed plug flow, velocity varies along the column	Full solution allowing for mass transfer	Ideal adsorbed solution theory	LDF	Isothermal	Orthogonal collocation
Ackley and Yang[24]	Separation of N_2/CH_4 mixture on a carbon molecular sieve for upgrading natural gas	Kinetically controlled; nitrogen is the faster component	Modified cycle (one-bed) with evacuation replacing blowdown and purge	Plug flow, velocity varies along the column	Frozen solid approximation and square wave change in column pressure	Binary Langmuir isothermal	LDF; $\Omega = 15$; micropore control	Isothermal	Method of characteristics
Ritter and Yang[25]	Purification of air by removing trace dimethyl methyl phosphate on activated carbon	Equilibrium controlled; DMMP is very strongly adsorbed	One-bed process, Skarstrom cycle using uncontaminated air as purge gas	Plug flow, constant velocity along the column	Frozen solid approximation and square wave change in column pressure	Langmuir isothermal	LDF; rate constant obtained by fitting the purity from a run with significant breakthrough	Isothermal	Finite difference
Farooq and Ruthven[26]	Air separation for N_2 production on a carbon molecular sieve	Kinetically controlled; oxygen is the faster component	Modified cycle with pressure equalization and no purge	Axial dispersed plug flow model, velocity varies along the column	Full solution allowing for mass transfer between fluid and solid (a square wave change in pressure)	Binary Langmuir isotherm	Pore model (micropore diffusion); variable diffusivity	Isothermal	Orthogonal collocation

[a] Two-bed process unless otherwise stated

[b] All the models assume negligible frictional pressure drop. Total column pressure remains constant during high pressure feed and desorption steps unless otherwise stated. To account for changing column pressure, unless any simplification is indicated, the actual pressure–time history (or appropriate best fit equation) was used.

Many of the models have been tested experimentally for particular systems, but no attempt is made here to review the results of such studies on an individual basis. Rather we have attempted to provide a conceptual summary in which the models are discussed in terms of their salient features. The models may be differentiated according to the following aspects:

1. The fluid flow pattern (generally plug flow or axially dispersed plug flow).
2. Constant or variable fluid velocity.
3. The form of the equilibrium relationship(s).
4. The form of the kinetic rate expression(s).
5. The inclusion of heat effects (isothermal/nonisothermal).
6. The numerical methods used to solve the system of equations.

5.1.1 Fluid Flow Models

Flow through an adsorption column is a PSA system is no different from flow through any fixed adsorbent bed. The flow pattern may therefore be adequately represented by the axial dispersed plug flow model. A mass balance for component i over a differential volume element yields:

$$-D_L \frac{\partial^2 c_i}{\partial z^2} + \frac{\partial}{\partial z}(v c_i) + \frac{\partial c_i}{\partial t} + \frac{1-\varepsilon}{\varepsilon} \frac{\partial \bar{q}_i}{\partial t} = 0 \tag{5.1}$$

In this model the effects of all mechanisms that contribute to axial mixing are lumped together into a single effective axial dispersion coefficient. More detailed models that include, for example, radial dispersion are generally not necessary. When mass transfer resistance is significantly greater than axial dispersion, one may neglect the axial dispersion term and assume plug flow. Axial dispersion is generally not important for large industrial units. In small laboratory units the axial mixing may be more significant due to the tendency of the smaller particles to stick together to form clusters that act effectively as single particles in their effect on the fluid flow. Subject to the plug flow approximation Eq. 5.1 reduces to:

$$\frac{\partial}{\partial z}(v c_i) + \frac{\partial c_i}{\partial t} + \frac{1-\varepsilon}{\varepsilon} \frac{\partial \bar{q}_i}{\partial t} = 0 \tag{5.2}$$

However, when the equations are to be solved numerically, it is generally advantageous to retain the form of Eq. 5.1 since inclusion of the axial dispersion term eliminates discontinuities in the slope of the concentration profile. The solution for the plug flow situation is then generated simply by assigning a very large value to the axial Peclet number (vL/D_L). This approach also allows easy investigation of the effect of axial dispersion on the cyclic steady-state performance.

Trace Systems

When the adsorbed component is present at low concentration in a large excess of an inert carrier (which is more or less true in purification processes

DYNAMIC MODELING OF A PSA SYSTEM 173

such as air drying and hydrogen purification), the change in the gas velocity through the bed due to adsorption/desorption can be neglected. Provided that the pressure drop through the bed is small, the interstitial velocity can therefore be considered as constant. In fact the frictional pressure drop in most actual systems is not very large and may usually be neglected.[28] For a trace system Eq. 5.1 therefore becomes:

$$-D_L \frac{\partial^2 c_i}{\partial z^2} + v\frac{\partial c_i}{\partial z} + \frac{\partial c_i}{\partial t} + \frac{1-\varepsilon}{\varepsilon}\frac{\partial \bar{q}_i}{\partial t} = 0 \tag{5.3}$$

Bulk Separation

When the mole fraction of the adsorbable component (or components) in the feed is large, the condition for constant velocity is no longer fulfilled and a more detailed analysis to account for the variation in velocity through the adsorbent bed is required, based on the continuity condition (assuming negligible pressure drop):

$$\sum_{i=1}^{n} c_i = C \ne f(z) \tag{5.4}$$

Constant Column Pressure

C in Eq. 5.4 is a constant when the adsorption column is operated at a constant total pressure. Therefore under constant column pressure condition the overall material balance equation, which gives the variation of fluid velocity through the column, takes the form:

$$C\frac{\partial v}{\partial z} + \frac{1-\varepsilon}{\varepsilon}\sum_{i=1}^{n}\frac{\partial \bar{q}_i}{\partial t} = 0 \tag{5.5}$$

Combining Eqs. 5.1 and 5.5, the component material balance equation for bulk separation at constant column pressure is obtained:

$$-D_L \frac{\partial^2 c_i}{\partial z^2} + v\frac{\partial c_i}{\partial z} + \frac{\partial c_i}{\partial t} + \frac{1-\varepsilon}{\varepsilon}\left(\frac{\partial \bar{q}_i}{\partial t} - y_i \sum_{i=1}^{n}\frac{\partial \bar{q}_i}{\partial t}\right) = 0 \tag{5.6}$$

Assuming that the ideal gas law holds [i.e., $c_i = Py_i/(R_g T_0)$], the component and material balance equations become:

$$-D_L \frac{\partial^2 y_i}{\partial z^2} + v\frac{\partial y_i}{\partial z} + \frac{\partial y_i}{\partial t} + \frac{1-\varepsilon}{\varepsilon}\frac{R_g T_0}{P}$$
$$\times \left(\frac{\partial \bar{q}_i}{\partial t} - y_i \sum_{i=1}^{n}\frac{\partial \bar{q}_i}{\partial t}\right) = 0 \tag{5.7}$$

$$\frac{\partial v}{\partial z} + \frac{R_g T_0}{P}\frac{1-\varepsilon}{\varepsilon}\sum_{i=1}^{n}\frac{\partial \bar{q}_i}{\partial t} = 0 \tag{5.8}$$

Variable Column Pressure

When the column pressure changes with the time, the overall material balance equation is:

$$C\frac{\partial v}{\partial z} + \frac{\partial C}{\partial t} + \frac{1-\varepsilon}{\varepsilon}\sum_{i=1}^{n}\frac{\partial \bar{q}_i}{\partial t} = 0 \tag{5.9}$$

which when substituted into Eq. 5.1 yields:

$$-D_L\frac{\partial^2 c_i}{\partial z^2} + v\frac{\partial c_i}{\partial z} + \frac{\partial c_i}{\partial t} + \frac{1-\varepsilon}{\varepsilon}\left(\frac{\partial \bar{q}_i}{\partial t} - y_i\sum_{i=1}^{n}\frac{\partial \bar{q}_i}{\partial t}\right)$$

$$- y_i\frac{\partial C}{\partial t} = 0 \tag{5.10}$$

Applying the ideal gas law, the component and overall material balance equations, under variable column pressure, assume the following form:

$$-D_L\frac{\partial^2 y_i}{\partial z^2} + v\frac{\partial y_i}{\partial z} + \frac{\partial y_i}{\partial t} + \frac{1-\varepsilon}{\varepsilon}\frac{R_g T_0}{P}$$

$$\times\left(\frac{\partial \bar{q}_i}{\partial t} - y_i\sum_{i=1}^{n}\frac{\partial \bar{q}_i}{\partial t}\right) = 0 \tag{5.11}$$

$$\frac{\partial v}{\partial z} + \frac{1}{P}\frac{\partial P}{\partial t} + \frac{R_g T_0}{P}\frac{1-\varepsilon}{\varepsilon}\sum_{i=1}^{n}\frac{\partial \bar{q}_i}{\partial t} = 0 \tag{5.12}$$

At this point it is important to recall that the elementary steps that constitute a PSA Skarstrom cycle are pressurization, purified product removal during the high-pressure feed step, and countercurrent blowdown to the low pressure followed by the low-pressure purge step. Modified versions of this cycle include various combinations of the following steps: pressure equalization or cocurrent depressurization (before the countercurrent blowdown step), vacuum desorption or low-pressure desorption without a purge stream, and partial pressurization with product stream before pressurization with feed (see Chapter 3). The form of the fluid flow model appropriate for each of the elementary steps will be determined by the nature of the separation (purification or bulk separation) and the pressure history of the column over a complete cycle. The standard (Danckwerts) inlet and exit boundary conditions for a dispersed plug flow system[29] apply for the component material balance in all the elementary steps. The velocity boundary conditions follow from the physical conditions controlling the cycle operation. Details of the flow boundary conditions are given in Table 5.2.

In an actual operation the column pressure changes continuously, and a model including the variable pressure condition should more closely represent the real situation. An interesting observation is that Eqs. 5.7 and 5.11, which are the component balance equations for constant and variable column pressure conditions, respectively, are of similar form. The minor difference arises from the fact that some of the coefficients in Eq. 5.11 are functions of time, but that does not introduce any additional complexity in the numerical

DYNAMIC MODELING OF A PSA SYSTEM

Table 5.2. Model Equations for PSA Simulation Using LDF Approximation[a]

The equations are written in general terms for component i ($= A$ for component A and $= B$ for component B) in bed j ($=1$ for bed 1 and $=2$ for bed 2). Although the subscript j should ideally appear with all the dependent variables, it is not shown here for simplicity.

Fluid-phase mass balance:

$$-D_L \frac{\partial^2 C_i}{\partial z^2} + v \frac{\partial c_i}{\partial z} + c_i \frac{\partial v}{\partial z} + \frac{\partial c_i}{\partial t} + \frac{1-\varepsilon}{\varepsilon} \frac{\partial \bar{q}_i}{\partial t} = 0 \qquad (1)$$

Continuity condition:

$$\sum_i c_i = C \neq f(z)$$

$$= f(t), \quad \text{pressurization and blowdown} \qquad (2a)$$
$$\neq f(t), \quad \text{high-pressure adsorption and purge} \qquad (2b)$$

Overall mass balance: high-pressure adsorption and purge (or a constant pressure step in general)

$$C \frac{\partial v}{\partial z} + \frac{1-\varepsilon}{\varepsilon} \sum_i \frac{\partial \bar{q}_i}{\partial t} = 0 \qquad (3)$$

pressurization and blowdown (or a variable pressure step in general)

$$C \frac{\partial v}{\partial z} + \frac{\partial C}{\partial t} + \frac{1-\varepsilon}{\varepsilon} \sum_i \frac{\partial \bar{q}_i}{\partial t} = 0 \qquad (4)$$

Mass transfer rates:

$$\frac{\partial \bar{q}_i}{\partial t} = k_i (q_i^* - \bar{q}_i) \qquad (5)$$

Adsorption equilibrium:

$$\frac{q_i^*}{q_{is}} = \frac{b_i c_i}{1 + \sum_i b_i c_i} \qquad (6)$$

Boundary conditions for fluid flow: pressurization, high-pressure adsorption, and purge

$$D_L \frac{\partial c_i}{\partial z}\bigg|_{z=0} = -v|_{z=0}(c_i|_{z=0^-} - c_i|_{z=0}); \quad \frac{\partial c_i}{\partial z}\bigg|_{z=L} = 0 \qquad (7)$$

$$(c_i|_{z=0^-})_{\text{purge}} = \frac{P_L}{P_H} (c_i|_{z=L})_{\text{adsorption}} \qquad (8)$$

blowdown

$$\frac{\partial c_i}{\partial z}\bigg|_{z=0} = 0; \quad \frac{\partial c_i}{\partial z}\bigg|_{z=L} = 0 \qquad (9)$$

Equation 7, which defines the standard (Danckwerts) inlet and exit boundary conditions for a dispersed plug flow system, reduces to Eq. 9 when the inlet velocity is set to zero. Similar boundary conditions apply for desorption without external purge. Equation (8) defines inlet gas

(Continued)

Table 5.2. (*Continued*)

concentration of the bed undergoing purge in terms of raffinate product concentration and is not applicable for a self-purging cycle.
Velocity boundary conditions:

$$v|_{z=0} = v_0 = f(P), \quad \text{pressurization} \tag{10a}$$

$$= v_{OH}, \quad \text{high-pressure adsorption} \tag{10b}$$

$$= 0, \quad \text{blowdown} \tag{10c}$$

$$= Gv_{OH}, \quad \text{purge;} \ G = 0 \ \text{for self-purging cycle} \tag{10d}$$

$$\left.\frac{\partial v}{\partial z}\right|_{z=L} = 0 \tag{11}$$

The additional velocity boundary condition at $z = L$ allows the convenience of using the same collocation coefficients for the velocity gradient as for the concentration gradient in the fluid phase.
Initial conditions: clean bed

$$c_i(z,0) = 0; \quad \bar{q}_i(z,0) = 0 \tag{12}$$

saturated bed

$$c_i(z,0) = c_0; \quad \bar{q}_i(z,0) = q_i^* \tag{13}$$

[a]The sum of the mole fractions of n components in the gas phase at every point in the bed is equal to one. Therefore solving for $(n-1)$ components in the gas phase is sufficient; the concentration of the remaining component in the gas phase is obtained by difference. This set of equations applies for flow from $z = 0$ to L. For flow from $z = L$ to 0 the $\partial/\partial z$ terms become negative, and the boundary conditions are interchanged.

---

integration. However, a full solution of the equations retaining the second term in Eq. 5.12 is very difficult and has not yet been attempted except in the work of Munkvold et al.[30] Several approximations have therefore been proposed to simplify the solution in an acceptable way.

The early PSA models[3-7] assume that the column pressure remains constant during the high- and low-pressure steps. It is further assumed that during pressurization and blowdown the solid phase remains frozen while the gas phase undergoes a square wave change in pressure. (During blowdown the mole fractions in the gas phase remains the same as at the end of the preceding high-pressure step while the pressure is reduced; during pressurization the residual gas profile is compressed so that it extends only through a fractional distance equal to the pressure ratio from the product end while the remainder of the bed is filled with feed.) These approximations are acceptable for purification processes operated on a Skarstrom cycle. The approximation of constant column pressure during the adsorption and desorption steps also holds for many bulk separation process cycles. However, the change in pressure is not instantaneous, and in a bulk separation process it becomes important to allow for mass transfer between fluid and solid during pressurization and blowdown, especially if the separation is equilibrium controlled. Yang and co-workers[8, 9, 12, 15, 17, 21] and Shin and Knaebel[20] have used the experimentally measured pressure–time history of the column (via a best fit equation) to account for the changing column pressure. The real situation is obviously best represented by this approach. In the absence of experimental data, however, an approximation is necessary. One approach is to consider that the column pressure varies either linearly or exponentially over the period of the pressurization or blowdown step.[22] An alternative approach is to assume that the column pressure changes instantaneously with the pressure change followed by mass transfer (at constant high or low pressure) between the gas and solid phases.[26] The former is a good approximation for an equilibrium-controlled separation, while the latter is more appropriate for kinetic separations. Experimentally measured pressure profiles for equilibrium-controlled air separation on 5A zeolite and kinetically controlled air separation on RS-10 (4A) molecular sieve are shown in Figure 5.1.

5.1.2 Equilibrium Isotherms

The pressure range of PSA operation often exceeds the pressure range over which experimental equilibrium data are available. Moreover, the multicomponent equilibria are commonly predicted from single-component isotherm data. Reliable models to represent both single and multicomponent adsorption equilibria are therefore an essential requirement.

Linear, Freundlich, and Langmuir isotherms have been used to define the single-component adsorption in PSA purification processes. Although the linear isotherm is the simplest equilibrium model, even a slight curvature of the isotherm influences the cyclic steady state of a PSA separation and should be considered. Since a PSA process involves both adsorption and desorption at the same temperature, simple qualitative reasoning suggests that the form of the isotherm should not deviate too greatly from linearity, otherwise either adsorption or desorption will become unacceptably slow. Moderate curvature of the isotherm (either type I or type II of Brunauer's classification) is acceptable, but it is obviously important that the portion of the isotherm over which the process operates should be completely reversible. Any hysteresis, as occurs for example in the alumina–water system,[33] (see Figure 2.5) will lead to an unacceptable buildup of the residual concentration in the adsorbed phase. In such a system PSA operation should be confined to the region below the point of inflection, where the isotherm is reversible.

The Langmuir model provides a reasonably good fit for most type I isotherms over a wide concentration range and for type II isotherms up to the inflection point. The Freundlich isotherm is also sometimes used, but, since it does not reduce to Henry's Law, it is likely to be less reliable in the low-concentration region. In the simulation of PSA purification processes

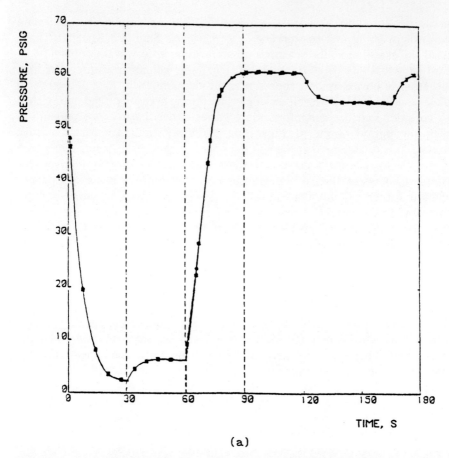

Figure 5.1 (a) Pressure profile in a small PSA column separating air on 5A zeolite for oxygen production. The steps in sequence are depressurization (30 s), purge (30 s), pressurization (30 s), and production (90 s). The pressure drop during the production step is due to additional product withdrawal for purging the other bed. Column size: 90 cm (length)×10 cm (i.d.), bed voidage = 0.35. (From Ref. 31; reprinted with permission.) (b) A representative pressure–time history in a small PSA column separating air on molecular sieve RS-10 for nitrogen production. The steps in sequence are pressurization (15 s), high-pressure feed (35 s), blowdown (6 s), and purge (3 s). Column size: 101.6 cm (length)×2.08 cm (i.d.), bed voidage = 0.6. (From Ref. 32; reprinted with permission.)

accurate representation of the Henry's Law region is obviously essential; so from this perspective the Langmuir equation is preferable.

The bulk PSA separation processes that have so far been modeled are for the most part binaries, together with a few ternary systems. To predict the mixture isotherms the extended Langmuir model, the loading ratio correlation (LRC) and ideal adsorbed solution theory (IAS) have been applied.

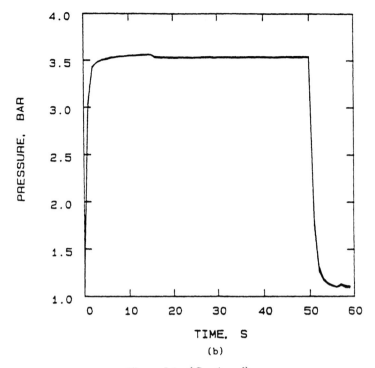

Figure 5.1 (*Continued*).

Extended Langmuir Model

$$\frac{q_i^*}{q_{is}} = \frac{b_i c_i}{1 + \sum_{i=1}^{n} b_i c_i} \qquad (5.13)$$

LRC Model

$$\frac{q_i^*}{q_{is}} = \frac{b_i c_i^{a_i}}{1 + \sum_{i=1}^{n} b_i c_i^{a_i}} \qquad (5.14)$$

The LRC model is essentially the multicomponent extension of the hybrid Langmuir–Freundlich equation. In the Langmuir–Freundlich equation the concentration terms are raised to arbitrary exponents to provide an improved empirical fit of the single-component data. Both extended Langmuir and LRC models provide explicit expressions for the adsorbed phase concentrations, so no iteration is needed during the simulation. By contrast, the IAS equations are implicit and an iterative subroutine is therefore needed to determine the composition of the equilibrium adsorbed phase. This increases the bulk of the computation so the simpler explicit equations are generally preferred, except in unusual situations.

Yang and co-workers[12] have reported that for the adsorption of various binary and ternary mixtures of CH_4, CO, CO_2, H_2, and H_2S on PCB activated carbon, the IAS and LRC methods give very similar results. It is also of interest to note that for these systems the exponent values used in the LRC model are close to unity. In a more recent study Yang and co-workers[34] have further shown that IAS and Langmuir models give very similar predictions for multicomponent adsorption of various mixtures of H_2, CH_4, CO, and CO_2 on 5A zeolite. The extended Langmuir model has also been successfully used to simulate the bulk PSA separation of methane–carbon dioxide[21] and nitrogen–methane[24] on carbon molecular sieve and oxygen–nitrogen on both 5A zeolite[22] and carbon molecular sieve.[13, 18, 26] One may conclude that for most practical systems there is little to be gained from using a more complex isotherm model.

5.1.3 Mass Transfer Models

The choice of an appropriate model to account for the resistance to mass transfer between the fluid and porous adsorbent particles is essential for any dynamic PSA simulation. The adsorbate gas must cross the external fluid film and penetrate into the porous structure during adsorption, and travel the same path in the reverse direction during desorption. The intraparticle transport by diffusion generally offers the controlling mass transfer resistance. The various mechanisms by which pore diffusion may occur have been discussed in Chapter 2. In an equilibrium-controlled PSA process macropore diffusion is often the major resistance to mass transfer. However, in the macropore control regime there is no significant kinetic selectivity. In a kinetically controlled process it is therefore desirable to operate under conditions such that all external mass transfer resistances are minimized, and the relative importance of the kinetically selective internal (micropore) diffusion process is maintained as large as possible.

Full simulations of PSA systems using pore diffusion models have been presented by Ruthven et al.[14] and by Shin and Knaebel.[16] The former study deals with macropore diffusion in a nonlinear trace system while the latter deals with micropore diffusion, with constant diffusivities, in a linear equilibrium system. Although the pore diffusion models are more realistic, the associated computations are very bulky. The linear driving force (LDF) model has therefore been widely used with varying degrees of success, regardless of the actual nature of the mass transfer resistance, since this approach offers a simpler and computationally faster alternative.

In the LDF model the mass transfer rate equation is represented as:

$$\frac{\partial q_i}{\partial t} = k_i(q_i^* - \bar{q}_i) \tag{5.15}$$

where

$$k_i = \Omega_i \frac{\epsilon_p D_p}{R_p^2} \frac{c_0}{q_0}, \quad \text{macropore control}$$

$$= \Omega_i \frac{D_c}{r_c^2}, \quad \text{micropore control} \tag{5.16}$$

In Eq. 5.15 q^* is the equilibrium value of the solid-phase concentration corresponding to fluid-phase concentration, c.

Nakao and Suzuki[35] have shown by solving the diffusion and LDF models independently for a single spherical particle subjected to alternate adsorption/desorption steps that, for cyclic processes, the value of Ω for macropore and micropore diffusion is not 15 (as suggested by Glueckauf and Coates[36]) but is in fact dependent on the frequency of the adsorption and desorption steps. They present a correlation from which the LDF constant, Ω may be estimated for any specified cycle time. Raghavan, Hassan, and Ruthven[14] in their study solved the pore diffusion model for a PSA system and by comparing the solutions derived from the simpler LDF model confirmed that Ω is indeed dependent on cycle time. The proposed correlation based on the full PSA simulation is, however, somewhat different from that proposed by Nakao and Suzuki based on a single-particle study, as may be seen from Figure 5.2. Farooq and Ruthven[26] ran a limited test to examine the validity of these correlations (based on a single-component study) for a binary system by comparing with constant-diffusivity pore model predictions. The results, shown in Figure 5.3, suggest that the LDF model with either correlation predicts the correct qualitative trends.

Alpay and Scott[37] addressed the same issue by a more fundamental approach using penetration theory. They assume that the dimensions of the adsorbent particle are sufficiently large that the concentration at the center is not significantly affected by the boundary condition at the particle surface and is therefore constant, even when the particle is subjected to a periodic change in surface concentration. Comparison of the LDF rate expression with the expression derived from the diffusion equation then yields $\Omega = 5.14/\sqrt{\theta_c}$, which, over the range $10^{-3} < \theta_c < 10^{-1}$ is very close to the correlation of Nakao and Suzuki.

Detailed studies of diffusion in microporous adsorbents reveal that, for both zeolites[38,39] and carbon molecular sieves,[40,41] the micropore diffusivity varies strongly with sorbate concentration. The concentration dependence of micropore diffusivity is even more pronounced in a binary system since,

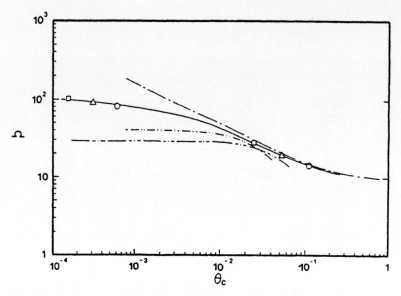

Figure 5.2 Dependence of Ω (constant in the LDF rate expression) upon cycle time. The correlations are given by Nakao and Suzuki[35] (— · — ·), Ruthven et al.[14] (— · · — for $\lambda = 0.05$ and Knudsen diffusion control, — · · — for $\lambda = 0.5$ and molecular diffusion control). Experimental data of Kapoor and Yang[21] (———). (Reprinted with permission.)

unlike single-component diffusion, binary diffusion is very sensitive to the concentration profile within the adsorbent particles. In a kinetically controlled PSA system the concentration profiles in the adsorbent particles are nonuniform and changing continuously. The concentration dependence of micropore diffusivity produces a more dramatic effect on the cyclic steady-state performance of a PSA separation than on the corresponding single breakthrough curve for a single column and must be considered when reliable extrapolation is required over a wide range of process conditions.

The constant-diffusivity model or the LDF model with an appropriately chosen value of Ω (Figure 5.2) can provide a qualitatively correct prediction of the effects of changes in process variables within a limited range, but such models do not predict correctly the effect of changes in the operating pressures. The deviations of the simplified models become more important at higher pressures where the effect of concentration dependence of the diffusivity is more pronounced.

The extension of the diffusion model to allow for concentration dependence of the diffusivities, however, adds considerably to the bulk of the numerical calculations. There is therefore a considerable incentive to adopt, where possible, the simpler LDF approach. A simple but practically useful way of minimizing quantitative disagreements at high operating pressures is to calibrate the Ω values by matching the model prediction of purity and

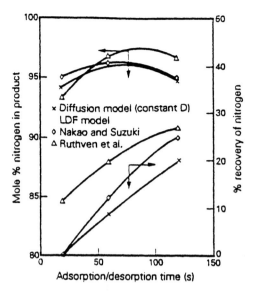

Figure 5.3 Comparison of the predictions of the diffusion and LDF models for a Skarstrom PSA cycle showing the performance of two available correlations for estimating the appropriate value of Ω as a function of cycle time. Pressurization/blowdown time = 20 s, $L/v_{OH} = 25$ s, $G = 1.0$, $D_A/r_c^2 = 3.73 \times 10^{-3}$ s^{-1}, $D_B/r_c^2 = 1.17 \times 10^{-4}$ s^{-1}, other parameters are same as in Table 5.8. (From Ref. 26.)

recovery with limited experimental data for the region of interest and then to use these Ω values to investigate the effects of the other operating variables. Kapoor and Yang[21] used this approach in their study of kinetic separation of methane from a mixture of methane and carbon dioxide over a carbon molecular sieve. Others[20] have used the constant-diffusivity pore diffusion model with the diffusivity values obtained by calibrating the model against limited number of experimental runs. The constant-diffusivity micropore diffusion model using calibrated effective diffusivities is, however, no better than a LDF model using the limiting diffusivity with the calibrated Ω values. In view of the computational efficiency, the latter approach appears preferable. The limitations of the LDF model discussed here are, however, not important for separations based on differences in adsorption equilibrium. For this class of separations, the LDF model approach is adequate in almost all situations.

5.1.4 Numerical Methods

Even the simplest PSA model including mass transfer resistance is not amenable to analytic solution, and efficient numerical methods for solving the coupled partial differential equations are therefore needed to solve the model. Commonly used techniques for solving partial differential equations are finite difference[42,43] and orthogonal collocation.[44-46] The dependent variables in the governing equations for a PSA system are functions of space and time. By applying either of the techniques the partial differential equations (PDEs) are discretized in space, thus reducing the PDEs into ordinary differential equations (ODEs), which are then integrated in the time domain using a standard numerical integration routine.[47,48] Algebraic equations (linear and/or nonlinear) may also appear together with the ODEs. In such cases Gaussian elimination for linear algebraic equations and Newton's method for nonlinear algebraic equations (or more sophisticated variations of these methods) are used simultaneously with the numerical integration routine for solving the coupled system of equations. Many standard computer programs[49,50] containing a variety of powerful integration routines and algebraic equation solvers are available. The software packages referred to here are those that are mentioned in the various reported studies of PSA simulation.

PDEs may also be reduced to ODEs by applying the method of characteristics. Whereas the method has been widely used to solve the equilibrium theory models, its application in the modeling of kinetic separations is quite limited.[3,24]

The main disadvantage of the finite difference method is the large number of segments needed to approximate the continuous system. This results in a large number of equations and therefore an increased integration load. On the other hand, one major advantage of orthogonal collocation is that it requires far fewer spatial discretization points to achieve the specified accuracy. From a comparative study of the two methods in relation to PSA simulation, Raghavan, Hassan, and Ruthven[7] concluded that orthogonal collocation is substantially more efficient in terms of computational CPU time.

5.2 Details of Numerical Simulations

The building blocks which constitute a dynamic simulation model for a PSA system have been discussed in a general way in the previous section. We now discuss the development of a complete simulation model. The simpler and computationally quicker LDF model approach as well as the more rigorous variable diffusivity micropore diffusion model approach are considered in detail with numerical examples.

5.2.1 LDF Model

PSA air separation for oxygen production on 5A zeolite is considered as an example of the LDF model approach. Detailed experimental and theoretical studies of this equilibrium-controlled separation have been presented by

DYNAMIC MODELING OF A PSA SYSTEM

Farooq, Ruthven, and Boniface,[22] using a simple two-bed process operated on a Skarstrom cycle (see Figure 3.4). The assumptions for the simulation model are:

1. The system is assumed to be isothermal.
2. Frictional pressure drop through the bed is negligible.
3. Mass transfer between the gas and the adsorbed phases is accounted for in all steps. The total pressure in the bed remains constant during the adsorption and the purge steps. During pressurization and blowdown the total pressure in the bed changes linearly with time.
4. The fluid velocity in the bed varies along the length of the column, as determined by the overall mass balance.
5. The flow pattern is described by the axial dispersed plug flow model.
6. Equilibrium relationships for the components are represented by extended Langmuir isotherms.
7. The mass transfer rates are represented by linear driving force rate expressions. Molecular diffusion controls the transport of oxygen and nitrogen in 5A zeolite.[51] For this particular operation, since the pressure is always greater than atmospheric, any contribution from Knudsen diffusion is neglected and the LDF constants are taken to be pressure dependent according to the correlation for molecular diffusion control given by Hassan et al.[10] This only affects the choice of mass transfer parameters and does not in any way alter the general form of the mass transfer rate equations.
8. The ideal gas law applies.
9. The presence of argon, which is adsorbed with almost the same affinity as oxygen and therefore appears with oxygen in the raffinate product, is ignored. This assumption reduces the mathematical model to a two-component system. Theoretically the model can be extended to any number of components, but the practicality is limited by the capability of the available integration routines.

The system of equations describing the cyclic operation subject to these assumptions is given in Table 5.2. Equations 1–11 in Table 5.2 are rearranged and written in dimensionless form. The dimensionless equations may then be solved by the method of orthogonal collocation to give the solid-phase concentrations of the two components as a function of the dimensionless bed length (z/L) for various values of time. Details of the collocation form are given in Appendix B. Computations are continued until cyclic steady state is achieved. In the study of air separation on 5A zeolite, depending on the parameter values, 15–25 cycles were required to approach the cyclic steady state.

The equilibrium and kinetic parameters used to simulate the experimental runs are summarized in Table 5.3, together with details of the adsorbent, the bed dimensions, and the cycle. The experimentally observed product purity and recovery for several operating conditions are summarized in Table 5.4.

Table 5.3. Kinetic and Equilibrium Data and Other Common Parameter values Used in the Simulations of PSA Air Separation for Oxygen Production

Feed composition	21% oxygen, 79% nitrogen
Adsorbent	Linde 5A zeolite
Bed length (cm)	35.0
Bed radius (cm)	1.75
Particle diameter (cm)	0.0707
Bed voidage	0.40
Ambient temperature (°C)	25.0
Blowdown pressure (atm)	1.0
Purge pressure (atm)	1.07 ± 0.05
Peclet number	500.0
Duration of step 1 or 3	0.3 of total cycle time
Duration of step 2 or 4	0.2 of total cycle time
Equilibrium constant for oxygen (K_A)	4.7^a
Equilibrium constant for nitrogen (K_B)	14.8^a
LDF constant for oxygen (k_A) (s^{-1})	62.0 (at 1 atm)^b
LDF constant for nitrogen (k_B) (s^{-1})	19.7 (at 1 atm)^b
Saturation constant for oxygen (q_{AS}) (mol/cm^3)	5.26×10^{-3c}
Saturation constant for nitrogen (q_{BS}) (mol/cm^3)	52.6×10^{-3d}

[a] Chromatographic data (dimensionless) (Boniface[52]).
[b] Molecular diffusion control, tortuosity factor = 3.0 and particle porosity = 0.33; all experimental conditions are within the large-cycle-time region, for which Ω approaches the Glueckauf limit of 15.
[c] Miller et al.[53]
[d] Since oxygen and nitrogen molecules are about the same size, their saturation capacities are assumed to be the same.

together with the theoretically predicted values from the numerical simulation. The mole fraction of oxygen in the product refers to the average oxygen concentration in the product at steady state. The theoretical oxygen concentration in the product at steady state was therefore computed at short intervals and was integrated to determine the average. Since the product rate rather than the purge rate was fixed, the recovery calculation was straightforward. The effects of cycle time, adsorption pressure, and product withdrawal rate on the purity and recovery are shown in Figure 5.4. It is evident that the theoretical model gives a reasonably accurate prediction of both the purity and recovery of the oxygen product over the range of experimental values examined.

The effect of varying the blowdown conditions was also investigated and the results are shown in Figure 5.4(a). There is clearly very little difference

Table 5.4. PSA Air Separation for Oxygen Production on 5A Zeolite: Summary of Experimental Conditions, Product Purity, and Recovery

Experiment No.	Feed flow rate[a] (cm^3/s)	Product flow rate[a] (cm^3/s)	Cycle time (s)	Adsorption pressure (atm)	Mole % O$_2$ in product		Recovery of O$_2$ (%)	
					Experiment	Theory	Experiment	Theory
1	25.0	1.13	100	1.48	80.0	93.4	17.0	20.1
						92.6[b]		19.9[b]
2	25.0	1.13	150	1.66	92.0	96.4	19.9	20.8
						96.2[b]		20.7[b]
3	25.0	1.13	200	1.73	86.0	78.2	18.5	16.8
						74.8[b]		16.1[b]
4	25.0	1.13	250	1.90	72.0	76.7	15.5	16.5
5	33.3	1.13	200	2.33	95.5	94.7	15.4	15.3
6	50.0	1.13	200	3.41	91.0	95.8	9.8	10.3
7	66.7	1.13	160	4.30	95.5	96.3	7.7	7.8
8	66.7	2.55	160	4.35	95.3	96.4	17.4	17.6
9	66.7	3.98	160	4.26	95.5	96.2	27.1	27.3

[a] 1 atm. 25°C.
[b] Instant pressure change assumed during blowdown. All other theoretical results correspond to linear pressure change during blowdown.
Source: From Ref. 22.

between the simulation results for an instantaneous pressure change or a linear pressure change during blowdown.

For two sets of operating conditions, represented by experiments 1 and 4 in Table 5.4, the effect of varying the mass transfer resistance was investigated theoretically. The results are summarized in Table 5.5. Under the conditions of experiment 1 a high-purity product is obtained, showing that the system must be operating without significant breakthrough. Reducing the mass transfer coefficient by a factor of 3 (case 2 of Table 5.5) gave very little change in either purity or recovery of the oxygen product, implying that under these conditions the system is operating close to equilibrium. Under the conditions of experiment 4 (Table 5.4) the effect of increasing the mass transfer resistance is more pronounced (case 3 and 4 of Table 5.5) since under these conditions there is significant breakthrough and any broadening of the concentration front as a result of increased mass transfer resistance leads to a lower-purity product. This simple investigation provides direct verification of the assumption that the dynamic LDF model can provide a reliable simulation of an equilibrium-controlled PSA system. Further direct support for this conclusion comes from the work of Cen, Cheng, and Yang.[8] For the separation of a H_2–CH_4–H_2S mixture on activated carbon, the concentration profiles calculated from both LDF and equilibrium theory models are practically identical (see Figure 5.5).

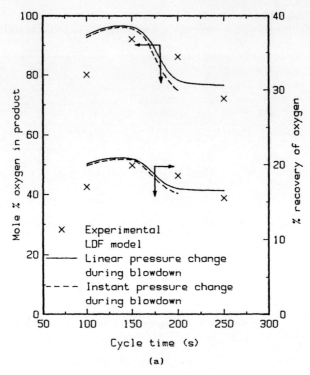

Figure 5.4 Effects of (a) cycle time (experiments 1–4 in Table 5.4), (b) adsorption pressure (experiments 3, 5, and 6 in Table 5.4) on purity and recovery of oxygen product in a dual-bed PSA air separation process operated on a Skarstrom cycle. Equilibrium and kinetic data and other common parameter values used for computing the LDF model predictions are given in Table 5.3. (From Ref. 22)

The simulation model discussed here may be applied to any other binary bulk separation using a two-bed process operated on a Skarstrom cycle. Perhaps a more important observation is that the model can handle mass transfer between fluid and solid adsorbent with forward and reverse flow under both constant and varying column pressure conditions. The linear pressure change approximation may be easily modified to include the actual pressure–time history either directly or through a best fit equation. The way in which the pressure equalization step is handled does not depend on the mass transfer model and is discussed in the context of the diffusion model. This simulation model therefore contains all the information necessary to simulate any other one- or two-bed PSA process operated on any of the simpler cycles. Although the computer code is written for a binary bulk separation, it is in fact possible to use the same computer code to simulate a purification process simply by assigning a zero value to the Langmuir constant for the second component and bypassing the subroutine that solves the

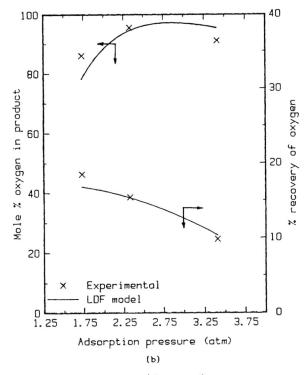

Figure 5.4 (*Continued*).

overall material balance equation. In this mode the computational load is only marginally greater than for the solution of the single-component, constant-velocity model. Further details of the modeling of trace PSA systems are given in papers of Raghavan et al.[7, 10]

For equilibrium-controlled separations the available Ω versus θ_c correlations (summarized in Figure 5.2) are directly applicable. These correlations cannot, however, account for the effect of the concentration dependence of micropore diffusivity, which, in a kinetically controlled separation, may be quite important. An empirical but practically useful way of overcoming this limitation was mentioned in Section 5.1. The effectiveness of such an approach is demonstrated by Kapoor and Yang[21] for the kinetic separation of methane and carbon dioxide on a carbon molecular sieve. They calibrated the Ω values with experiments conducted by varying the adsorption/desorption time and product withdrawal rate at 3.72 atm (high pressure). For a fixed cycle time the purity versus recovery profile obtained by varying product withdrawal rates was matched with LDF model predictions, and the pair of Ω values that gave the best fit was chosen. The same procedure was repeated for runs with different cycle times. The experimental and best fit profiles are shown in Figure 5.6, and the resultant empirical Ω versus θ_c correlation is

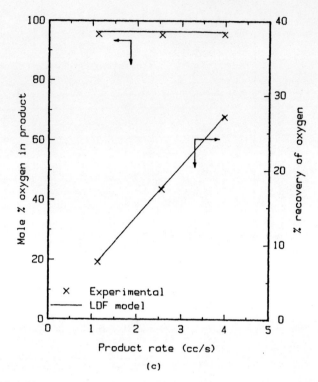

Figure 5.4 (c) Effect of product withdrawal rate (experiments 7–9 in Table 5.4) on purity and recovery of oxygen product in a dual-bed PSA air separation process operated on a Skarstrom cycle. Equilibrium and kinetic data and other common parameter values used for computing the LDF model predictions are given in Table 5.3. (From Ref. 22)

Table 5.5. Effect of Mass Transfer Resistance on Product Purity and Recovery (PSA Oxygen)[a]

No.	Feed rate (cm³/s)	Product rate (cm³/s)	k_A (s⁻¹)	k_B (s⁻¹)	Mole % O_2 in product	Recovery of O_2 (%)
1	25.0	1.13	62.0	19.7	93.40	20.10
2	25.0	1.13	20.0	6.6	93.29	20.08
3	25.0	1.13	62.0	19.7	76.70	16.50
4	25.0	1.13	20.0	6.6	75.78	16.30

[a] For 1 and 2: adsorption pressure = 1.48 atm, blowdown pressure = 1.0, purge pressure = 1.07 atm, cycle time = 100 s; for 3 and 4: adsorption pressure = 1.90 atm, blowdown pressure = 1.0, purge pressure = 1.1 atm, cycle time = 250 s. Other parameters as in Table 5.3.
Source: From Ref. 22.

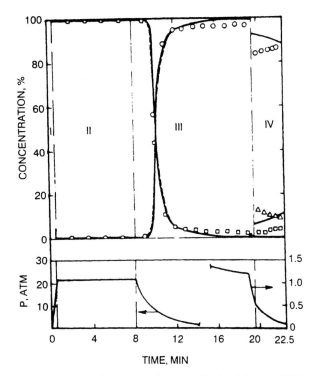

Figure 5.5 Effluent concentration at steady state obtained from a PSA cycle separating 49.5:59.5:1.0 $H_2:CH_4:H_2S$ mixture on activated carbon showing the close agreement between numerically solved equilibrium model (———) and LDF model (- - -). (○), (□) and (△) represent experimental data for CH_4, H_2, and H_2S, respectively. (From Ref. 8; reprinted with permission.)

presented in Figure 5.2 together with the other correlations. It is evident from Figure 5.7 that the LDF model using the Ω values derived from this calibration can correctly predict the effects of varying the product withdrawal rate and the high operating pressure in the range 2.36–3.72 atm. The calibrated Ω values approach the limiting value of 15 for large cycle times. It is important to note that this particular observation is specific to the system and the operating conditions and therefore should not be taken as a universal upper limit for the calibrated Ω values in the long-cycle-time region.

5.2.2 Variable-Diffusivity Micropore Diffusion Model

In most of the studies in which the LDF model has been successfully applied to PSA separations based on kinetic selectivity, key parameter values have been chosen empirically by matching the model predictions to experimental data. In this situation the model predictions can be considered reliable only

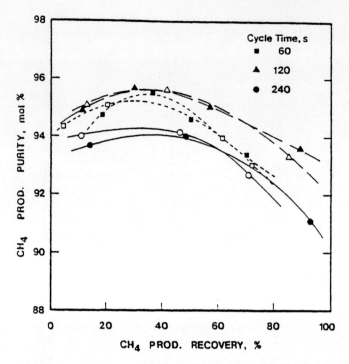

Figure 5.6 Performance curves for PSA separation of CH_4–CO_2 mixture on a carbon molecular sieve. The cycle consists of four steps of equal duration: pressurization, product withdrawal accompanied by cocurrent depressurization, cocurrent blowdown, and countercurrent evacuation. Adsorption pressure = 3.72 atm, evacuation pressure = 0.34 atm, feed rate = 2.7 1 STP/cycle, product withdrawal rate varied from 0.21 to 1.35 1 STP/cycle; adsorbent diameter = 0.318 cm, bed length = 60.6 cm, bed radius = 2.05 cm (inner), bed voidage = 0.3. Solid symbols are experimental, open symbols are the best fit LDF model results obtained by adjusting Ω values. For equilibrium and kinetic data see Table 3.2. (From Ref. 21; reprinted with permission.)

over the range covered by the experiments. Such an approach is clearly unsuitable for the a priori prediction of system performance. A reliable and complete a priori estimate of PSA performance, based on independently measured single-component equilibrium and kinetic data over a wide range of operating conditions is a major target of PSA modeling. For a kinetically controlled PSA separation this requires the full micropore diffusion model including the concentration dependence of micropore diffusivity.[26] A simple two-bed process operated on a modified cycle with pressure equalization and no external purge (Figure 3.16) is considered, and the variable diffusivity micropore diffusion model is developed for a binary bulk separation. It is important to note that the extension of this model, even to three components, is not straightforward.

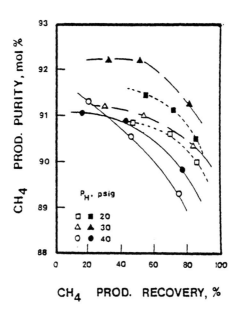

Figure 5.7 Experimental product purity and recovery (solid symbols) for PSA separation of CH_4–CO_2 mixture on a carbon molecular sieve are compared with the LDF model predictions (open symbols) using calibrated Ω values. The cycle consists of five steps of equal duration: pressurization, high-pressure feed, cocurrent depressurization, countercurrent blowdown, and countercurrent evacuation. Total cycle time = 150 s and feed rate = 2.8 l STP/cycle; other details are same as in the caption of Figure 5.6. (From Ref. 21; reprinted with permission.)

The assumptions 1, 2, 4, 5, and 6 made for the LDF model are retained here. Other approximations for the present model are:

1. Mass transfer between gas and adsorbed phases is accounted for in all steps except the pressure equalization, which is assumed to be instantaneous. During pressurization and blowdown steps a square wave change in pressure is assumed, with a constant pressure maintained throughout the adsorption or desorption step. These are good approximations for small-scale laboratory units, although in large industrial-scale operations the column pressure is never really constant, since the product withdrawal valve is normally opened before pressurization is complete. The modeling of varying column pressure does not depend on the kinetic model and has been discussed in greater detail in Section 5.1.
2. The pressure equalization step is difficult to handle in a rigorous manner. The approximate representation adopted by Hassan et al.[18] is therefore retained. It is assumed that the adsorbed phase concentration in both columns remains frozen during pressure equalization. Since the columns are either pressurized or depressurized simultaneously from both ends, it

is assumed that there is no mixing in the midsections of the columns. The feed and the product ends of the high-pressure column are connected to the product and feed ends, respectively, of the low-pressure column. The gas from the corresponding halves is assumed to be uniformly mixed, with due allowance for the difference in initial pressures and therefore in the number of moles of gas initially present in each half-column. The resulting gas mixtures are then uniformly distributed through the relevant halves of the two columns.
3. In the particle mass balance it is assumed that the adsorbent consists of uniform microporous spheres; any macropore diffusional resistance is neglected. This is a good approximation, since, in a kinetically selective adsorbent such as carbon molecular sieve, the diffusional resistance of the micropores is much larger than that of the macropores. (The relevant particle radius in the time constant is that of the microparticles.)
4. The gradient of chemical potential is taken as the driving force for micropore diffusion with a constant intrinsic mobility. This leads to a Fickian diffusion equation in which the diffusivity is a function only of the adsorbed-phase concentrations. Ideal Fickian diffusion with constant diffusivity is also investigated for comparison.
5. The fluid and the solid phases are linked through an external film resistance, even though a large value is usually assigned to the external film mass transfer coefficient to approximate equilibrium at the particle surface. In conjunction with the collocation method this proves simpler than the alternative approach involving the direct application of the equilibrium boundary condition at the particle surface.

The model equations subject to these assumptions are summarized in Table 5.6. Equations that are similar to those in Table 5.2 are not repeated in Table 5.6. Equations 1, 2a, 3, and 7–11 in Table 5.2 together with Eqs. 1, 2, 7 and the appropriate set of diffusion equations (Eqs. 4, 5, and 8 for the constant-diffusivity case, and Eqs. 11–14 and 4 for the concentration-dependent diffusivity case) in Table 5.6 are rearranged and written in dimensionless form. The dimensionless equations may then be solved by the method of orthogonal collocation to obtain the gas-phase composition as a function of dimensionless bed length (z/L), and the solid-phase composition as a function of both the dimensionless bed length and the dimensionless particle radius (r/R_p) for various values of time. Details of the collocation form are given in Appendix B. Starting from a given initial condition the computations are continued as usual until cyclic steady state is reached.

The air separation data for nitrogen production on a carbon molecular sieve reported by Hassan et al.[18] are chosen to illustrate the importance of the concentration dependence of the micropore diffusivity on the performance of the kinetically controlled PSA separation. The experiments, carried out in a two-bed PSA unit using the modified cycle with pressure equalization and no purge, were conducted over a wide range of high operating

Table 5.6. Equations for PSA Simulation Using Pore Diffusion Model

---

Except for the following changes all other equations in Table 5.2 apply. The subscript i has the same meaning as in Table 5.2.

Mass transfer rate across the external film:

$$\frac{\partial \bar{q}_i}{\partial t} = \frac{3}{R_p} k_f \left(c_i - c_i|_{r=R_p} \right) \tag{1}$$

Velocity boundary condition for pressurization:

$$v|_{z=L} = 0 \tag{2}$$

Equation 2, which replaces Eq. 10a in Table 5.2, is a more appropriate velocity boundary condition for the pressurization step. Moreover, with this boundary condition it is no longer necessary to specify pressurization gas quantity as an input.

Particle balance:

$$\frac{\partial q_i}{\partial t} = \frac{1}{r^2} \left[\frac{\partial}{\partial r} \left(D_i^2 \frac{\partial q_i}{\partial r} \right) \right] \tag{3}$$

Boundary conditions:

$$\left.\frac{\partial q_i}{\partial r}\right|_{r=0} = 0 \tag{4}$$

$$\left. D_i \frac{\partial q_i}{\partial r} \right|_{r=R_p} = k_f \left(c_i - c_i|_{r=R_p} \right) \tag{5}$$

$c_i|_{r=R_p}$ in Eq. 5 is related to $q_i|_{r=R_p}$ through the equilibrium isotherm:

$$\frac{q_i|_{r=R_p}}{q_{is}} = \theta_i = b_i c_i|_{r=R_p} \bigg/ \left(1 + \sum_i b_i c_i|_{r=R_p} \right) \tag{6}$$

Equation 6, written for the two components and then solved simultaneously, yields

$$c_i|_{r=R_p} = \frac{1}{b_i} \theta_i \bigg/ \left(1 - \sum_i \theta_i \right) \tag{7}$$

Constant Diffusivity. If the micropore diffusivity (D_i) is independent of concentration, Eq. 3 becomes:

$$\frac{\partial q_i}{\partial t} = D_i \left(\frac{\partial^2 q_i}{\partial r^2} + \frac{2}{r} \frac{\partial q_i}{\partial r} \right) \tag{8}$$

The associated boundary conditions, Eqs. 4 and 5, remain the same.

Concentration-dependent diffusivity. The expressions for the diffusivities in a binary Langmuir system with constant intrinsic mobilities (D_{A0}, D_{B0}) have been given by Habgood[54] and Round et al.[55a]:

$$D_A = \frac{D_{A0}}{1 - \theta_A - \theta_B} \left((1 - \theta_B) + \theta_A \frac{\partial q_B / \partial r}{\partial q_A / \partial r} \right) \tag{9}$$

$$D_B = \frac{D_{B0}}{1 - \theta_A - \theta_B} \left((1 - \theta_A) + \theta_B \frac{\partial q_A / \partial r}{\partial q_B / \partial r} \right) \tag{10}$$

(*Continued*)

Table 5.6. (*Continued*)

Except for the following changes all other equations in Table 5.2 apply. The subscript i has the same meaning as in Table 5.2.
The appropriate forms for the diffusion equations are obtained by substituting Eqs. 9 and 10 in the particle balance equations for component A ($i = A$ in Eq. 3) and component B ($i = B$ in Eq. 3):

$$\frac{\partial q_A}{\partial t} = \frac{D_{A0}}{1-\theta_A-\theta_B}\left[(1-\theta_B)\left(\frac{\partial^2 q_A}{\partial r^2} + \frac{2}{r}\frac{\partial q_A}{\partial r}\right) + \theta_A\left(\frac{\partial^2 q_B}{\partial r^2} + \frac{2}{r}\frac{\partial q_B}{\partial r}\right)\right]$$
$$+ \frac{D_{A0}}{(1-\theta_A-\theta_B)^2}\left((1-\theta_B)\frac{\partial \theta_A}{\partial r} + \theta_A\frac{\partial \theta_B}{\partial r}\right)\left(\frac{\partial q_A}{\partial r} + \frac{\partial q_B}{\partial r}\right) \quad (11)$$

$$\frac{\partial q_B}{\partial t} = \frac{D_{B0}}{1-\theta_A-\theta_B}\left[(1-\theta_A)\left(\frac{\partial^2 q_B}{\partial r^2} + \frac{2}{r}\frac{\partial q_B}{\partial r}\right) + \theta_B\left(\frac{\partial^2 q_A}{\partial r^2} + \frac{2}{r}\frac{\partial q_A}{\partial r}\right)\right]$$
$$+ \frac{D_{B0}}{(1-\theta_A-\theta_B)^2}\left[(1-\theta_A)\frac{\partial \theta_B}{\partial r} + \theta_B\frac{\partial \theta_A}{\partial r}\right]\left(\frac{\partial q_B}{\partial r} + \frac{\partial q_A}{\partial r}\right) \quad (12)$$

Similarly the appropriate boundary conditions for the two components at the particle surface are obtained by substituting Eqs. 9 and 10 in Eq. 5 written for components A and B and solving simultaneously for $(\partial q_A/\partial r)|_{r=R_p}$ and $(\partial q_B/\partial r)|_{r=R_p}$:

$$\left.\frac{\partial q_A}{\partial r}\right|_{r=R_p} = \frac{k_f}{D_{A0}}(1-\theta_A)(c_A - c_A|_{r=R_p}) - \frac{k_f}{D_{B0}}\theta_A(c_B - c_B|_{r=R_p}) \quad (13)$$

$$\left.\frac{\partial q_B}{\partial r}\right|_{r=R_p} = \frac{k_f}{D_{B0}}(1-\theta_B)(c_B - c_B|_{r=R_p}) - \frac{k_f}{D_{A0}}\theta_B(c_A - c_A|_{r=R_p}) \quad (14)$$

[a] Equations 9 and 10 are true for $q_{AS} = q_{BS}$; if this is not true, the expressions will contain additional terms.

Table 5.7. PSA Air Separation for Nitrogen Production on a Carbon Molecular Sieve: Summary of Experimental Conditions, Product Purity, Recovery, and Productivity[a]

Run No.	L/v_{OH} ratio(s)	Adsorption pressure (atm)	Mole % oxygen in product	% Recovery of nitrogen[b]	Productivity $\left(\dfrac{\text{cm}^3\,N_2}{\text{hr. cm}^3 \text{ adsorbent}}\right)$
1	25	3.0	10.5	56.4	81
2	25	4.4	7.5	53.7	106.2
3	25	5.8	6.0	49.2	137.4
4	25	6.8	4.4	42.1	135.75
5	37	3.0	4.0	29.2	27.4
6	37	4.4	1.8	21.6	30.76
7	37	5.8	0.7, 0.75	11.1	21.4
8	37	6.8	0.6, 0.7	7.7	17.3

[a] Feed composition = 21% oxygen, 79% nitrogen, blowdown/desorption pressure = 1 atm pressurization/blowdown time = 2 s, adsorption/desorption time = 60 s, pressure equalization = 2 s.
[b] Corrected for pressurization gas quantity.
Source: Hassan et al.[18]

Table 5.8. Kinetic and Equilibrium Data and Other Common Parameter values Used in the Simulations of PSA Air Separation for Nitrogen Production[a]

Adsorbent	Carbon molecular sieve (Bergbau–Forschung)
Bed length (cm)	35.0
Bed radius (cm)	1.75
Particle size (cm)	0.3175 (pellet)
Particle density (g/cm^3)	0.9877
Bed voidage	0.40
Ambient temperature (°C)	25.0
Peclet number	1000.0
Equilibrium constant for oxygen (K_A)	9.25[b]
Equilibrium constant for nitrogen (K_B)	8.9[b]
Saturation constant for oxygen (q_{AS}) (mol/cm^3)	2.64 × 10^{-3}[b]
Saturation constant for nitrogen (q_{BS}) (mol/cm^3)	2.64 × 10^{-3}[b]
Limiting diffusional time constant for oxygen (D_{A0}/r_c^2) (s^{-1})	2.7 × 10^{-3}[c]
Limiting diffusional time constant for nitrogen (D_{B0}/r_c^2) (s^{-1})	5.9 × 10^{-5}[c]

[a] Ω values of 14 and 85 for oxygen and nitrogen, respectively, taken from the correlation of Nakao and Suzuki have been used in computing the LDF model predictions.
[b] Ruthven et al.[56]
[c] Farooq and Ruthven.[26]

pressures (3–7 atm) and therefore provide a suitable database. The experimental conditions and the observed product purity, recovery, and productivity are summarized in Table 5.7. The equilibrium and kinetic parameters taken from independent, single-component measurements and used in simulating these experimental runs are given in Table 5.8. The diffusion model (with constant and variable diffusivity) and the LDF model predictions are compared with the experimental results in Figure 5.8. The predictions of the constant-diffusivity pore diffusion model and the LDF model, with Ω values adjusted for cycle time according to the correlation of Nakao and Suzuki,[35] are very close. (This agreement provides additional confirmation of the results shown in Figure 5.3.) It is clear (from Figure 5.8) that the concentration-dependent diffusivity model predicts the correct qualitative trends for purity and recovery over the range of experimental variables examined. The constant-diffusivity model (and the LDF model), on the other hand, cannot predict the correct trend of recovery with operating pressure, and, even though this model predicts the correct trend for the variation of product purity, the qualitative disagreement at the higher pressures is too large.

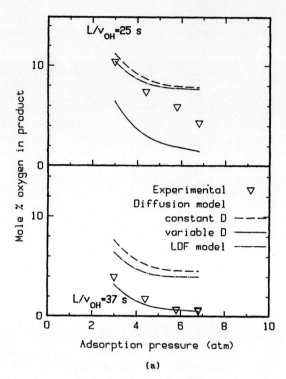

Figure 5.8 Effect of adsorption pressure on (a) nitrogen product purity and (b) nitrogen recovery at two different L/v_{OH} ratios in a kinetically controlled PSA air separation process (modified cycle) showing the comparison among the experimental data, the LDF model, and the diffusion model with constant and variable diffusivity. Experimental conditions are given in Table 5.7 and other parameters in Table 5.8. (From Ref. 26.)

Industrial PSA nitrogen units operate at pressures between 7 and 10 atm (with blowdown to atmospheric pressure) and at a relatively low L/v_{OH} ratio (< 25 s). A fairly pure nitrogen product ($> 99\%$) is produced. This level of purity is not predicted by either the LDF model or by the constant-diffusivity pore diffusion model. In fact at $L/v_{OH} = 25$ s the nitrogen product purity profiles from the LDF model as well as the constant-diffusivity pore diffusion model become asymptotic at an oxygen concentration of about 8%. It is therefore evident that the formation of nitrogen product containing less than 1% oxygen, which is routinely observed in large-scale industrial units operating at relatively low L/v_{OH} ratios, is correctly predicted only when allowance is made for the concentration dependence of the diffusivity. The concentration dependence of micropore diffusivity evidently has a strong effect on the steady-state performance of a kinetically controlled PSA separation. The variable-diffusivity model provides a reliable a priori estimate of such perfor-

DYNAMIC MODELING OF A PSA SYSTEM

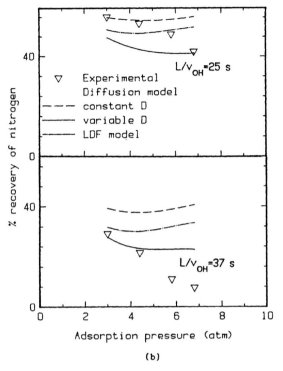

Figure 5.8 (*Continued*).

mance based on independently measured single-component equilibrium and kinetic data.

The flexibility of a PSA simulation model for accommodating various cyclic operations is determined mainly by the versatility of the fluid flow model. (The flexibility of the fluid flow model used here has been discussed in Section 5.1.) There is therefore no reason to prevent the application of the pore diffusion model to other cycle configurations.

Effects of Process Variables

Having established the validity of the concentration-dependent diffusivity model for air separation over carbon molecular sieve, the model may now be used to investigate the effects of some important operating parameters on the performance of the system. Some results are summarized in Figure 5.9. For a Skarstrom cycle operation with no purge, increasing adsorption pressure improves product purity at the expense of diminishing recovery. It is shown that dual-ended pressure equalization improves both purity and recovery over those obtained by Skarstrom cycle operation without purge. Desorption without purge does not produce a high-purity product when the feed pres-

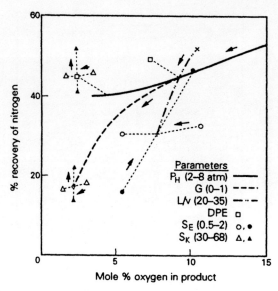

Figure 5.9 Effects of some important operating parameters on purity and recovery of nitrogen in a kinetically controlled PSA air separation process. Adsorption/desorption time = 60 s, pressurization/blowdown time = 15 s, kinetic and equilibrium parameters are given in Table 5.8. The solid line shows the effect of increasing the adsorption pressure (in the direction of the arrow) for a Skarstrom cycle with no purge. For two different operating pressures the effect of introducing a double-ended pressure equalization step, as in the modified cycle is shown by dotted lines leading to the points ($\square$). The effect of increasing purge/feed is shown by chain dotted line leading to point ($\diamond$) at purge/feed = 1.0, and the effect of changing the L/v_{OH} ratio from 20 s at point ($\times$) to 35 s at point ($+$) is indicated by dot–dash line. The effects of increasing and decreasing the kinetic and equilibrium parameters for oxygen and nitrogen (by factors of 1.5 and 2.0, respectively, relative to the experimental values) are also shown. The open symbols show the effect of changing the values for oxygen, and the closed symbols show the effect of changing the values for nitrogen. The directions of the increasing variables (L/v_{OH}, P_H, and G) and the kinetic and equilibrium selectivities (S_K and S_E) are indicated by arrows.

sure is low. The product purity may be significantly improved in a low-feed pressure operation by regenerating the adsorbent bed with product purge. However, the recovery is reduced by introducing purge. Another way of increasing the product purity when the feed pressure is low is to increase the L/v_{OH} ratio, but product purge appears to be more efficient. Of course, when operated at high feed pressure a very high-purity product may be achieved without resorting to any external purge and consequently, at comparable purity, the recovery is much higher than that from a Skarstrom cycle. The effects of varying the kinetic selectivity ($S_K = D_{O_2}/D_{N_2}$) and the equilibrium selectivity ($S_E = K_{O_2}/K_{N_2}$) about their experimental values are also illustrated. It is evident that varying the oxygen diffusivity or equilibrium mainly affects the purity. Varying the nitrogen diffusivity affects mainly the recovery, and changing the nitrogen equilibrium affects both purity and recovery. An improved molecular sieve for nitrogen from air by pressure swing adsorption would therefore require stronger oxygen equilibrium and/or slower nitrogen diffusion. It should be recalled here that there are limits up to which such improvements may be effectively exploited in this type of cycle (see Section 3.4).

5.3 Continuous Countercurrent Models

In order to understand the continuous countercurrent flow model for a PSA separation process, it is necessary to recall once again that the steps involved

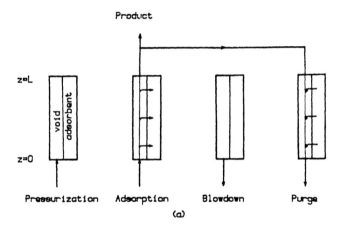

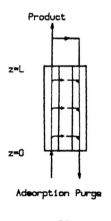

Figure 5.10 (a) Skarstrom PSA cycle. (b) Continuous countercurrent flow model representation of a Skarstrom PSA cycle.

in a basic Skarstrom PSA cycle are pressurization, high-pressure adsorption, countercurrent blowdown, and countercurrent purge, as represented in Figure 5.10(a). If mass transfer between solid and gas phases during the pressurization and blowdown steps is assumed to be negligible, then at cyclic steady state the amount adsorbed during the adsorption step should be equal to the amount desorbed during the purge step. Transient PSA simulation with a frozen solid approximation during pressurization and blowdown has been validated for both purification processes[4-7,10,25] and kinetically controlled bulk separation.[13,57] Therefore, for these processes, the operation of a Skarstrom cycle at steady state can be viewed as a continuous countercurrent flow (CCF) system in which the immobile solid phase adsorbs from the high-pressure stream and desorbs to the purge stream with zero net accumulation in the solid phase. This representation is shown schematically in Figure 5.10(b).

The idea of representing the PSA system as a continuous countercurrent flow operation was first proposed by Suzuki.[58] He developed the CCF model for a trace component system and compared the steady-state concentration profiles from this model with those from the transient simulation. In addition to using the frozen solid approximation during pressurization and blowdown for the transient simulation, Suzuki also adopted rapid cycling to attain a

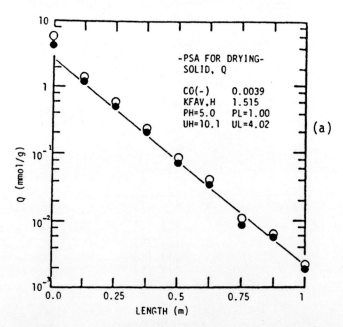

Figure 5.11 (a) Solid-phase and (b) gas-phase concentration profiles from the CCF model (——— adsorption, --- desorption) and the transient simulation model (O end of adsorption, ● end of desorption) are compared for air drying on activated alumina. (From Ref. 58; reprinted with permission.)

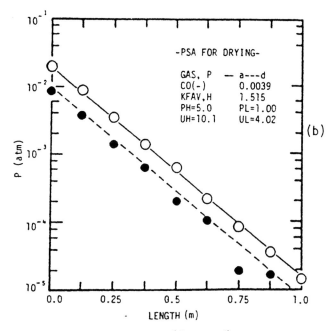

Figure 5.11 (*Continued*).

small throughput ratio so that solid- and gas-phase profiles also remained nearly frozen during the adsorption and the desorption steps. Under these conditions, the steady-state profiles from the CCF model were found to be in good agreement with those from the transient simulation, as may be seen from Figure 5.11. While these extreme assumptions may be realistic for a purification process, in a bulk separation process there will generally be significant excursions of the concentration profiles during the adsorption and the desorption steps. Farooq and Ruthven[59] extended the CCF model for a bulk separation process (the model equations are given in Table 5.9) and showed that, even without a very small throughput ratio, the CCF model still correctly predicts the qualitative trends of experimental purity and recovery data for a PSA nitrogen unit using carbon molecular sieve. The results, including transient model predictions, are shown in Figure 5.12. This study also showed that mass transfer during pressurization and blowdown steps will not impair the predictions of the CCF model. Although the transient model is quantitatively superior, the simplicity and computational efficiency make the CCF model useful at least for initial selection of the range of conditions within which more detailed studies should be concentrated. There are, however, some limitations of this approach. The CCF model solution is invariant to changes in cycle time as long as the durations of the adsorption and description steps are kept equal. The CCF model is applicable in

Table 5.9. Continuous Countercurrent Flow Model Equations for a Bulk PSA Separation Process[a]

Assumptions 1–7 discussed in connection with the LDF model apply.
Component mass balance:

$$m\frac{d(v_j c_{ij})}{dz} + \frac{1-\varepsilon}{\varepsilon}\frac{d\bar{q}_{ij}}{dt} = 0 \qquad (1)$$

Continuity condition:

$$\sum_i c_i = C_j \ (\text{constant}) \qquad (2)$$

Overall mass balance:

$$mC_j\frac{dv_j}{dz} + \frac{1-\varepsilon}{\varepsilon}\sum_i \frac{d\bar{q}_{ij}}{dt} = 0 \qquad (3)$$

Mass transfer rates:

$$\frac{d\bar{q}_{ij}}{dt} = k_{ij}(q_{ij}^* - \bar{q}_i) \qquad (4)$$

Adsorption equilibrium:

$$\frac{q_{ij}^*}{q_{is}} = \frac{b_i c_{ij}}{1 + \sum_i b_i c_{ij}} \qquad (5)$$

In these equations $i = A$ for component A and B for component B, $j = $ H or L, and $m = +1$ or -1. The values $j = $ H and $m = +1$ represent high-pressure flow, $j = $ L and $m = -1$ represent purge flow.
Boundary conditions: high-pressure flow

$$c_{AH}|_{z=0} = c_{A0}; \qquad v_H|_{z=0} = v_{OH} \qquad (6)$$

purge flow

$$c_{AL}|_{z=L} = \left(\frac{P_L}{P_H}\right) c_{AH}|_{z=L}; \qquad v_L|_{z=L} = v_{OL} = Gv_{OH} \qquad (7)$$

The concentration boundary condition for purge flow represents the fact that part of the high-pressure product is expanded to low pressure and used for purge. v_{OL} and v_{OH} are related by the purge to feed velocity ratio, G.

The assumption of zero net accumulation in the solid phase leads to:

$$\alpha\frac{d\bar{q}_{iH}}{dt} + (1-\alpha)\frac{d\bar{q}_{iL}}{dt} = 0 \qquad (8)$$

In this equation $\alpha = t_H/(t_H + t_L)$. The factors α and $(1-\alpha)$ with the solid phase accumulation terms for the high- and low-pressure steps, respectively, have been introduced to maintain consistency in mass transfer rates between the CCF approximation and the transient simulation.

(Continued)

Table 5.9. (*Continued*)

From Eqs. 4 and 8 we get:

$$\bar{q}_i = \frac{\alpha k_{iH} q^*_{iH} + (1-\alpha) k_{iL} q^*_{iL}}{\alpha k_{iH} + (1-\alpha) k_{iL}} \qquad (9)$$

Substituting Eq. 9 into 4 and rearranging yields:

$$\frac{d\bar{q}_{ij}}{dt} = \frac{m}{(n-1+m\alpha)} k_{ei}(q^*_{iH} - q^*_{iL}) \qquad (10)$$

where $n = 1$ for high-pressure flow and 2 for purge flow; and

$$k_{ei} = \frac{1}{1/\alpha k_{iH} + 1/(1-\alpha) k_{iL}} \qquad (11)$$

Combining Eqs. 1, 3, and 10 gives:

$$v_j \frac{dc_{ij}}{dz} + \frac{1-\varepsilon}{\varepsilon} \frac{1}{(n-1+m\alpha)} (k_{ei}(q^*_{iH} - q^*_{iL})$$

$$- X_{ij} \sum_i k_{ei}(q^*_{iH} - q^*_{iL}) \Big) = 0 \qquad (12)$$

$$C_j \frac{dv_j}{dz} + \frac{1-\varepsilon}{\varepsilon} \frac{1}{(n-1+m\alpha)} \sum_i k_{ei}(q^*_{iH} - q^*_{iL}) = 0 \qquad (13)$$

Equation 12 for $i = A$ and Eq. 13 are simultaneously integrated for high- and low-pressure flow using the fourth-order Runge–Kutta method to obtain c_{Aj} and v_j at different axial locations in the bed. The boundary conditions for high-pressure flow are known at $z = 0$ and those for purge are given at $z = L$; therefore, the solution procedure requires iteration. Values of the dependent variables for purge flow are guessed at $z = 0$. The integration is then performed from $z = 0$ to L, which is repeated until known boundary conditions for purge flow at $z = L$ are satisfied. Jacobian analysis is performed to update the trial values, which accelerates the speed of convergence. Values of c_{Bj} at different locations are obtained by difference from total concentration. Steady-state adsorbed-phase concentrations are then calculated from Eq. 9.

---

[a] The material balance Eqs. 1 and 3 should actually be written in terms of equivalent velocities in order to ensure that the total volume of feed and purge used in actual operation are the same in the steady-state representation. However, such a restriction on the mass balance leads to an equivalent purge to feed velocity ratio given by $G[(1-\alpha)/\alpha]$. For $\alpha = 0.5$ $G[(1-\alpha)/\alpha] = G$, but when $\alpha \to 1$, $G \to 0$, and when $\alpha \to 0$, $G \to \infty$, and therefore the CCF model fails for cycles in which the adsorption and desorption steps are unequal (i.e., $\alpha \neq 0.5$). The present form of material balance equations, although less accurate from the point of mass balance between the actual operation and steady-state approximation (except when $\alpha = 0.5$), extends the usefulness of the CCF model to the cycles in which $\alpha \neq 0.5$.

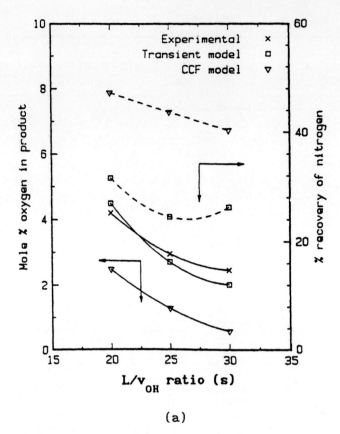

(a)

Figure 5.12 Effects of (a) L/v_{OH} ratio ($G = 1.0$) and (b) purge to feed velocity ratio ($L/v_{OH} = 25$ s) on the purity and recovery of nitrogen product in a PSA air separation process (Skarstrom cycle) showing comparison among the experimental results, the CCF model, and the transient model predictions. Adsorption pressure = 3 atm, blowdown/purge pressure = 1.0 atm, adsorption/desorption time = 60 s, pressurization/blowdown time = 15 s, diffusional time constants used for oxygen and nitrogen were 3.73×10^{-3} s^{-1} and 1.17×10^{-4} s^{-1}, respectively. The CCF model results were computed with $\Omega = 15$ for both oxygen and nitrogen; cycle-time-dependent Ω values according to the correlation of Nakao and Suzuki were used in the transient model simulation. (From Ref. 59.)

principle to both kinetic and equilibrium-based separations. Its applicability to the latter class is restricted to trace systems with significant mass transfer resistance. The fact that one solid-phase concentration profile cannot be in equilibrium with two different gas-phase profiles precludes the use of this model for equilibrium-controlled separations with negligible mass transfer resistance. Similar reasoning also precludes the extension of the CCF model to account for heat effects.

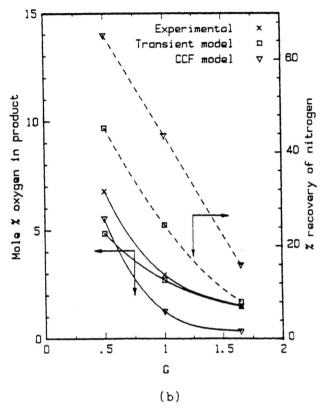

(b)

Figure 5.12 (*Continued*).

5.4 Heat Effects in PSA Systems

One factor that has been ignored in our discussion on the dynamic modeling of PSA cycles is the heat effect (see Section 4.8) and, so far, the basic assumption of isothermal behavior has been retained. As a result of the heat of sorption, there is always some temperature excursion in a PSA cycle, and depending on the magnitude, this may significantly reduce the efficiency of separation. The amplitude of the temperature swing depends primarily on the heat of adsorption, the throughput, and the heat transfer characteristics of the packed adsorbed column. In laboratory scale units, small-diameter thick-walled metal columns are generally used, and the high heat capacity and thermal conductivity of the column walls serve to minimize the temperature variation. Even then it is found that the isothermal assumption usually predicts a better separation than is actually achieved. In order to approach true isothermal operation, one has to go to a very small column diameter and a small throughput ratio, as in the air drying experiments (on silica gel) reported by Chihara and Suzuki.[5] These authors were the first to investigate the effect of nonisothermality in a trace component PSA system.[4] Later, Yang and co-workers[9,12,15] showed that heat effects are even more detrimental to the performance of equilibrium-controlled bulk separation processes. Nonisothermal studies have further revealed, as may be intuitively deduced, that the adiabatic condition, which is approached in a large commercial operation, generally results in the worst separation. Whereas for many equilibrium-controlled separations, the isothermal approximation may mean a major departure from physical reality, this is usually a good approximation for separations based on kinetic selectivity since mass transfer rates are generally much slower in such systems.

The LDF model discussed in Section 5.2 is extended here to allow for nonisothermal PSA operation. The additional assumptions are summarized and the heat balance equations are given in Table 5.10.

1. The equilibrium constants are the most sensitive temperature-dependent terms, and it is assumed that the Langmuir constants show the normal exponential temperature dependence ($b = b_0 e^{-\Delta H/RT}$).
2. The temperature dependence of the gas and solid properties and the transport parameters is assumed negligible.
3. Effective thermal conductivities of the commercial adsorbent particles are relatively high, and therefore intraparticle temperature gradients can be neglected. Thermal equilibrium is assumed between the fluid and the adsorbent particles, which is also a very common assumption in adsorber calculations.[60]
4. Bulk flow of heat and conduction in the axial direction are considered in the heat balance equation. For heat conduction we consider the contribution from axial dispersion only. The contribution from the solid phase becomes important at low Reynolds number, which is, seldom, if ever, approached in a PSA operation. An overall heat transfer coefficient is used to account for heat loss from the system. The temperature of the column wall is taken to be equal to that of the feed. Farooq and Ruthven,[61,62] in their studies on heat effects in adsorption column dynamics, have shown that the major resistance to heat transfer in an adsorption column is at the column wall, and a simple one-dimensional model with all heat transfer resistance concentrated at the column wall provides a good representation of the experimentally observed behavior. The advantage of using the one-dimensional model is that the system behavior at the isothermal and adiabatic limits may be very easily investigated by assigning a large value and zero, respectively, to the heat transfer coefficient. The isothermal condition is also approached if the simulation is carried out with $\Delta H = 0$.
5. The boundary conditions for the heat balance equation are written assuming the heat–mass transfer analogy for a dispersed plug flow system. The

Table 5.10. Additional Equations for Nonisothermal PSA Simulation Using LDF Approximation

The set of equations in Table 5.2 together with the following equations constitute the model equations for nonisothermal PSA simulation. The subscripts have the same meanings as in Table 5.2.

Fluid phase heat balance[a]:

$$-K_L \frac{\partial^2 T}{\partial z^2} + C_g \left(v \frac{\partial T}{\partial z} + T \frac{\partial v}{\partial z} \right) + \left(C_g + \frac{1-\varepsilon}{\varepsilon} C_s \right) \frac{\partial T}{\partial t}$$

$$-\sum_i (-\Delta H_i) \frac{1-\varepsilon}{\varepsilon} \frac{\partial \bar{q}_i}{\partial t} + \frac{2h}{\varepsilon r_{in}} (T - T_w) = 0 \qquad (1)$$

Temperature dependency of Langmuir constant:

$$\ln \frac{b_i}{b_{i0}} = -\frac{\Delta H_i}{R_g} \left(\frac{1}{T} - \frac{1}{T_0} \right) \qquad (2)$$

Boundary conditions: pressurization, high-pressure adsorption, and purge

$$K_L \frac{\partial T}{\partial z}\bigg|_{z=0} = -v|_{z=0} \rho_g C_{pg} (T|_{z=0^-} - T|_{z=0}); \quad Q \frac{\partial T}{\partial z}\bigg|_{z=L} = 0 \qquad (3)$$

$$(T|_{z=0^-})_{\text{purge}} = (T|_{z=L})_{\text{adsorption}} \qquad (4)$$

Equation 4 defines inlet gas temperature of the bed undergoing purge in terms of raffinate product temperature from the high-pressure bed. This is not applicable when the beds are not coupled through a purge stream. Blowdown:

$$\frac{\partial T}{\partial z}\bigg|_{z=0} = 0; \quad \frac{\partial T}{\partial z}\bigg|_{z=L} = 0 \qquad (5)$$

Initial condition (same for both clean and saturated bed conditions)

$$T(z,0) = T_0 \qquad (6)$$

---

[a] When the small laboratory columns are insulated from outside the fluid phase heat balance Eq. 1 is modified as follows to account for the heat capacity of and conduction through the column wall:

$$-\left(K_L \frac{A_{cs}}{A} + \frac{K_{steel}}{\varepsilon} \frac{A'}{A} \right) \frac{\partial^2 T}{\partial z^2} + \left(v \frac{\partial T}{\partial z} + T \frac{\partial v}{\partial z} \right)$$

$$+ \left(\frac{A_{cs}}{A} C_g + \frac{1-\varepsilon}{\varepsilon} \frac{A_{cs}}{A} C_s + \frac{A'}{\varepsilon A} \rho_{steel} C_{psteel} \right) \frac{\partial T}{\partial t}$$

$$- \sum_i (-\Delta H_i) \frac{1-\varepsilon}{\varepsilon} \frac{\partial \bar{q}_i}{\partial t} + \frac{2h}{\varepsilon r_{out}} (T - T_w) = 0$$

---

effective axial thermal conductivity of the fluid also follows from the assumed similarity between mechanisms of fluid phase mass and heat transfer.
6. Weighted average values for gas density and heat capacity based on feed composition are used in a multicomponent system.

The numerical solution of the set of coupled nonlinear equations 1–11 in Table 5.2 and 1–6 in Table 5.10 gives gas- and solid-phase concentration and the bed temperature at several locations in the column for various values of

Table 5.11. Parameters Used in Computing the Theoretical Curves in Figures 5.14–5.16 Showing the Existence of Two Different Cyclic Steady States

	5A zeolite[a]	Activated alumina[b]
Feed gas composition	1% Ethylene in helium	0.39% Moisture in air
Particle diameter (mm)	0.7	4–5
Bed length (cm)	35.0	100.0
Bed radius, i.d. (cm)	1.75	10.0
o.d. (cm)	2.1	—
Bed voidage	0.4	0.4
Feed temperature (K)	298.0 (ambient)	303.0 (ambient)
Gas density (g/cm^3)	1.5×10^{-4} (1 atm)	1.2×10^{-3} (1 atm)
Adsorbent density (g/cm^3)	1.14[c]	1.2
Heat of adsorption (cal/mol)	-8000.0[d]	-12404.0
Gas heat capacity (cal/g °C)	1.2376	0.238
Adsorbent heat capacity (cal/g °C)	0.206[c]	0.3
Heat transfer coefficient (cal/cm^2 s °C)	0.0	0.0
Effective axial thermal conductivity of fluid (cal/cm s °C)	Deduced from analogy of mass and heat transfer	0.0018
Adsorption pressure (atm)	3.0	5.0
Blowdown/purge pressure (atm)	1.36	1.0
Peclet number	110[a]	10^6 (plug flow)
L/v_{OH} ratio (s)	5.91	4.0
Purge to feed velocity ratio	1.54	2.0
q_0/c_0	837.0[e]	8993.16
q_0/q_s	0.92[e]	0.0 (linear)
LDF mass transfer coefficient (s^{-1})		
high pressure	0.19[e]	2.78×10^{-4}
low pressure	0.18[e]	1.39×10^{-3}
Adsorption/purge time (s)	80.0	540.0
Pressurization/blowdown time (s)	20.0	140.0

[a] Heat capacity and thermal conductivity of steel were used to account for the wall effect.
[b] Chihara and Suzuki.[4]
[c] Ruthven et al.[64]
[d] Farooq and Ruthven.[62]
[e] Hassan et al.[10]

time. Convergence to cyclic steady state may be very slow under adiabatic conditions, requiring in some cases up to 100 cycles. A more detailed account of nonisothermal PSA simulation is given in Ref. 19.

Yang and co-workers used similar nonisothermal models to study several equilibrium-controlled separation processes (see Table 5.1). They neglected the axial thermal conduction but considered the temperature dependence of the gas concentration according to the ideal gas law. The temperature variation in the bed at any given point was 20–40°C. The experiments were carried out in small-diameter columns (4.1 cm i.d.) and therefore suffered from heat loss by wall conduction and exchange with the surroundings. (Under true adiabatic conditions the temperature variation could reach 100°C.) In some studies they replaced the LDF rate equation with more detailed diffusion models. In the introductory section of this chapter it was pointed out that in an equilibrium-controlled separation the detailed form of the kinetic model is of only secondary importance. This is also true in the

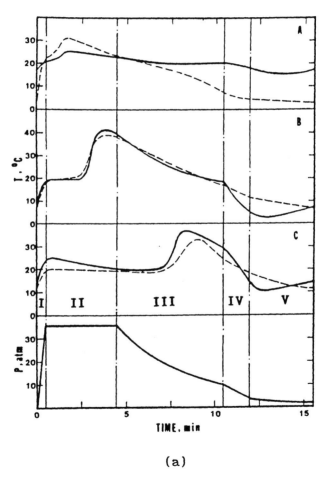

(a)

Figure 5.13 (a) Steady-state temperature–time histories measured at three locations (feed end A, middle B, and product end C) for the equilibrium controlled PSA bulk separation of an equimolar H_2-CO mixture on activated carbon in a single bed, five-step cycle (I-V indicate pressurization, adsorpation, cocurrent blowdown, countercurrent blowdown and purge). ——, experimental, ---, numerically solved equilibrium model (equivalent to LDF model with large mass transfer coefficients. (From Cen and Yang[15], with permission.)

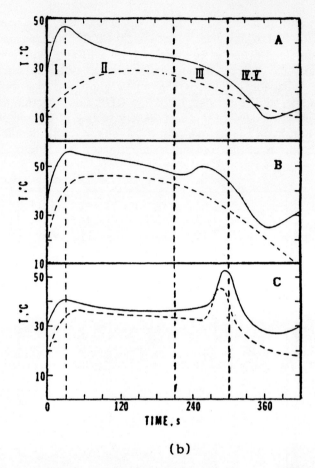

Figure 5.13 (b) Steady-State temperature histories for an equilibruim PSA bulk separtion (equimolar mixture of H_2-CH_4 on activated carbon, ---, experimental; ——, macropore diffusion model). Other details as in part (a). (From Doong and Yang[12], with permission.)

prediction of nonisothermal behavior. Representative temperature profiles from the work of Yang et al. are shown in Figure 5.13. Since direct measurement of concentration profiles in the bed is not easy, the temperature vs. time profiles measured at various positions in the column provide a practically useful way of locating the advancing mass transfer zone in the column. Espitalier-Noel[31] observed a sharp rise in the bed temperature at the product end prior to breakthrough of nitrogen during the high pressure production step of a PSA air separation cycle for oxygen production on 5A zeolite. The temperature variation at various points along the bed is shown in Figure 5.14. The temperature profiles are also useful for understanding the improved performance of a PSA separation that may be achieved by allowing

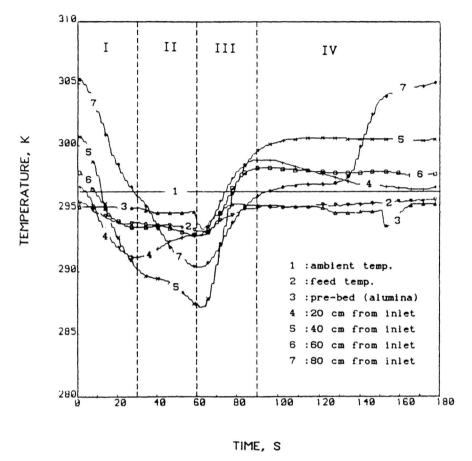

Figure 5.14 Variation of temperature with time at various points along the bed for an oxygen PSA system, at cyclic steady state, operating on a modified Skarstrom cycle. I–IV indicate, respectively depressurization, purge, repressurization, and high-pressure production. (From Ref. 31; reprinted with permission.)

heat exchange between adsorbers or by introducing high heat capacity inert additives, as proposed by Yang and Cen.[63]

5.4.1 Two Different Cyclic Steady States in PSA Systems

A detailed study of a PSA purification system under both isothermal and nonisothermal conditions was conducted by Farooq, Hassan, and Ruthven.[19] The simulation results reveal that if the system is adiabatic there are at least two different solutions to the model equations so that, depending on the initial condition of the beds, two different cyclic steady states are obtained. The desirable steady state, giving a pure high-pressure product, is achieved

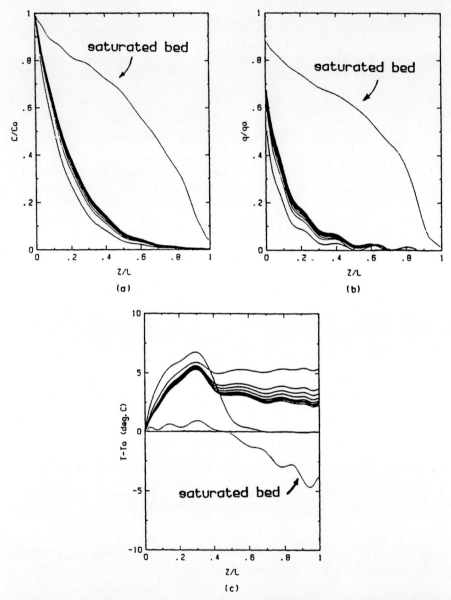

Figure 5.15 Computed profiles for PSA air drying on activated alumina showing approach to cyclic steady state from clean bed initial conditions. Steady-state profiles with both beds initially equilibrated with feed at high pressure are also shown. The profiles represent the end of the high-pressure adsorption step and, starting for a clean bed, are shown after 19, 39, 49, 59, 69, 79, 89, 99, and 109 half-cycles. (a) Gas-phase concentration profile, (b) adsorbed-phase concentration profile, (c) temperature profile. Parameters used in the numerical simulation are given in Table 5.11. (From Ref. 19.)

when the beds are initially clean. Starting the system from a saturated condition leads to a different steady state with significantly different profiles of both concentration and temperature and a less pure final product. Steady-state bed profiles for clean and saturated bed initial conditions are shown in Figure 5.15. This is equally true for both linear and nonlinear equilibrium isotherms. Different cyclic steady states corresponding to clean and saturated bed initial conditions may also be obtained for an isothermal system when the equilibrium relationship is nonlinear.[19, 25] In their study Farooq, Hassan, and Ruthven further showed that for a linear isothermal system the steady state is unique and the solution of the model equations (using a large value of h or $\Delta H = 0$) converges to the same final cyclic steady state from all initial conditions. It is clear that multiplicity can arise only when the equations contain a significant nonlinearity. In the isothermal case the nonlinearity comes from curvature of the equilibrium isotherm, but in a nonisothermal linear equilibrium system the same type of behavior arises from the temperature dependence of the adsorption equilibrium constant.

Limited experiments, employing a dual-bed system operated on a Skarstrom cycle, were conducted with the ethylene–helium–5A system to confirm the existence of more than one cyclic steady state. The columns were insulated as much as possible to attain a near-adiabatic condition. When insulated from outside, the heat capacity of the column wall and conduction

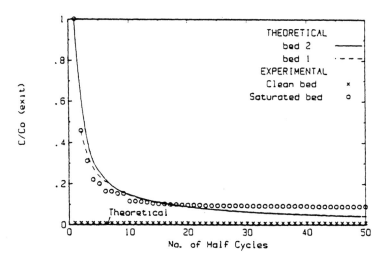

Figure 5.16 Comparison of experimental product composition with prediction of theoretical model for PSA separation of ethylene–helium on 5A zeolite showing the difference in behavior for the clean and saturated bed initial conditions (bed 1 saturated with feed at low pressure and bed 2 equilibrated with feed at high pressure). Note that for the clean bed case theory predicts a perfectly pure product. Parameters used in computing the model predictions are given in Table 5.11. (From Ref. 19.)

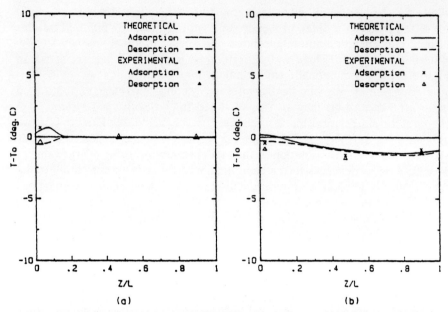

Figure 5.17 Comparison of theoretical and experimental temperature profiles for PSA separation of ethylene–helium on 5A zeolite at cyclic steady state. (a) Clean bed initial condition, (b) saturated bed initial condition (bed 1 equilibrated with feed at low pressure and bed 2 equilibrated with feed at high pressure). Parameters used in computing the model predictions are given in Table 5.11. (From Ref. 19.)

of heat through the column wall become important factors that must be accounted for in the heat balance equation. The approach of the product concentration to cyclic steady state from two different initial conditions is shown in Figure 5.16. The temperature profiles at the end of adsorption and desorption steps after the cyclic steady state was reached are given in Figure 5.17. Both the product concentration and the experimental temperature profiles agree well with the model predictions and the existence of two distinct steady states, depending on the initial condition of the beds, is confirmed. Clearly, the performance of the system is markedly superior when starting from initially clean beds in that a purer high-pressure product is obtained.

The existence of multiple steady state has certain practical implications, since it means that under certain conditions the operation may be unstable to a perturbation in feed composition or flow rate. Furthermore, if a unit is shut down during operation, it may be necessary to purge the beds before the operation can be successfully resumed with the desired product purity. An

---

Note: A novel mathematical anaylsis that provides for the direct determination and stability analysis of cyclic steady states has recently been developed: D. T. Croft and M. D. LeVan, *Chem. Eng. Sci.* (in press). See also Proceedings of Fourth International Conference on Adsorption, Kyoto, May 1992.

attempt to restart the system after it had been standing in a loaded condition for some time could well lead to the *saturated bed* steady-state operation rather than the more desirable *clean bed* steady-state operation, which gives a purer product. Of course the change of steady state would not show in the product concentration if the system is operated such that the concentration front is maintained well inside the column with a large margin of safety.

References

1. J. C. Kayser and K. S. Knaebel, *Chem. Eng. Sci.* **44**(1), 1 (1989).
2. G. Flores-Fernandez and C. N. Kenney, *Chem. Eng. Sci.* **38**(6), 827 (1983).
3. J. E. Mitchell and L. H. Shendalman, *AIChE Symp. Ser.* **69**(134), 25 (1973).
4. K. Chihara and M. Suzuki, *J. Chem. Eng. Japan* **16**(1), 53 (1983).
5. K. Chihara and M. Suzuki, *J. Chem. Eng. Japan* **16**(4), 293 (1983).
6. J. W. Carter and M. L. Wyszynski, *Chem. Eng. Soc.* **38**(7), 1093 (1983).
7. N. S. Raghavan, M. M. Hassan, and D. M. Ruthven, *AIChE J.* **31**(3), 385 (1895).
8. P. L. Cen, W. N. Chen, and R. T. Yang, *Ind. Eng. Chem. Process Des. Dev.* **24**(4), 1201 (1985).
9. R. T. Yang and S. J. Doong, *AIChE J.* **31**(11), 1829 (1985).
10. M. M. Hassan, N. S. Raghavan, D. M. Ruthven, and H. A. Boniface, *AIChE J.* **31**(12), 2008 (1985).
11. N. S. Raghavan and D. M. Ruthven, *AIChE J.* **31**(12), 2017 (1985).
12. S. J. Doong and R. T. Yang, *AIChE J.* **32**(3), 397 (1986).
13. M. M. Hassan, D. M. Ruthven, and N. S. Raghavan, *Chem. Eng. Sci.* **41**(5), 1333 (1986).
14. N. S. Raghavan, M. M. Hassan, and D. M. Ruthven, *Chem. Eng. Sci.* **41**(11), 2787 (1986).
15. P. L. Cen and R. T. Yang, *Ind. Eng. Chem. Fundam.* **25**(4), 758 (1986).
16. H. S. Shin and K. S. Knaebel, *AIChE J.* **33**, 654 (1987).
17. S. J. Doong and R. T. Yang, *AIChE J.* **33**(6), 1045 (1987).
18. M. M. Hassan, N. S. Raghavan, and D. M. Ruthven, *Chem. Eng. Sci.* **42**(8), 2037 (1987).
19. S. Farooq, M. M. Hassan, and D. M. Ruthven, *Chem. Eng. Sci.* **43**(5), 1017 (1988).
20. H. S. Shin and K. S. Knaebel, *AIChE J.* **34**(9), 1409 (1988).
21. A. Kapoor and R. T. Yang, *Chem. Eng. Sci.* **44**(8), 1723 (1989).
22. S. Farooq, D. M. Ruthven, and H. A. Boniface, *Chem. Eng. Sci.* **44**(12), 2809 (1989).
23. J. L. L. Liow and C. N. Kenney, *AIChE J.* **36**(1), 53 (1990).
24. M. W. Ackley and R. T. Yang, *AIChE J.* **36**(8), 1229 (1990).
25. J. A. Ritter and R. T. Yang, *Ind. Eng. Chem. Res.* **30**(5), 1023 (1991).
26. S. Farooq and D. M. Ruthven, *Chem. Eng. Sci.* **46**(9), 2213 (1991).

27. M. Chlendi, J. Granger, E. Carcin, and D. Tondeur, *Hydrogen Purification by Multicolumn, Multisorbent PSA: Modelling, Design, Optimization*. Fourth International Conference on Adsorption, Kyoto, Japan, May 17-22, 1992.
28. S. J. Doong and R. T. Yang, *AIChE Symp. Ser.* **264**(84), 145 (1988).
29. J. F. Wehner and R. A. Wilhelm, *Chem. Eng. Sci.* **6**, 89 (1956).
30. G. Munkvold, K. Teague, T. F. Edgar, J. J. Beaman, *Gas Separation & Purification* (in press).
31. P. M. Espitalier-Noel, *Waste Recycle Pressure Swing Adsorption to Enrich Oxygen from Air*. PhD. thesis, University of Surrey, Guildford, UK, 1988.
32. H. S. Shin, *Pressure Swing Adsorption: A Study of Diffusion Induced Separation*. Ph.D. thesis, The Ohio State University, Columbus, Ohio, 1988.
33. R. Desai, M. Hussain, and D. M. Ruthven, *Can. J. Chem. Eng.* (in press)
34. Y. D. Chen, J. A. Ritter, and R. T. Yang, *Chem. Eng. Sci.* **45**(9), 2877 (1990).
35. S. Nakao and M. Suzuki, *J. Chem. Eng. Japan* **16**(2), 114 (1983).
36. E. Glueckauf and J. J. Coates, *J. Chem. Soc.* 1315 (1947).
37. E. Alpay and D. M. Scott *Chem Eng. Sci.* **47**, 499 (1992).
38. D. M. Ruthven and K. F. Loughlin, *Chem. Eng. Sci.* **26**, 1145 (1971).
39. I. H. Doetsch, D. M. Ruthven, and K. F. Loughlin, *Can. J. Chem.* **52**, 2717 (1974).
40. K. Kawazoe, M. Suzuki, and K. Chihara, *J. Chem. Eng. Japan* **7**(3), 151 (1974).
41. K. Chihara, M. Suzuki, and K. Kawazoe, *J. Chem. Eng. Japan* **11**(2), 153 (1978).
42. D. U. von Rosenberg, *Methods for the Numerical Solution of Partial Differential Equations*. Gerald L. Farrar and Associates, Inc., Tulsa, Oklahoma, 1977 (fourth print).
43. L. M. Sun and F. Meunier, *AIChE J.* **37**(2), 244 (1991).
44. B. A. Finlayson, *Method of Weighted Residuals and Variational Principles*. Academic Press, New York, 1972.
45. J. V. Villadsen and W. E. Stewart, *Chem. Eng. Sci.* **22**, 1483 (1967).
46. J. V. Villadsen and M. L. Michelsen, *Solution of Differential Equation Models by Polynomial Approximation*. Prentice-Hall, Englewood Cliffs, New Jersey, 1978.
47. L. Lapidus and J. H. Seinfeld, *Numerical Solution of Ordinary Differential Equations*. Academic Press, New York, 1971.
48. J. H. Seinfeld, L. Lapidus, and M. Hwang, *Ind. Eng. Chem. Fundam.* **9**(2), 266 (1970).
49. *FORSIM, A Fortran Package for the Automated Solution of Coupled Partial and/or Ordinary Differential Equation Systems*. Atomic Energy of Canada Limited, 1976.
50. *IMSL Library User's Manual*, IMSL, Inc., Houston, Texas, 1982.
51. N. Haq and D. M. Ruthven, *J. Colloidal Interface Sci.* **112**(1), 164 (1986).
52. H. Boniface, *Separation of Argon from Air using Zeolites*. Ph.D. thesis, University of New Brunswick, Fredericton, Canada, 1983.
53. G. W. Miller, K. S. Knaebel, and K. G. Ikels, *AIChE J.* **33**, 194 (1987).
54. H. W. Habgood, *Can. J. Chem.* **36**, 1384 (1958).
55. G. F. Round, H. W. Habgood, and R. Newton, *Separation Sci.* **1**, 219 (1966).

56. D. M. Ruthven, N. S. Ragavan, and M. M. Hassan, *Chem. Eng. Sci.* **41**, 1325 (1986).
57. S. Farooq and D. M. Ruthven, *Chem. Eng. Sci.* **45**(1), 107 (1990).
58. M. Suzuki, *AIChE Symp. Ser.* **81**(242), 67 (1985).
59. S. Farooq and D. M. Ruthven, *AIChE J.* **36**(2), 310 (1990).
60. R. T. Yang, *Gas Separation by Adsorption Processes.* Butterworths, Stoneham, MA (1987), p. 211.
61. S. Farooq and D. M. Ruthven, *Ind. Eng. Chem. Res.* **29**(6), 1076 (1990).
62. S. Farooq and D. M. Ruthven, *Ind. Eng. Chem. Res.* **29**(6), 1084 (1990).
63. R. T. Yang and P. L. Cen, *Ind. Eng. Chem. Process Des. Dev.* **25**(1), 54 (1986).
64. D. M. Ruthven, D. R. Garg, and R. M. Crawford, *Chem. Eng. Sci.* **30**, 803 (1975).

CHAPTER

6

PSA Processes

In previous chapters we have described the principles underlying the operation of a PSA process and shown how the process may be represented in terms of simplified mathematical models. However, apart from the discussion in Chapter 3 concerning the factors governing the choice of cycle, the more practical aspects of PSA process design have been ignored and few details of comparative performance have been given. An account of several representative PSA processes that have been developed to the industrial scale is presented in this chapter together with comments on some of the more practical design aspects and the thermodynamic efficiency. Detailed information on practical design and system performance is not widely available in the open literature and the comments presented here therefore represent only a partial account. Further information can be obtained from the extensive patent literature, although the details are often confusing.

6.1 Air Drying

The "heatless drier" was the earliest practical PSA process, and the factors governing the design and performance were studied in detail by Skarstrom.[1] The process, which is widely used for small-scale applications such as the drying of instrument air operates on the simple two-bed Skarstrom cycle (see Section 3.2). Essentially bone-dry product air (1 ppm H_2O) can be readily achieved with either alumina or zeolite (4A or 5A) as the desiccant, but beaded alumina is the usual choice.

The equilibrium isotherm for water vapor on alumina is less strongly curved than the corresponding isotherm for zeolite adsorbents (see Figure 2.5) and as a result, the working capacity in a PSA system is higher. However, a more important consideration is that alumina beads are physically more robust than most other desiccants and do not fracture or suffer attrition under the rather harsh operating conditions of a PSA process. Indeed, provided that the feed air is clean, the life of a PSA drier packed with alumina beads is very long; continuous operation without changing the adsorbent over a period of 20 years has been reported.

6.1.1 Design Considerations

Product purity is normally the primary requirement and, since the water vapor concentration in the feed is generally quite low, a pressure equalization step is not normally included in the cycle. Minimization of the purge, subject to the requirement of a pure product, is therefore the major objective in optimization of the operating cycle. The process operates under essentially adiabatic conditions, and conservation of the heat of adsorption is therefore a major consideration determining the selection of operating conditions. In essence the desiccant bed must be sufficiently long that, during the adsorption step, the thermal wave (which travels faster than the mass transfer zone) does not escape from the bed. The heat retained in the bed heats the purge gas during the desorption step, thus reducing the volume of purge required to desorb the water. If the bed is too short so that the thermal front is not contained, some of the heat will be lost (as sensible heat in the product stream). Under these conditions a greater volume of purge will be needed to clean the bed with consequent loss of process efficiency.

Skarstrom enunciated three general principles for the design of a "heatless drier":

1. Conservation of the heat of adsorption (as discussed). This requires relatively short cycles with low throughput per cycle.
2. Regeneration at low pressure using a fraction of the product stream as a reverse-flow purge. In order to produce a pure product the actual purge volume should exceed the actual feed volume at all points in the bed.
3. For a pure product the absolute pressure ratio (P_H/P_L) should be greater than the reciprocal of the mole ratio of the product to the feed.

It is clear that, under cyclic steady-state conditions, all water vapor entering with the feed must be removed in the purge (apart from the small loss in the blowdown). The maximum water vapor content of the purge gas will be the same as that of the high-pressure feed. Cyclic steady-state operation is therefore only possible with a volumetric purge-to-feed ratio greater than unity. The third principle also follows directly from the overall mass balance. A useful summary of the procedure that is normally followed in the design of a PSA drier has been given by White.[2]

6.1.2 Bed Diameter

The bed diameter is chosen in the normal way based on the design throughput requirement. The maximum velocity in upflow is normally limited to 75% of the minimum fluidization velocity to avoid the increased attrition resulting from particle vibration, which becomes serious, even in a well packed bed, as the fluidization velocity is approached. A somewhat higher velocity, perhaps double the fluidization velocity, can be tolerated in downflow.[3]

6.1.3 Bed Length

The bed length is determined primarily by the requirement to contain the temperature front, which, in an air drier, travels at a higher velocity than the concentration front (the mass transfer zone). The situation is as sketched in Figure 6.1. The concentration front is confined to the entry region of the bed and oscillates during the cycle over only a relatively small distance. The precise form of the profiles and the degree of penetration into the bed depend of course on the humidity level and cycle time as well as on the nature of the desiccant. Experimental concentration profiles from an operating unit are shown in Figure 6.2; the corresponding effluent concentrations are shown in Figure 6.3. The temperature front extends towards the product end of the bed, and the amplitude of its movement during the cycle is much greater than that of the concentration front. The area between the two limiting temperature fronts is proportional to the latent heat of the water adsorbed (and desorbed) during each cycle, while the area between the two concentration fronts is proportional to the mass of water adsorbed and desorbed in each cycle.

More than half the bed generally operates simply as a gas–solid heat exchanger, and it would indeed be possible to replace the adsorbent in this

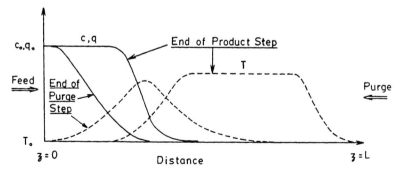

Figure 6.1 Sketch showing qualitative form of concentration and temperature profiles in a "heatless drier" at cyclic steady state. The precise form of the profiles depends on many factors, including the properties of the adsorbent and the duration of the cycle.

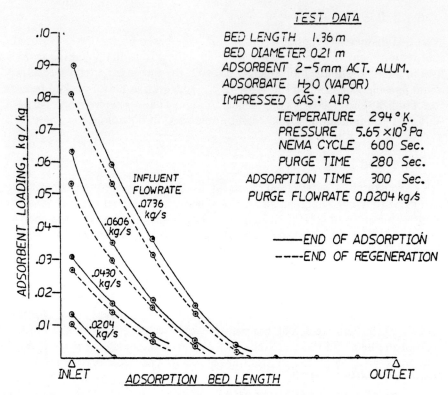

Figure 6.2 Experimental concentration profiles for a heatless air drier packed with an activated alumina adsorbent. (From D. H. White,[2] with permission.)

region by an inert solid with a high heat capacity. Since the heat capacity of the inert solid can be higher than that of the adsorbent, a reduction in overall bed volume can be achieved. The required bed length (typically 1–2 m) is normally determined from a classical heat transfer calculation, following the method of Anzelius,[4] although a full dynamic simulation of the nonisothermal PSA cycle, as described in Section 5.4, is preferable, since such an approach provides more detailed and reliable information concerning the effects of the process variables.

6.1.4 Purge Flow Rate

Under ideal equilibrium conditions the partial pressure of moisture in the purge stream leaving the bed will be the same as that of the entering feed. Consequently, the stoichiometric minimum purge volume (measured at purge pressure) is equal to the actual feed volume (measured at feed pressure). If the purge flow is reduced below this level, the steady-state mass balance will

PSA PROCESSES

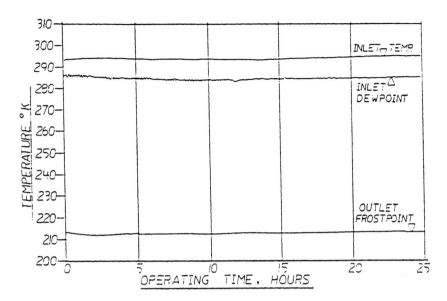

Figure 6.3 Performance test data showing constancy of effluent dew point over a 25 hour period for a "heatless drier" packed with an activated alumina adsorbent. (From D. H. White,[2] with permission.)

not be satisfied and the moisture front will slowly advance through the bed. To allow for the obvious deviation from the ideal equilibrium situation, a margin of at least 15% over the theoretical minimum purge is normally desirable:

$$\frac{\text{purge rate}}{\text{feed rate}} = 1.15 \left(\frac{t_f}{t_p}\right)\left(\frac{P_L}{P_H}\right) \tag{6.1}$$

where the flow rates are expressed on a molar basis and t_f, t_p refer to the feed and purge times.

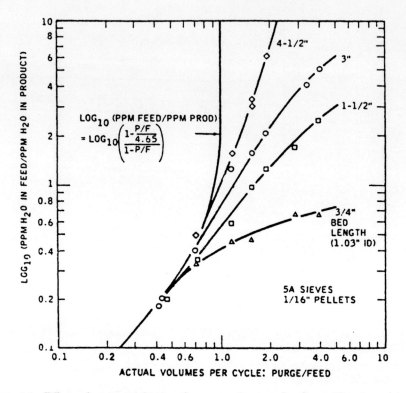

Figure 6.4 Effect of purge-to-feed ratio on product purity for a "heatless drier" packed with 5A zeolite. Feed 26 SCFH at 80° F, 4.7 atm, 6000 ppm H$_2$O. (From Skarstrom,[5] with permission.)

6.1.5 Product Purity

The product purity is determined by the bed length (or the dimensionless ratio L/vt_c) and the purge-to-feed ratio. Experimental results obtained with a column of 5A sieve of different lengths at various purge-to-feed ratios are summarized in Figure 6.4. The asymptotic line is simply calculated from the mass balance:

$$\frac{c_F}{c_p} = \frac{F - (P_L/P_H)P}{F - P} \tag{6.2}$$

With adequate purge and adequate bed length the product air is essentially bone dry. If the purge is insufficient or if the bed length is not long enough to contain the temperature wave, the performance deteriorates.

6.2 Production of Oxygen

Air separation to produce oxygen was one of the processes described in the early PSA patents of Skarstrom.[5] It has been commercialized at scales ranging from a few liters per minute, for medical oxygen, to tens of tons per day for industrial systems. A zeolite adsorbent, generally 5A or 13X, is normally used. When thoroughly dehydrated, such adsorbents show a selectivity towards nitrogen with a separation factor of about 3.0–3.5.[6] Oxygen and argon are adsorbed with almost the same affinity; so the separation is in effect between oxygen plus argon and nitrogen. The maximum attainable oxygen purity is therefore about 95–96%. The presence of argon as an impurity is of little consequence for medical purposes, but it is a significant disadvantage for welding and cutting, since the flame temperature, and therefore the cutting speed, are significantly reduced. Substantially higher separation factors ($\alpha_{N_2/O_2} \sim 8$–10, with correspondingly higher nitrogencapacity) have been reported for thoroughly dehydrated CaX and, various ionic forms of chabazite.[7,8] The Henry constant for nitrogen on these "second generation" adsorbents is too high to permit effective desorption at atmospheric pressure, making them unsuitable for use in a conventional PSA cycle operating at pressures above atmospheric. These adsorbents are, however, used in many of the more modern large-scale vacuum swing or pressure/vacuum swing processes. The development of these processes provides an excellent example of the need to tailor the process cycle to the properties of the adsorbent.

6.2.1 Small-Scale Medical Oxygen Units

At the scale of domestic medical oxygen units the cost of power is a less significant consideration than process simplicity and reliability. Most small-scale units use a two-bed system, operated on a Skarstrom cycle, sometimes with the addition of a pressure equalization step (see Figure 3.6). Typical performance data are shown in Figure 6.5 (see also Figure 3.7). Although a fairly high product purity is attainable, the recovery is relatively low so that the power consumption is high. A significant improvement in recovery can be achieved by the inclusion of a pressure equalization step (see Figure 3.8), but in small-scale units the additional complexity may not be justified.

6.2.2 Industrial-Scale Units

At larger scales of operation proper optimization of a PSA system is essential in order to compete with alternative processes such as cryogenic distillation. Recent trends have been reviewed by Smolarek and Campbell.[9] Capital and operating costs contribute almost equally to the overall cost of a PSA process, and the cost of power is by far the most important component of the operating cost. These considerations lead to competing requirements in optimization. Process efficiency and therefore power cost can be reduced by introducing additional pressure equalization steps, but the increased process complexity, requiring an increased number of beds and with the associated valves and piping, increases the capital cost.

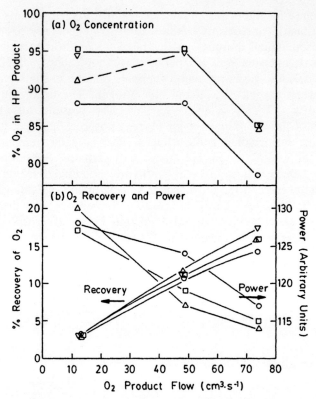

Figure 6.5 Experimental performance data for a simple two-bed PSA oxygen generator operated on a Skarstrom cycle. Note that for high product purity the recovery is less than 10%. (Previously unpublished experimental data.)

The "Lindox" process,[10] shown schematically in Figures 3.10 and 3.11, is typical of the first generation of large-scale PSA oxygen systems. The process operates on a modified Skarstrom cycle with two or three pressure equalization steps (depending on the number of beds). Both three- and four-bed versions were developed at scales of up to 40 tons/day of oxygen product. The process is normally operated between three and one atmosphere pressure and produces a 90% oxygen product gas (dry and free of CO_2) with an oxygen recovery of about 38% and a power requirement of about 1.7 kWh per 100 SCF oxygen product (equivalent to 48,000 J/mole product gas). Productivity is about 0.018 moles of oxygen product gas per kg of zeolite per cycle. The overall mass balance is shown in Figure 6.6, and a more detailed description of the cycle is given in Section 3.2.

Since the original commercialization of the Lindox process in the early 1970s a good deal of further development has occurred, leading to major improvements in process economics. To take full advantage of the higher

PSA PROCESSES

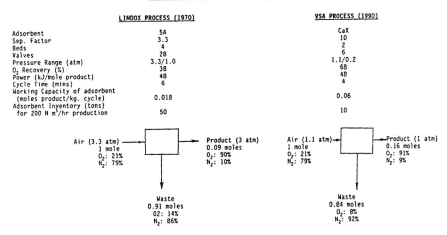

Figure 6.6 Comparison of LINDOX PSA oxygen process with a modern two-bed VSA process. Figures are approximate estimates based on information from a variety of sources.

selectivity and capacity of the "second generation" adsorbents requires desorption at subatmospheric pressure, necessitating the use of a vacuum swing or pressure/vacuum swing cycle. This leads to a substantial reduction in the adsorbent inventory relative to the traditional pressure swing process but at the cost of a somewhat more complex cycle with both a feed compressor and a vacuum pump. Since the valves and piping must be larger for vacuum operation, there is some penalty in capital cost, but this is more than offset by the reduction in adsorbent inventory and power cost.

Modern VSA oxygen systems generally operate between about 1.5–2.5 atmospheres on the high-pressure side with desorption at 0.25–0.35 atmospheres so the pressure ratio is substantially greater than for the earlier supra-atmospheric pressure processes. The cycle is basically similar to the original Air Liquide cycle (Figure 3.12) but with the addition of a pressure equalization step. The adsorbent (CaX) is poisoned by traces of water or carbon dioxide so pre-beds are often included to remove these components. The use of vacuum desorption eliminates the requirement for multiple-bed systems to recover the energy of compression. As a result the more modern VPSA oxygen processes generally use only two beds rather than the three or four beds of the earlier large-scale pressure swing processes, with consequent reduction of capital costs. An approximate performance comparison between a modern VSA oxygen process and the original LINDOX process is given in Figure 6.6. Power requirements are similar, but the adsorbent inventory has been reduced to about 20% of that required in the first generation processes as a result of improved adsorbent capacity, reduced cycle time, and the

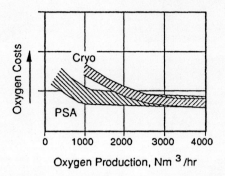

Figure 6.7 Comparative costs of oxygen produced by VSA and cryogenic processes as function of product rate. (From Smolarek and Campbell,[9] with permission.)

elimination of the need for multiple beds. Such processes can also produce the nitrogen product in relatively pure form, which is of course an advantage where both oxygen and nitrogen are required. The VSA process is claimed to be competitive with cryogenic distillation for product rates of up to 3000 N m^3/m or about 100 tons/day (see Figure 6.7).

6.3 Production of Nitrogen

The most common PSA nitrogen process depends on the use of a kinetically selective carbon molecular sieve adsorbent in which oxygen diffuses faster than nitrogen. This difference in diffusion rates makes oxygen the preferentially adsorbed component, even though there is very little equilibrium selectivity (see Figure 2.11). The choice of contact time is critical since if the contact time is too short there will be no significant adsorption, while, if the time is too long, equilibrium will be approached and there will be no selectivity. In this type of process the argon goes with the nitrogen product since the diffusivities are similar.

Most modern nitrogen PSA processes use a two-bed configuration operated on the cycle shown in Figure 3.16.[11] Descriptions of the process have been given by Schröter and Jüntgen,[12] Pilarczyk and Knoblauch,[13] and by Nitrotec.[14] This system can produce 98–99% pure nitrogen (+ Ar) which is adequate for most purging and inerting operations. Although a higher-purity nitrogen product can be obtained directly from this process, it is generally more economic to use a final ("DEOXO") polishing step. The stoichiometric quantity of hydrogen required to oxidize the residual oxygen is introduced, and the gas stream is then passed over a catalyst bed in which essentially all the oxygen is oxidized to water, which is then removed by adsorption on a zeolite desiccant.

Figure 6.8 Small-scale skid-mounted PSA nitrogen generator. (Courtesy of Nitrotec, Inc.[14])

A schematic diagram of the Bergbau Forschung process is shown in Figure 3.15, and a typical small-scale unit, produced and marketed by Nitrotec, is shown in Figure 6.8. The process operates between 6-8 and 1 atmospheres pressure. Standard units are produced in varying sizes with nitrogen product rates from 60 to 60,000 SCFH. A typical overall mass balance is shown in Figure 6.9, and representative performance data are summarized in Table 6.1. The reduction in throughput and the corresponding reduction in recovery and the increase in the specific energy requirement with increasing purity of the nitrogen product are clearly apparent. The economics of this type of process are at their best in the 98-99% purity range at product rates of 200-800 Nm^3/h, although much larger units (up to 4800 Nm^3/hr or about 150 tons/day) have been built.[15] Typical operating costs (assuming electric power at $0.05 per kwh) are about $0.30-0.40 per 1000 SCF (28 Nm^3) for 98% and 99.5% purity respectively (1992 figures)*. At this level costs are comparable with a cryogenic unit over a fairly wide range and only at the

---

*These cost estimates were kindly provided by Mr. Herbert Reinhold of Nitrotec.

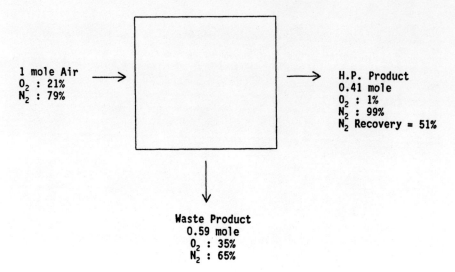

Figure 6.9 Overall mass balance for PSA nitrogen production process. See Table 6.1.

Table 6.1. Performance Characteristics of Nitrotec Nitrogen Generator[a]

Product rate (N m³/h)	O₂ impurity (%)	N₂ recovery (%)	Power (kW)	Specific energy (kWh/N m³ product)
96	3	84	26	0.27
57	1.0	51	23	0.40
39	0.5	35	20	0.51
21	0.1	15	16	0.76

[a] Feed rate: 140 N m³/h, half-cycle time 2 mins; adsorbent beds (2): 76 cm diam × 150 cm height (approx); working pressure: 8 atm/1 atm. Data are from Nitrotec brochure,[14] courtesy Nitrotec Corporation, Glen Burnie, MD, and other sources.

highest purity levels (> 99.5%) and production rates (> 200 tons/day) does the cryogenic system gain a clear economic advantage (see figure 8.10).

6.4 PSA Process for Simultaneous Production of O_2 and N_2

Most PSA processes produce one pure product and an impure byproduct, but by proper design of the operating cycle it is in fact possible to produce two reasonably pure products, subject of course to the limitations imposed by the overall mass balance. An example of such a process is the Air Products

PSA PROCESSES

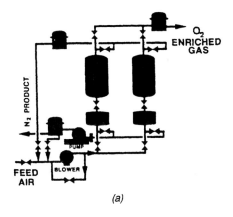

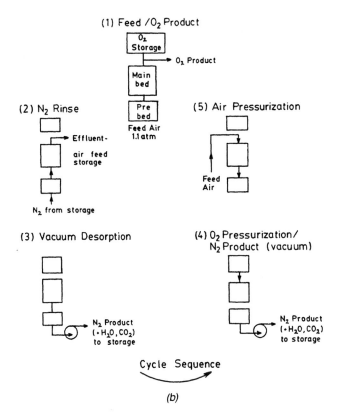

Figure 6.10 Schematic diagrams showing (a) the flowsheet for the Air Products vacuum swing process for simultaneous production of O_2 and N_2 and (b) the sequence of the operating cycle.[17]

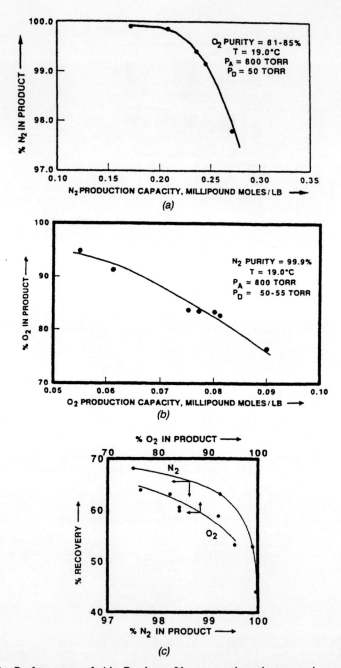

Figure 6.11 Performance of Air Products Vacuum swing air separation process. Adsorbent, Na mordenite; feed pressure, 1.05 atm; desorption pressure (main bed), 50–55 Torr; prebed, 25–30 Torr. (a) Nitrogen product purity; (b) oxygen product purity as a function of product rate; and (c) recovery–purity profile for both products. (From Sircar[17] with permission.)

vacuum swing adsorption process (VSA),[16,17] which is described here. Another similar cycle is discussed in Sections 4.4.5 and 4.5. The process operates with what is basically a two-bed system, but each "bed" actually consists of a prebed for impurity removal in series with the main adsorption bed. The sequence of operations, which involves five distinct steps, is shown schematically in Figure 6.10(b). The following description is taken from Sircar.[17]

Feed / O_2 Product. Air at near-ambient pressure is passed through the prebed and main bed, which have been previously raised to ambient pressure in steps (d) and (e) of the cycle. H_2O and CO_2 are removed in the prebed, and the dry CO_2 free air then passes to the main bed, where N_2 is selectively desorbed to yield the oxygen-rich product stream, some of which is stored in a gas tank for use as the pressurization gas in step (c). This step is terminated at or before the breakthrough of N_2.

N_2 Rinse. A stream of the N_2-rich product is passed through both the prebed and the main bed in the cocurrent direction. The effluent is a dry CO_2 free gas with a composition close to that of air. A part of this gas is therefore recycled as feed air to reduce the load on the prebed. This step is continued until both the prebed and main bed are essentially saturated with nitrogen.

Vacuum Desorption. Both adsorbers are now evacuated from the O_2 product end (countercurrent direction), producing the nitrogen-enriched product stream. This stream, however, contains essentially all the CO_2 and H_2O desorbed from the prebed. A fraction of this gas is stored for use as the nitrogen rinse (step b) in the other pair of beds, while the remainder is withdrawn as the N_2-rich product. Evacuation is continued in this manner until the pressure reaches a preset value at which the valve between the prebed and the main bed is closed.

Evacuation / O_2 Pressurization. Evacuation of the prebed is continued with the desorbate being added to the nitrogen-rich product. Meanwhile the main bed is pressurized with part of the oxygen-rich product from the storage tank.

Air Pressurization. Finally the interconnecting valve between the pre and main beds is opened, and the prebed is brought up to feed pressure with oxygen from the storage tank through the main bed, thus completing the cycle. The performance is shown in Figure 6.11. (See also Figures 4.14 and 4.15.)

6.5 Hydrogen Recovery

The increasing demand for hydrogen, particularly in petroleum refining and petrochemical processing, has provided a strong economic motivation to develop processes to recover hydrogen from refinery fuel gases, coke oven gases, and other similar sources as well as from more traditional sources such

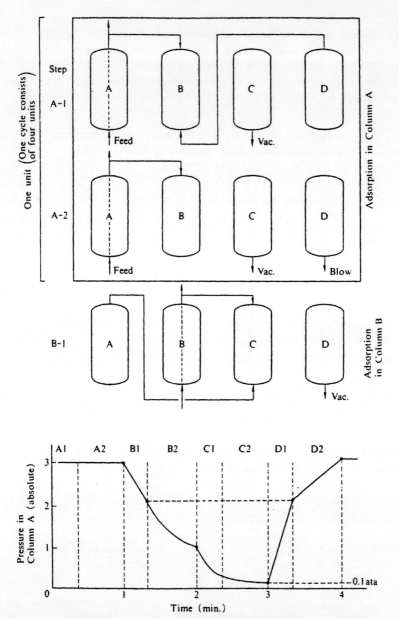

Figure 6.12 Sequence of bed switching and pressure variation in a four-bed hydrogen PSA process. (From Suzuki,[29] with permission.)

as reformer off gas. For most applications a high-purity hydrogen product is required (at least 99.99% and often 99.999%). Since hydrogen is adsorbed much less strongly than almost any other species, a well-designed pressure swing system can meet this challenge. Indeed this is one application for which PSA has a clear advantage over almost all other possible approaches, for many of which these purity levels are unattainable.

Commercial implementation of PSA processes for hydrogen recovery dates from the late 1960s, but the earlier processes were small-scale three- or four-bed units with relatively modest performance ($\sim 70\%$ recovery) de-

Table 6.2. Details of Test Conditions and Performance Data for a Four-Bed Hydrogen PSA Purification System

Adsorbent	5A zeolite
Size	1–4 mm (4–18 mesh)
Form	spherical beads
Bulk density	0.74 kg/l
Feed conc.	
H_2	69.2 vol %
N_2	26.8
CO	2.2
CH_4	1.8
Pressure	
Adsorption P_a	8.6 kg/cm^2
Purge P_p	1.5 kg/cm^2
First equilibration P_1	5.2 kg/cm^2
Second equilibration P_3	1.9 kg/cm^2

Run no.	1	2	3	4	5	6	7	8	9	10	11
Feed rate (Nl/min)^a	14.9	14.6	14.6	21.0	20.9	20.9	8.3	8.3	14.6	14.6	14.5
Cycle Time (min)	12.0	24.0	48.0	8.4	16.8	33.6	21.0	42.0	4.7	6.0	8.0
Column size	4.3 cm i.d. × 200 cm L.								4.3 cm i.d. × 100 cm L.		
Adsorbent (kg)	2.06 × 4 bed								1.0 × 4 bed		
Product conc.											
H_2 (vol %)	99.73	85.61	76.19	99.38	84.22	76.55	99.93	83.48	99.87	97.93	88.41
N_2	0.27	13.41	21.15	0.62	14.52	20.86	0.07	15.19	0.13	2.06	10.9
CO	0.0	0.52	1.48	0.0	0.68	1.45	0.0	0.68	0.0	0.0	0.39
CH_4	0.0	0.47	1.18	0.0	0.58	1.14	0.0	0.65	0.0	0.01	0.30
H_2 Recovery (%)	77.7	88.4	94.9	77.2	90.3	94.6	77.7	90.5	70.0	78.6	83.9
Adsorbent productivity Q (Nl/kg ads., cycle)	21.7	42.7	84.9	21.4	42.6	85.2	21.3	42.5	17.0	21.8	29.0
Space velocity (1/min)	0.66	0.63	0.62	0.90	0.87	0.84	0.37	0.37	1.30	1.29	1.26

^{a} Nl = liters at 273 K 1 atm.
From Tomita et al.,[23] with permission.

signed to provide local sources of high-purity hydrogen.[18-23] Large-scale multiple-bed processes were developed in the late 1970s, the first being the 41 MMSCFD plant at the Wintershall AG Lingen refinery in Germany.[24] These processes, which employ up to twelve beds with several pressure equalization steps, can achieve hydrogen recoveries in the 85–90% range at 99.99% purity from a typical feed containing 75% H_2.[25-28]

6.5.1 The Four-Bed Process

The system for the four-bed process is essentially similar to the four-bed oxygen process (Lindox) shown in Figures 3.9 and 3.10. A mixed adsorbent of activated carbon and 5A zeolite is commonly used, although, since the selectivity for most impurities relative to hydrogen is high, almost any adsorbent can be employed. Production of a hydrogen product with 99.99% purity at 75% recovery from a feed containing 70–80% hydrogen is claimed for the four-bed process with two pressure equalization steps. The process operates between 20–30 atmospheres and 1–2 atmospheres. The cycle is

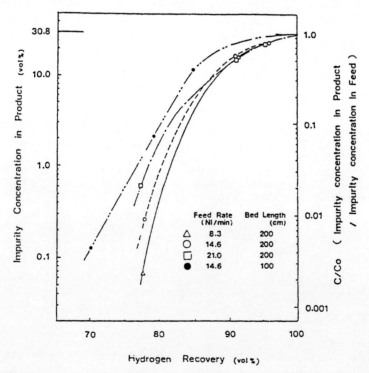

Figure 6.13 Experimental recovery–purity profile for four-bed H_2 PSA system. (From Tomita et al.,[23] with permission.)

PSA PROCESSES

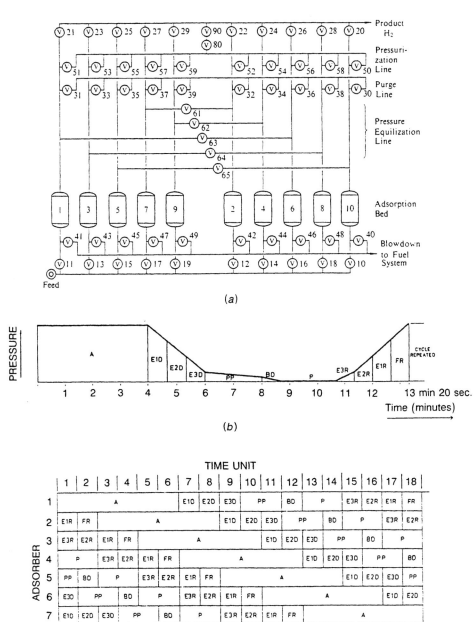

Figure 6.14 (a) Flowsheet, (b) pressure history, and (c) switching sequence for polybed (ten-bed) PSA hydrogen purification process. (Courtesy of Union Carbide.)

shown in Figure 6.12. Details concerning the optimization have been given by Doshi et al.[22] Results of an experimental pilot scale study of the four-bed hydrogen PSA process have been reported by Tomita et al.[23] Some of their results are presented in Table 6.2 and Figure 6.13.

A detailed theoretical optimization of the design of a zeolite-based hydrogen PSA unit, for feed and product specifications typical of industrial practice, has been carried out by Smith and Westerberg.[30] Their results illustrate clearly the economy obtained from an optimal choice of operating pressures and the number of equalization steps. For smaller-scale plants a single equalization step is preferable, but as the throughput increases the optimum shifts towards two or three equalization steps as a result of the proportionately greater importance of operating versus capital costs (see Figure 1.3). The optimal operating pressure, for a system with two equalizations, is about 18 atm. These conclusions are in line with current industrial practice.

6.5.2 Polybed Process

The polybed process operates on a similar principle but with seven to ten beds and at least three pressure equalization steps. This increases the hydrogen recovery to the 85–90% range, but the improvement in performance must be weighed against the increase in capital cost. The process flowsheet is shown in Figure 6.14. Both four-bed and polybed processes operate typically with a pressure ratio 20–30 atm/1 atm. The overall mass balance is shown in Figure 6.15. Purities as high as 99.9999% are achievable with the polybed system, although 99.999% is more common.

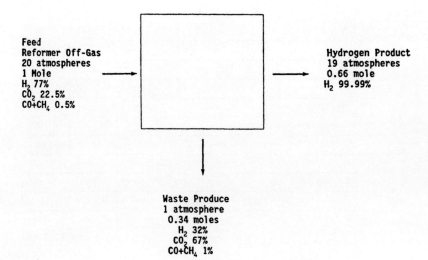

Figure 6.15 Mass balance for polybed PSA hydrogen purification process.

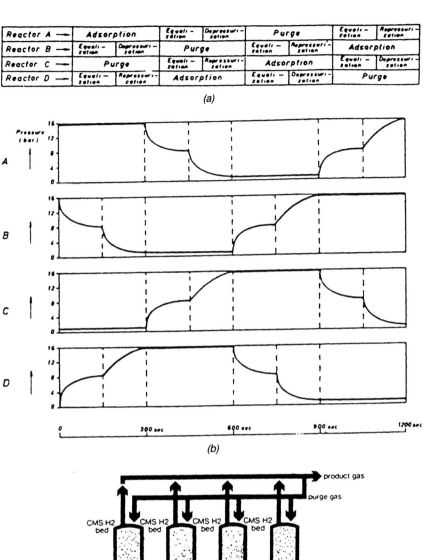

Figure 6.16 Switching sequence, pressure history, and schematic flowsheet for four-bed carbon molecular sieve hydrogen purification process. (From Pilarczyk and Knoblauch,[13] with permission.)

Table 6.3. Composition of Typical Coke Oven Gas

Components	Concentration (% by vol.)
H_2	59.99
CH_4	22.76
N_2	6.68
CO	5.45
C_2H_4	1.66
O_2	1.40
CO_2	1.26
C_2H_6	0.53
C_3H_6	0.11
C_2H_2	0.09
C_3H_8	0.02
C_4H_6	0.02
C_6H_6	0.02
C_4H_6	0.01
C_4H_{10}	0.01
C_5H_{12}	0.01
C_6H_{14}	0.01
C_7H_{16}	0.01
C_7H_8	0.01

Source: Ref. 30.

6.5.3 Bergbau–Forschung Process

As an alternative to the zeolite-based hydrogen recovery processes developed by Union Carbide, Bergbau–Forschung has developed equivalent processes using a wide-pore carbon molecular sieve as the adsorbent. A four-bed system is used with five smaller preadsorption beds containing activated carbon and operating between atm and 1 atm. The process sequence, which is basically similar to that used in the four-bed Union Carbide system, is shown in Figure 6.16. This system has been used primarily to recover hydrogen from coke oven gas containing about 60% hydrogen (see Table 6.3). Hydrogen product purities as high as 99.999% at a recovery of 85% are claimed, and the largest units have a product rate of 10^4 Nm^3/h. The performance thus appears to be broadly similar to that of the Union Carbide polybed system, but since only four beds are used, there should be a capital cost advantage.

6.6 Recovery of CO_2

Carbon dioxide is present at relatively high concentrations (15–35%) in the flue gases from many industries such as steel and lime production. Since CO_2 is strongly adsorbed on many adsorbents, including both zeolites and carbon

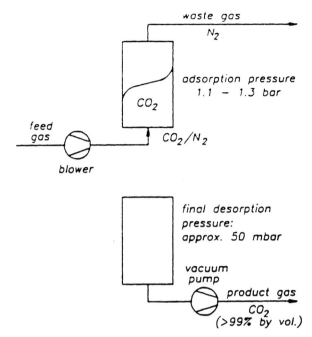

Figure 6.17 Schematic of vacuum swing process for CO_2 recovery from the effluent gas from a steel works. (From Schröter and Jüntgen,[12] with permission.)

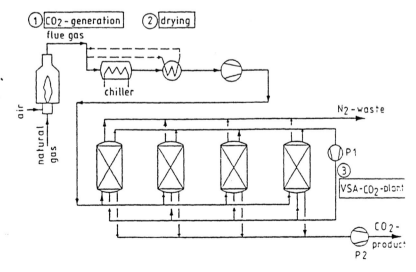

Figure 6.18 Schematic diagram of a four-bed vacuum swing system for recovery of CO_2 from flue gas. (From Pilarczyk and Schröter,[13] with permission.)

Table 6.4. Performance of Bergbau VSA Process for Recovery of CO_2 from Flue Gas

	Run 1	Run 2	Run 3
Compressor P1 (m³/h)	65	65	65
Compressor P2 (m³/h)	350	160	100
CO_2 purity (%)	98.8	98.6	98.7
Product rate (kg/h)	5.9	4.8	4.1
Recovery (%)	72	62	53
Total energy of P1 + P2 kW/h kg CO_2	3.0	1.0	0.8

[a] Feed: flue gas containing ~11 vol% CO_2 at 36 m³/h (STP); adsorber: four beds, each 96 liters packed with wide-pore CMS.
Source: Ref. 30.

molecular sieves, a vacuum desorption system is necessary. A high selectivity for CO_2 is achieved by the use of a narrow-pore carbon molecular sieve adsorbent, similar to that used for nitrogen production. However, the process differs from the nitrogen process in that the product (CO_2) is the more strongly adsorbed component. A simple two-bed vacuum swing system operating between about 1.2 and 0.05 atm has been developed and built at a Japanese steel works.[31] The process schematic is shown in Figure 6.17. 99% product purity and a recovery of 80% at a product rate of 3000 N m³/h are claimed. A more complex four-bed version of this process has also been developed to the pilot plant stage.[32] The schematic of the process is shown in Figure 6.18 and performance data are given in Table 6.4. As a result of environmental pressures the possibility of extracting CO_2 from the stack gases of coal-fired power stations is under active study at the pilot plant stage,[31] although, with current technology, the power costs are too high to make such a process economically attractive.

6.7 Recovery of Methane from Landfill Gases

A somewhat similar process has been developed for recovery of methane from landfill gases.[12,13] These gases contain mainly methane (50–65%) and carbon dioxide (35–50%) as well as small amounts of nitrogen, with many different hydrocarbons and sulfur compounds in trace concentrations. A two-stage initial purification process is employed. In the first stage hydrogen sulfide is removed at a temperature of 40–50° C using a bed of iodine-impregnated activated carbon. This acts as an efficient catalyst for conversion to elemental sulfur, and residual H_2S levels as low as 1 ppm can be achieved. In the second stage halocarbons and heavier hydrocarbons are removed in a conventional activated carbon adsorber. The final stage of the process utilizes a four-bed vacuum swing system to recover methane from the purified landfill gas using a narrow-pore carbon molecular sieve. The remaining impurities

PSA PROCESSES

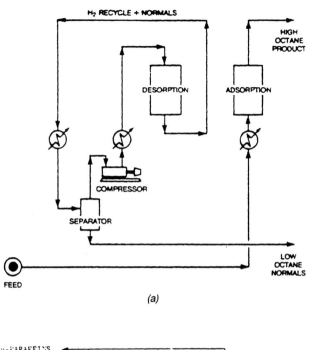

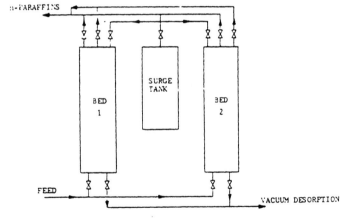

Figure 6.19 (a) Simplified flow diagram and (b) cycle sequence for the Isosiv paraffin separation process. (From Cassidy and Holmes,[27] with permission.)

(CO_2, N_2, O_2) diffuse into the adsorbent much faster than methane, which is therefore produced as the raffinate product. A high recovery (over 90%) is reported, with the product containing 87–89% methane with the balance $CO_2 + N_2$.

6.8 Hydrocarbon Separations

The separation of linear from branched and cyclic hydrocarbons using 5A zeolite as the adsorbent is one of the earliest examples of a molecular sieve separation process. In the low-molecular-weight range (up to C_{10}) a pressure swing version of the "Isosiv" process is commonly used. The system, which is in essence a standard pressure/vacuum swing system, is shown schematically in Figure 6.19. High-purity linear paraffins are produced as the desorbate during the evacuation step, while during the adsorption step a raffinate stream depleted of normals is produced. The operating temperature and pressures depend on the molecular weight range, but for the C_6 feedstock, 350°C with a pressure swing from 20 to 0.2 atm is typical.[27] A somewhat similar process has been developed by B.P.[32]

6.9 Process for Simultaneous Production of H_2 and CO_2 from Reformer Off-Gas

The crude hydrogen stream from a steam reformer contains significant quantities of carbon dioxide, which, by appropriate design of the separation

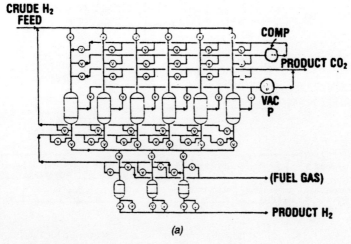

Figure 6.20 (a) Simplified process flow sheet, (b) flow schedule for the Gemini-9 process for simultaneous production of H_2 and CO_2 from reformer gases. (From Sircar[17] and Kumar et al.,[34] with permission.)

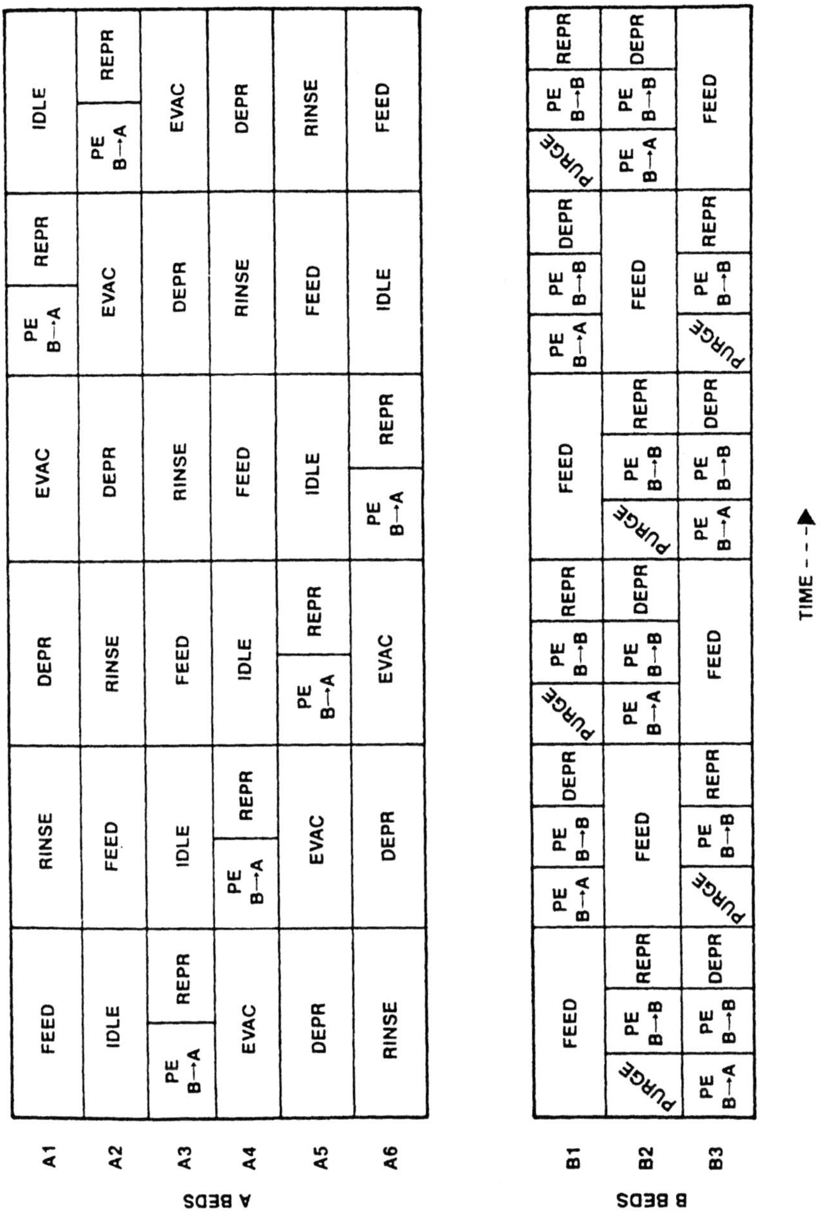

(b)

Figure 6.20 (*Continued*).

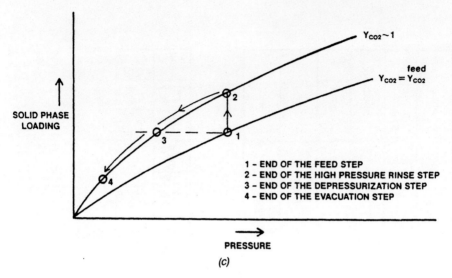

Figure 6.20(c) Variation in CO_2 loading during the cycle for the Gemini-9 Process. (From Kumar et al.,[34] with permission.)

process, can be recovered as a valuable byproduct. A PSA process for simultaneous production of pure hydrogen and carbon dioxide from such a feed gas has recently been developed by Air Products and is described here as an example of the third generation of PSA processes designed to achieve both energy efficiency and dual product recovery. Two variants of this process have been developed. The original version (Gemini-9)[17,33] used nine adsorbent beds in a series-parallel arrangement but in the later version improved performance was obtained, at a somewhat lower capital cost, by using a modified cycle with only eight beds (Gemini-8).[34]

The flowsheet for Gemini-9 is shown schematically in Figure 6.20. There are six parallel beds containing a zeolite adsorbent (NaX) that selectively removes H_2O and CO_2 from the feed gas (the A beds) and three parallel beds packed with a second zeolite adsorbent (a mixture of NaX and 5A) that selectively removes CO_2, CO, CH_4, and N_2 impurities from the hydrogen product (the B beds). During the adsorption steps one A and one B bed are connected in series, but the desorption steps for the A and B beds are different, as will be described.

6.9.1 Cycle for the A Beds

A1. **Adsorption.** Following pressurization with the hydrogen-rich product gas the feed gas is passed through the bed at the highest pressure of the cycle (P_H). CO_2 and water vapor are removed, and the effluent passes to a B bed for removal of the trace impurities (CO, CH_4, and N_2) together

PSA PROCESSES

with any CO_2 that has passed through A bed. This step is terminated just prior to breakthrough of the C_2 mass transfer front.

A2. CO_2 Rinse. At the termination of the adsorption step high-purity CO_2 at the feed gas pressure is passed through the bed in the cocurrent direction. The effluent has a composition similar to that of the feed, and it is recycled as feed to another of the A beds. This step is continued until the bed is essentially saturated with high-purity CO_2.

A3. Countercurrent Depressurization. The A bed is blown down countercurrently to atmospheric pressure, and the effluent from this step, which consists of high-purity CO_2, is collected as byproduct. Part of this gas is recompressed to P_H for use in the CO_2 rinse step A2.

A4. Countercurrent Evacuation. The bed is evacuated from the feed end to the lowest pressure of the cycle (P_L). The residual CO_2 from this step is added to the byproduct stream.

A5. Countercurrent Pressurization I. The evacuated A bed is connected with a B bed undergoing B2 (see Section 6.9.2) in order to transfer a part of the residual gases from the B bed to the A bed (product end), thus raising the pressure in the A bed to an intermediate level P_1. ($P_L < P_1 < P_H$).

A6. Countercurrent Pressurization II. To complete the cycle is pressurized to P_H using the hydrogen product gas introduced at the product end. The pressurizing gas is in fact the recycled effluent from a B bed undergoing step B7.

The variation in CO_2 loading during the cycle (for the A beds) is shown schematically in Figure 20(c). At the end of the feedstep (A1) the beds are at point (1) on the feed isotherm. At the end of the rinse step (A2) the loading corresponds to point (2) on the pure CO_2 isotherm. At the end of the blowdown step (A3) the CO_2 loading has fallen to point (3) and the gas released is recompressed for use as the high-pressure rinse gas. During the evacuation (step A4) the loading falls to point 4 and the CO_2 desorbed in this step constitutes the product stream.

6.9.2 Cycle for the B Beds

B1. Adsorption. Prior to this step the B bed is pressurized to P_H with hydrogen product gas. The B bed in series with an A bed during step A1 receives the CO_2 depleted gas from the A bed and removes the remaining CO_2 and other impurities to yield highly pure hydrogen as effluent. Part of the hydrogen product is used to purge another B bed (step B5) and to pressurize both B and A beds (steps B7 and A6). The adsorption/product withdrawal step is terminated just before the leading impurity breaks through.

B2. Countercurrent Depressurization I. The B bed is connected with an A bed, which is undergoing step A5 and a portion of the desorbed and

Table 6.5. Performance of Gemini-9 Process for Simultaneous Production of H_2 and CO_2

Run	Feed[a] P (psig)	Relative amounts of gases					H_2 Product		CO_2 Product		Fuel gas[b]	
		CO_2 rinse	A bed desorption	A bed evacuation	H_2 purge to B bed	Amount	% H_2	Recovery	Amount	% CO_2	Recovery	Amount
A	200	0.252	0.275	0.160	0.078	0.645	99.999+	85.9%	0.183	99.4	90.1%	0.180
B	250	0.259	0.295	0.152	0.061	0.657	99.999+	87.1%	0.188	99.4	94.0%	0.172
C	300	0.283	0.310	0.144	0.050	0.656	99.999+	86.7%	0.171	99.4	86.9%	0.165

Performance Comparison

	Gemini-8	Gemini-9
H_2 purity (mol %)	99.999	99.999
H_2 recovery (%)	86	87
CO_2 purity (mol %)	97.0+	99.0+
CO_2 recovery (%)	86	91
Relative compressor size	0.43	1.00
Relative compressor power	0.57	1.00
Relative vacuum blower power	0.86	1.00
Number of adsorbent beds	8	9
Number of switch valves	508w.	68

[a] Feed gas contains 75.4% H_2, 19.9% CO_2, 0.96% CO, and 3.73% CH_4.
[b] Fuel gas composition: 9.5% CO_2, 6.1% CO, 22.5% CH_4, and 61.9% H_2.

interstitial gas is transferred through the feed end to the A bed (countercurrent), thus reducing the pressure from P_H to P_1.

B3. **Countercurrent Depressurization II.** The bed is connected to another B bed, which is undergoing step B6 and more of the desorbate and interstitial gas is removed (countercurrently) through the feed end, reducing the pressure to P_2.

B4. **Countercurrent Depressurization III.** The bed is blown down from P_2 to atmospheric pressure, and the effluent gas, which contains a proportion of the feed impurities together with some hydrogen, is rejected.

B5. **Countercurrent Purge.** The bed is purged at atmospheric pressure with high-purity hydrogen product to desorb any impurities further. The effluent is rejected.

B6. **Cocurrent Pressurization.** The pressure in the bed is raised to P_2 by connecting with another B bed undergoing step B5.

B7. **Countercurrent Pressurization.** Final pressurization of the B bed to P_H is accomplished with hydrogen product, introduced from the product end. During the later part of this step the B bed is connected with another A bed undergoing step A6 and both beds are then pressurized to P_H.

The B beds pass through two complete cycles (steps B1 to B7) while the A beds go through one cycle so that each B bed handles the gas from two A beds during the complete cycle. This approach reduces significantly the size of the B beds. A key feature of this cycle is that the A and B beds are connected in series during the adsorption step but they are regenerated by two entirely different sequences. The overall performance is summarized in Table 6.5. It is evident that the hydrogen product has a purity greater than 99.999% and the fractional recovery is about 86–87%, while the CO_2 product is produced at a purity of about 99.4% with about 90% recovery. The Gemini-8 process gives slightly lower purity and recovery of CO_2, but there is a significant reduction in the size of the compressors and the power consumption.

6.10 PSA Process for Concentrating a Trace Component

In the processes described so far in this chapter the objective has generally been to produce a pure raffinate product, although in some cases the more strongly adsorbed species (the extract product) is also recovered in concentrated form. However, particularly in environmental applications, it is often necessary to concentrate a trace component for disposal or further processing. Provided that a sufficiently selective adsorbent is available, PSA appears to be well suited to such applications, although to date few, if any, processes of this kind have been commercialized. Examples of two such processes that have been developed to the pilot plant scale are described in this section.

The basic principle of a PSA "concentration process" may be understood from equilibrium theory (see Section 4.2 and Figure 4.25). Consider an adsorbent bed equilibrated at pressure P_0 and mole fraction (of the more strongly adsorbed species) y_0. If the equilibria are linear:

$$q_A^* = K_A c_A = K_A y \frac{P}{RT}; \qquad q_B^* = K_B c_B = K_B (1-y) \frac{P}{RT} \qquad (6.3)$$

By combining the differential mass balance expressions for both components (Eq. 4.4), we obtain:

$$\frac{1}{\beta_B} \frac{\partial P}{\partial t} + \left(\frac{1}{\beta_A} - \frac{1}{\beta_B} \right) \frac{\partial (Py)}{\partial t} + \frac{\partial (Pv)}{\partial z} = 0 \qquad (6.4)$$

and by eliminating the term $\partial v/\partial z$ between Eqs. 4.4 and 6.3:

$$\left(\frac{\partial y}{\partial t} \right)_z + \frac{\beta_A v}{1 + (\beta - 1)y} \left(\frac{\partial y}{\partial z} \right)_t = \frac{(\beta - 1)(1-y)y}{1 + (\beta - 1)y} \frac{d \ln P}{dt} \qquad (6.5)$$

in which $\beta = \beta_A/\beta_B$ and the total pressure differential arises from the assumption of negligible pressure drop across the column $[P = P(t)]$. Since $y = y(z,t)$, the left-hand side of Eq. 6.5 is simply the total time derivative of y, and the variation of composition during pressurization or blowdown steps is given by Eq. 4.8. It follows by direct integration that the variation in composition during pressurization or blowdown will be given by:

$$\frac{y}{y_0} = \left(\frac{1-y}{1-y_0} \right)^\beta \left(\frac{P}{P_0} \right)^{\beta - 1} \qquad (6.6)$$

Equations 4.4 and 6.4 yield:

$$\frac{1}{\beta_B} \frac{\partial P}{\partial t} + (\beta - 1) \frac{\partial (vPy)}{\partial z} + \frac{\partial (vP)}{\partial z} = 0 \qquad (6.7)$$

Neglecting the axial variation of pressure leaves only the time dependence; so the velocity during a pressurization or blowdown step can be found simply integrating from the closed end:

$$v = \frac{-z}{\beta_B [1 + (\beta - 1)y]} \frac{d \ln P}{dt} \qquad (6.8)$$

The fraction of A desorbed during blowdown from pressure P_0 to P is given by:

$$F_A = \int_{P_0}^{P} \beta_A \frac{vyP \, dt}{z y_0 P_0} \qquad (6.9)$$

and substituting for v and y from Eqs. 6.8 and 6.6:

$$F_A = \frac{\beta}{1 - \beta} \left(\frac{1 - y_0}{y_0} \right)^{\beta/(\beta - 1)} \int_{P_0}^{P} \frac{y^{1/(\beta - 1)}}{(1-y)^{(2\beta - 1)/(\beta - 1)}} dy \qquad (6.10)$$

Similarly, for the fraction of B desorbed:

$$F_B = \frac{1}{1 - \beta} \left(\frac{y_0}{1 - y_0} \right)^{1/(1-\beta)} \int_{P}^{P_0} \frac{y^{(2-\beta)/(\beta - 1)}}{(1-y)^{\beta/(\beta - 1)}} dy \qquad (6.11)$$

PSA PROCESSES

In the limit of $y_0 \to 0$ and $\beta \to 0$ (high selectivity) these expressions reduce to:

$$\frac{y}{y_0} = \frac{P_0}{P} \qquad (6.12)$$

$$F_A = \beta \ln\left(\frac{y/y_0}{1-y}\right)$$

$$F_B = 1 - P/P_0$$

Figure 6.21 shows F_A and F_B plotted against the pressure ratio for different values of y_0. The curves are nonlinear, but, when the selectivity is

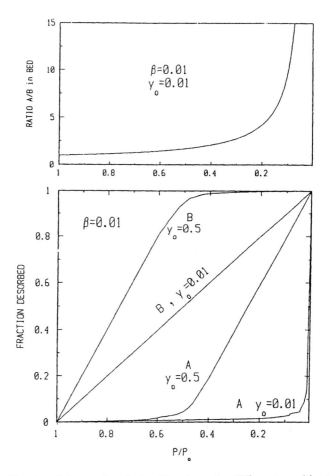

Figure 6.21 Fractional desorption during blowdown for different combinations of y_0 and β, calculated according to Eqs. 6.10 and 6.11.

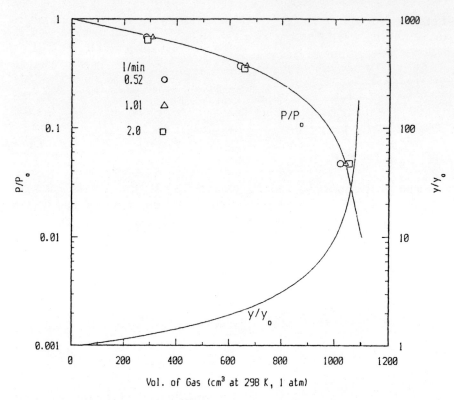

Figure 6.22 Experimental data showing concentration of hydrogen from a H_2–He mixture during blowdown of a small experimental column of 5A zeolite, equilibrated initially at 77 K with a stream containing 0.16% H_2 in He at 21.4 atm. Column 15.6 cm × 0.77 cm i.d. packed with 20–40 mesh 5A mol sieve particles. (From Ruthven and Farooq.[36])

Table 6.6. Recovery of H_2 from the He by Cryogenic PSA[36a]

Adsorbent	5.6 g 5A zeolite (pelleted)
Feed	0.18% H_2 in He at 21.4 atm
Hydrogen uptake	240 cm³ STP
Exhaust gas	220 cm³ STP
Purity	93% H_2
Recovery	85%

[a] Bed was equilibrated with feed, blown down to atmospheric pressure, and evacuated to 0.01 atm. Exhaust gas was collected from vacuum pump.

modest ($\beta = 0.5$) and the initial mole fraction is large ($y_0 = 0.5$), the deviations from linearity, for both F_A and F_B, are small. However, when the selectivity is high and the initial mole fraction of A is small, the curve for F_B becomes essentially linear, but the curve for F_A assumes a highly nonlinear form. It is clear that, in this situation, by a sufficiently large reduction in total pressure almost all of component B can be desorbed with very little desorption of A. A further deep blowdown or evacuation step then allows A to be removed in concentrated form.

This analysis is for a linear equilibrium system, but the effect of isotherm nonlinearity is actually to enhance the degree of separation and concentration that can be achieved in this type of process. Since the isotherm for the more strongly adsorbed species will generally have the higher curvature, even less of this component is desorbed during the initial blowdown compared with the equivalent linear equilibrium system.

A process of this kind has recently been developed as a means of concentrating and removing the traces of tritium from the helium purge stream of a lithium breeder reactor.[35] To achieve a high concentration ratio ($\sim 10^3$) requires a high selectivity ratio (as well as a high pressure ratio), and for H_2 (or tritium) this can be achieved only by operating at cryogenic temperatures with vacuum desorption at a very low pressure. Laboratory data showing the feasibility of recovering hydrogen at greater than 90% purity and with a similarily high fractional recovery from a stream containing traces of H_2 in He are summarized in Figure 6.22 and Table 6.6. The process schematic is shown in Figure 6.23.

The same principle was used by Yang and co-workers[37,38] in recent studies of the possibility of removing and concentrating trace organics from air and SO_2 from flue gas. It is also utilized in the Air Products fractional vacuum swing adsorption process (FVSA), which produces 90% oxygen together with 98–99% nitrogen.[39] The cycle, which is essentially similar to that used in the hydrogen recovery process, involves four steps:

- Adsorption with feed air at 1.1 atm abs. with simultaneous withdrawal of oxygen product.
- Reverse flow blowdown with discharge of the blowdown gas (impure nitrogen) to waste.
- Evacuation to 0.1 atm with collection and recompression of the nitrogen product.
- Pressurization with product oxygen.

With CaX zeolite as the adsorbent the selectivity ($\alpha \sim 10$) is high enough that most of the oxygen is eliminated in the blowdown step. About half the adsorbed nitrogen can be recovered at 98–99% purity during the evacuation step which run at about 0.1 atm abs. The schematic diagram together with performance data are shown in Figure 6.24.

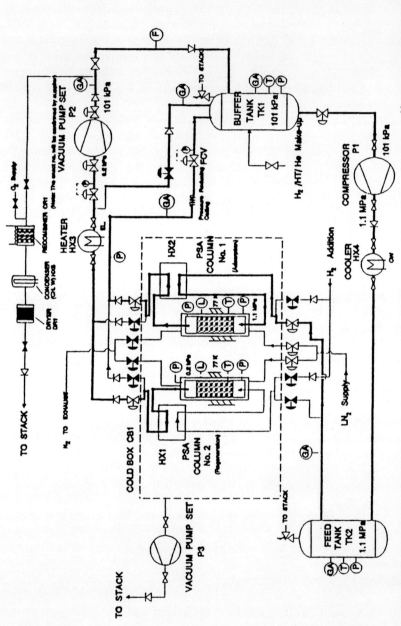

Figure 6.23 Schematic diagram of pilot plant for tritium concentration by cryogenic PSA. (From Sood et al.[35]) Feed: 2.2 mole/sec, 99.9% He, 0.1% hydrogen isotopes at 12 ats. abs. Blowdown to atmosphere; evacuation to 0.01 atm.

PSA PROCESSES

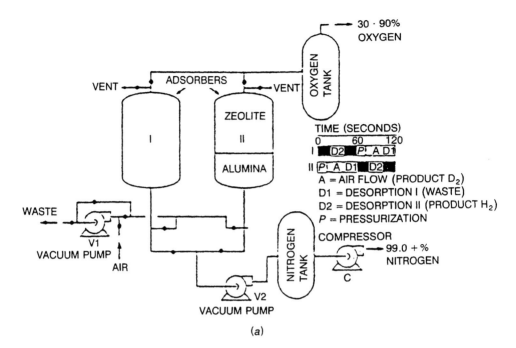

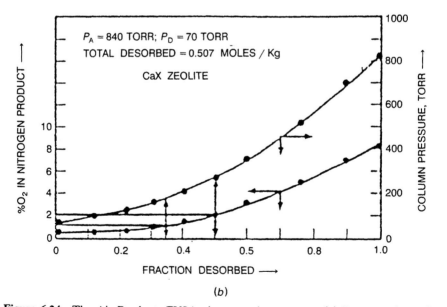

Figure 6.24 The Air Products FVSA air separation process. (a) Process schematic; (b) performance data. (From Sircar,[39] with permission.)

6.11 Efficiency of PSA Processes

The thermodynamic efficiency of any separation process (the First Law efficiency) can be defined simply as the ratio of the minimum work of separation (the negative free energy of mixing) to the actual work required to drive the separation process. Such efficiencies are generally less than 15% and even lower than this for most PSA systems. The values for two representative air separation processes are given in Table 6.7. The nitrogen production process is markedly less efficient than oxygen production, reflecting the irreversibility inherent in a kinetically based separation. Although thermodynamic efficiency provides a rational basis for comparing the efficiency of different PSA processes based on the same type of cycle, it is much less useful for comparing different types of separation process or even radically different PSA cycles. Furthermore, the thermodynamic efficiency gives only an overall measure of performance and provides no information as to the sources of efficiency. An exergy analysis provides far greater insight.

The exergy of a substance is the maximum useful work that can be obtained by interaction with the environment. It is in essence the free energy relative to the normal environment as standard state. For a nonreacting system in which potential and kinetic energies are insignificant:

$$\text{Ex} = (h - h_0) - T_0(s - s_0) \tag{6.13}$$

The exergetic efficiency is then defined as:

$$\frac{\text{moles product}}{\text{moles feed}} \times \frac{\text{Ex}_{product}}{\text{Ex}_{feed}} \tag{6.14}$$

The feed exergy (Ex_{feed}) includes the work of compression while the product exergy ($\text{Ex}_{product}$) includes the energy of compression (or expansion) to reduce the product to atmospheric pressure. For comparing different PSA processes, operated over different pressure ranges, a more useful definition is the overall efficiency (η) defined by:

$$\eta = \frac{\text{molar exergy of product, corrected to 1 atm}}{\text{net energy input}} \tag{6.15}$$

where the net energy input is the energy of feed compression less the energy

Table 6.7. Thermodynamic (First Law) Efficiencies for PSA Air Separation Processes

Process	Principal product	Process Energy (J/g mole product)	Separative work (J/g mole product)	Efficiency (%)
"Lindox" (Figure 6.8)	90% O_2	4.8×10^4	3055	6.3
"Nitrotec" (Figure 6.11 and Table 6.1)	99% N_2	3.2×10^4	660	2.1

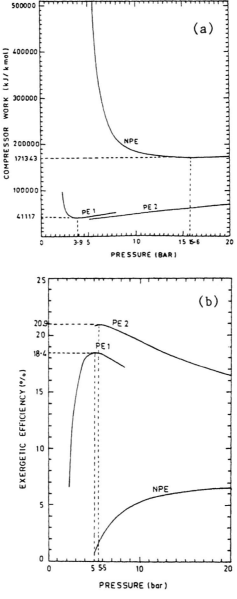

Figure 6.25 Variation of (a) compressor work and (b) exergetic efficiency with operating pressure for a two-bed Skarstrom cycle for oxygen production with and without pressure equalization. (From Banerjee et al.,[40] with permission.)

of expansion of the product to atmospheric pressure. For a vacuum swing process in which the product is produced at subatmospheric pressure, the latter quantity will be negative. Equation 6.15 differs from the normal definition of the First Law efficiency in that the separative work associated with the waste product is excluded and the energy of compression (or expansion) of the product to atmospheric pressure is allowed for.

A detailed exergy analysis of PSA air separation processes has been reported by Banerjee et al.[40,41] For oxygen production two process configurations were considered: the two-bed Skarstrom cycle with and without a pressure equalization step. The variation of compressor work and exergetic efficiency with operating pressure is shown in Figure 6.25. For the cycle

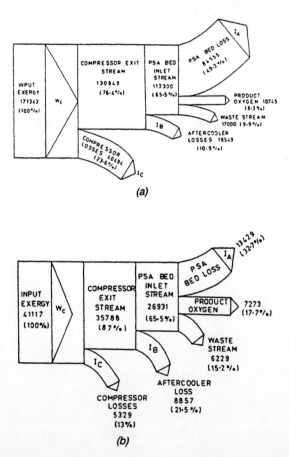

Figure 6.26 Grassman diagrams showing losses of exergy for Skarstrom cycle (blowdown to atmospheric pressure) for oxygen production (a) without and (b) with pressure equalization. Operating pressures are, respectively, 15.6 and 3.9 bar, the values for which the compressor work is minimized. (From Banerjee et al.,[39] with permission.)

PSA PROCESSES

without pressure equalization there is a shallow minimum in the compressor work curve at about 15 atm, while for the process with pressure equalization the compressor work is greatly reduced and there is a sharper minimum at about 4 atm. The exergetic efficiency reaches a maximum at about 5 atm for the process with pressure equalization but increases monotonically for the process without pressure equalization. The corresponding Grassman diagrams for operation at their optimal pressures (in terms of compressor work) are shown in Figure 6.26. The exergetic efficiency is about 17% for the

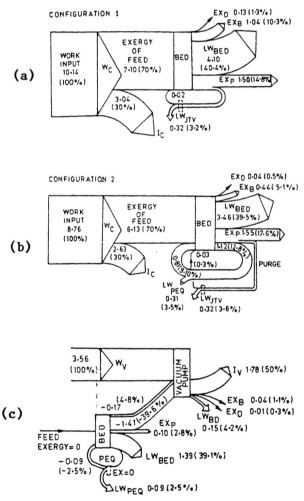

Figure 6.27 Grassman diagrams showing exergy losses for nitrogen production with three different cycles. (a) Skarstrom cycle and (b) Skarstrom cycle with pressure equalization (P_H) = 8 atm, P_L = 1 atm); (c) vacuum swing cycle (P_H = 1 atm, P_1 = 0.2 atm). (From Banerjee et al.,[41] with permission.)

Table 6.8. Exergy Analysis of Three PSA Air Separation Processes[a]

	(1) PSA O_2 process		(2) PSA N_2 process		(3) VSA N_2 process	
Adsorbent	5A zeolite		CMS		CMS	
Cycle	Skarstrom + P.E.		Self-purging + P.E.		Vac. swing	
P_H/P_1 (atm)	3.9/1.0		8.0/1.0		1.0/0.2	
Recovery (%)	60		33		63	
Energy input						
(a) Compressor/vac.	41,120		31,500		12,800	
(b) Product compression /expression	−2,800	(%)	−4,800	(%)	+5,700	(%)
Total input	38,320	100	26,700	100	18,500	100
Product exergy	7,270		5,600		360	
Product exergy at 1 atm	4,470	11.7	800	3.0	4,360	23.5
Waste product	6,230	16.3	1,730	6.5	720	3.9
Bed loss	13,430	35.0	12,450	46.5	5,100	27.6
Compressor/cooler losses	14,200	37	9,470	35.5	8,050	43.43
Other losses (e.g., valves, etc.)			2,270	8.5	320	1.7
Process efficiency (%)		11.7		3.0		23.5

[a] Energy and exergy expressed as J/mole product. Product exergy is corrected to 1 atm, in all cases, by allowing for work of expansion or compression. Process efficiency is defined by Eq. 6.15.

process with pressure equalization when operated under optimal conditions in terms of the power requirement.

A similar analysis has also been made for the two-bed nitrogen production process including both the self-purging cycle and the vacuum swing cycle (Figures 3.12 and 3.17). The corresponding Grassman diagrams are shown in Figure 6.27. The exergetic efficiency of the self-purging process is about 17.6% at an operating pressure of 8 atm, and the corresponding energy requirement is about 8.7 kWh per kmole of product. For the vacuum swing cycle the exergetic efficiency is much lower ($\sim 2.8\%$) since the product is delivered at subatmospheric pressure. However, the energy requirement is also lower (about 3.6 kWh per kmole of product). If the product nitrogen were compressed to 8 atm, the total work requirement would be increased to 5.6 kWh/kmole product, but this is still substantially lower than for the pressure swing process.

A detailed thermodynamic comparison of the three air separation processes, based on Banerjee's figures, is given in Table 6.8. For all three processes the major sources of inefficiency are the losses in the feed compressor or the vacuum pump and in the adsorbent bed. Comparing the two supra-atmospheric pressure processes (1 and 2), the efficiency of the kinetically controlled nitrogen process is substantially lower than that of the equilibrium-based oxygen process, reflecting the inherent irreversibility of the kinetic process. More significantly, however, the comparison between the vacuum swing process (3) and the supra-atmospheric processes (1 and 2) shows clearly the thermodynamic advantage of vacuum swing. This advantage stems from the large reduction in the energy input so that, even when the energy required to compress the product is allowed for, the net energy requirement is substantially reduced. However, this advantage, which translates directly into a reduction of the process operating cost, must be offset against the increased capital costs associated with vacuum swing operation, which requires both a compressor and a vacuum pump as well as much larger ducts and valves.

References

1. C. W. Skarstrom, "Heatless Fractionation of Gases over Solid Adsorbents," in *Recent Developments in Separation Science*, pp. 95–106, Vol. 2, N. N. Li, ed., CRC Press, Cleveland (1972).

2. D. H. White, *The Pressure Swing Adsorption Process*, AIChE National Meeting, paper 87b, New Orleans, LA, March 8 (1988). See also D. H. White and G. Barclay, *Chem. Eng. Prog.* **85**(1), 25 (1989).

3. D. M. Ruthven, *Principles of Adsorption and Adsorption Processes*, Chap. 7, Wiley, New York (1984).

4. A. Anzelius, *Zeit. Angew. Math. Mech.* **6**, 291–94 (1926).

5. C. W. Skarstrom, U.S. Patents 2,944,627 (1958) and 3,237,377 (1966) to Esso Research and Engineering.

6. G. A. Sorial, W. H. Granville, and W. O. Daley, *Chem. Eng. Sci.* **38**, 1517 (1983).

7. C. G. Coe, G. E. Parris, R. Srinivasan, and S. R. Auvil, in *Proceedings of Seventh International Zeolite Conference*, Tokyo, p. 1033, Y. Murakami, A. Lijima, and J. W. Ward, eds., Kodansha–Elsevier, Tokyo (1986).

8. C. G. Coe, in *Gas Separation Technology*, pp. 149–59, E. F. Vansant and R. Dewolfs, eds., Elsevier, Amsterdam (1990).

9. J. Smolarek and M. J. Campbell, in *Gas Separation Technology*, p. 281, E. F. Vansant and R. Dewolfs, eds., Elsevier, Amsterdam (1990).

10. L. B. Batta, U.S. Patent 3,636,679 to Union Carbide (1972).

11. K. Knoblauch, H. Heinback, and B. Harder, U.S. Patent 4,548,799 (1985), to Bergbau Forschung.

12. H. J. Schröter and H. Jüntgen, in *Adsorption: Science and Technology*, p. 269, NATO ASI 158, A. E. Rodrigues, M. D. LeVan, and D. Tondeur, eds., Kluwer, Dordrecht (1989).

13. E. Pilarczyk and K. Knoblauch, *Separation Technology*, p. 522, N. Li and H. Strathmann, eds., Eng. Foundation, NY (1988).

14. Nitrotec brochure, Nitrotec Engineering Co., Glen Burnie, MD (1988).

15. Anon., "Pressure Swing Adsorption Picks Up Steam," *Chem. Eng.*, **95**, Sept. 26, 1988, p. 26.

16. S. Sircar and J. W. Zondlo, U.S. Patent 4,013,429 (1977) to Air Products.
17. S. Sircar, in *Adsorption: Science and Technology*, p. 285, NATO ASI E158 A. E. Rodrigues, M. D. LeVan, and D. Tondeur, eds., Kluwer, Dordrecht (1989).
18. J. L. Wagner, U.S. Patent 3,430,418 to Union Carbide (1969).
19. L. B. Batta, U.S. Patent 3,564,816 to Union Carbide (1971).
20. R. W. Alexis, *Chem. Eng. Prog. Symp. Ser.* **63**(74), 50 (1968).
21. H. A. Stewart and J. L. Heck, *Chem. Eng. Prog.* **65**(9), 78 (1969).
22. K. J. Doshi, C. H. Katiro, and H. A. Stewart, *AIChE Symp. Ser.* **67**(117) (1971).
23. T. Tomita, T. Sakamoto, U. Ohkamo, and M. Suzuki, in *Fundamentals of Adsorption II*, p. 89, A. I. Liapis, ed., Eng. Foundation, NY (1987).
24. J. L. Heck and T. Johansen, *Hydrocarbon Processing*, p. 175 (Jan. 1978).
25. A. Fuderer and E. Rudelstorfer, U.S. Patent 3,846,849 to Union Carbide (1976).
26. R. T. Cassidy, "Polybed Pressure Swing Hydrogen Processes," in *Adsorption and Ion Exchange with Synthetic Zeolites*, W. H. Flanck, ed., *ACS Symp. Ser.* **135**, p. 247, Am. Chem. Soc., Washington D.C. (1980).
27. R. T. Cassidy and E. S. Holmes, *AIChE Symp. Ser.* **80**(233), 68 (1984).
28. G. Keller, "Gas Adsorption Processes: State of the Art in Industrial Gas Separations," *Am. Chem. Soc. Symp. Ser.* (223) (1983).
29. M. Suzuki, *Adsorption Engineering*, p. 247, Kodansha Elsevier, Tokyo (1990).
30. O. J. Smith and A. W. Westerberg, *Chem. Eng. Sci.* **46**, 2967 (1991).
31. J. Izumi, Mitsubishi Heavy Industries Ltd., personal communication (1992).
32. J. Grebbell, *Oil and Gas Journal*, p. 85, April 14 (1985).
33. W. C. Kratz, D. L. Rarig, and J. M. Pietrantonio, *AIChE Symp. Ser.* **84**(264) (1988).
34. R. Kumar, W. C. Kratz, D. E. Guro, D. L. Rarig, and W. P. Schmidt, "Gas Mixture Fractionation to Produce Two High Purity Products by PSA," Sep. Sci. and Technol. **27**, 509 (1992).
35. S. K. Sood, C. Fong, K. M. Kalyanam, A. Busigin, O. V. Kveton, and D. M. Ruthven, *Fusion Technology* **24**, 299 (1992).
36. D. M. Ruthven and S. Farooq, *Chem. Eng. Sci.* (in press).
37. J. A. Ritter and R. T. Yang, *I and E.C. Research* **30**, 1023 (1991).
38. E. S. Rikkinides and R. T. Yang, *I and E.C. Research* **30**, 1981 (1991).
39. S. Sircar, Fourth International Conferences on Adsorption, Kyoto, Japan, May, 1992, plenary lecture, "Novel Applications of Adsorption Technology."
40. R. Banerjee, K. G. Narayankhedkar, and S. P. Sukhatine, *Chem. Eng. Sci.* **45**, 467 (1990).
41. R. Banerjee, K. G. Narayankhedkar, and S. P. Sukhatine, *Chem. Eng. Sci.* **47**, 1307 (1992).

CHAPTER

7

Extensions of the PSA Concept

The basic pressure and vacuum swing processes have been developed in a variety of ways by making use of ingenious multiple-bed cycles to conserve energy and separative work. The processes described in Chapter 6 give some indication of the range of such solutions. In all these processes the relationship with the original PSA concept is quite clear. However, the PSA concept has also been developed in other ways, leading to processes in which the relationship to the parent process is less obvious. Three such developments, none of which has so far been developed on an industrial scale, are described in this chapter.

7.1 The Pressure Swing Parametric Pump

The term *parametric pumping* was coined by Wilhelm in the 1960s to describe a novel class of liquid-phase separation processes in which separation is achieved in an oscillating flow system subjected to a periodic change in temperature and other intensive thermodynamic variable.[1] He and his co-workers focused on temperature swings, but they contemplated also the synchronous cycle of pressure, pH, and electrical and magnetic fields. In fact, he cited the earliest patent of Skarstrom (PSA air dryer) as an example of a pressure parametric pump.

The essential features of a thermally driven system are shown in Figure 7.1. During the heating half-cycle liquid flows upwards, while the flow is reversed during the cooling half-cycle. The basis of the separation can be

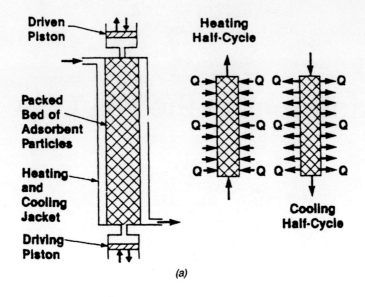

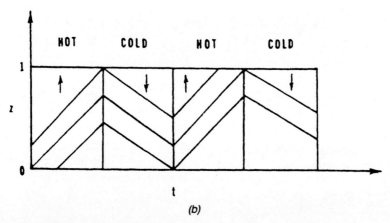

Figure 7.1 Thermal parametric pump showing (a) the principle of operation and (b) the zigzag progression of the concentration profile in successive cycles.

easily understood from equilibrium theory. The wave velocity is given by (Eq. 2.50):

$$w_c \equiv \left(\frac{\partial z}{\partial t}\right)_c = \frac{v}{1 + ((1-\epsilon)/\epsilon)\, dq^*/dc} \tag{7.1}$$

If adsorption is exothermic:

$$\left(\frac{dq^*}{dc}\right)_{hot} < \left(\frac{dq^*}{dc}\right)_{cold} \tag{7.2}$$

so:

$$(w_c)_{hot} > (w_c)_{cold} \tag{7.3}$$

Each cycle therefore leads to a net upward movement of the solute, as illustrated in Figure 7.1(b). Over a large number of cycles a very high degree of separation can be achieved between the two reservoirs.

The main drawback of parametric pumping is that it was originally envisioned and reduced to practice as a batch process. In fact, published accounts of experimental attempts indicate that perhaps 50 or more cycles are necessary to approach cyclic steady state. Pigford and co-workers[2] attempted to remedy that by suggesting a similar process, which they called "cycling zone adsorption." Their concept was to admit feed to the first of several zones (fixed beds of adsorbent) connected in series, rather than to drive it back and forth through a single fixed bed, as in parametric pumping. They were able to relate the number of zones for continuous operation to the corresponding number of batchwise cycles. Nevertheless, to achieve a good separation in a continuous system requiring several fixed beds has since been shown to be just as impractical as a batchwise system that required enormous time (i.e., several days).

It is often said that PSA separations are similar to parametric pumping with pressure rather than temperature as the controlling thermodynamic variable. However, the relationship between parametric pumping and a

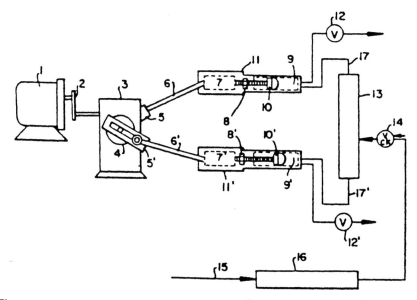

Figure 7.2 A pressure swing parametric pump (the "molecular gate"). 1–6, motor and drive; 7–11, adjustable stroke pistons; 13, adsorbent bed; 12,12', product drawoff points; 15,16, feed and pre-drier. (From Keller and Kuo,[3] with permission.)

conventional PSA system is somewhat remote, since independent control of pressure and flow, which is a key feature of the parametric pump, is lacking in PSA. A true pressure swing parametric pump has, however, been demonstrated by Keller and Kuo,[3] who called their process the *molecular gate*. The essential components of such a system are shown schematically in Figure 7.2. The pistons, which are of unequal displacement, are coupled so that a constant phase angle is maintained. The synchronized movement of the pistons is adjusted so that the gas flows upwards through the bed at high pressure and downwards through the bed at low pressure but, since the displacments are unequal, there is a net flow towards the smaller piston. Feed is introduced near the center of the adsorbent bed (the optimal point depends on the feed and product compositions), while a fraction of the gas is discharged as product from each cylinder at each stroke. The velocity of the concentration wavefront in the adsorbent bed is governed by Eq. 7.1. The isotherm slope dq^*/dc may be expressed in terms of the mole fraction in the

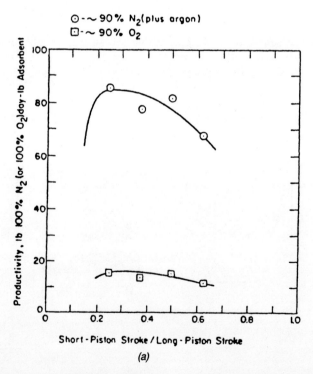

Figure 7.3 Performance of the molecular gate in air separation showing (a) the effect of piston stoke and (b) the effect of phase angle on productivity. Both products at ~ 90% purity; in (a) the phase cycle is 45°, and in (b) the stroke ratio is 4:1. (From Keller and Kuo[3] with permission.)

gas phase (y).

$$\frac{dq^*}{dc} = \frac{RT}{P}\frac{dq^*}{dy} \tag{7.4}$$

Thus at high pressure dq^*/dc is relatively smaller and w_c is correspondingly larger. The analogy between the thermal parametric pump driven by temperature variation and the gas-phase system driven by a pressure variation is therefore clear.

The maximum variation in the volume between the pistons occurs when the difference in phase angle is 45°. If kinetic effects are unimportant this condition should give the maximum variation in pressure and therefore the best performance. For air separation over a zeolite adsorbent the separation factor (α_{N_2/O_2}) is about 3.5. Assuming linear isotherms, this means that the capacity of the bed is about three and one-half times as great for nitrogen as for oxygen. For any given pressure change the ratio of the nitrogen and oxygen volumes required to pressurize the bed must lie in this ratio. The experimental data, shown in Figure 7.3, are in approximate accordance with

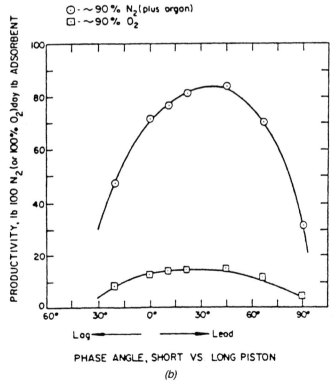

Figure 7.3 (*Continued*).

these simple arguments; optimal performance is obtained when the phase angle difference is about 45° and the ratio of the piston strokes is about 3.5:1.

As a separation process the pressure swing parametric pump has two major advantages:

1. It can produce two pure products; so complete resolution of a binary mixture may be achieved without the complexity of the purge and rinse steps that are required to accomplish this in the normal PSA mode (see Section 6.3).
2. The system can be easily designed to provide efficient energy recovery, since, on the expansion stroke, the pistons are driven by the pressure of the gas. Conservation of this energy for use in the next compression stroke can be easily accomplished either using a flywheel or by coupling together two units operating out of phase.

There is, however, one serious disadvantage: the pistons and cylinders must be large enough to accommodate virtually all the gas desorbed from the bed at the lowest pressure of the cycle. For a bench-scale unit this is not a serious problem, but it does present a serious obstacle to scaleup.

7.2 Thermally Coupled PSA

In the previous section we considered the molecular gate as a pressure-driven parametric pump. This system is also closely related to the Stirling engine and thus to a novel class of processes that utilize periodic variations in both pressure and temperature together with an oscillating gas flow to effect an energetically efficient separation. The basic elements of a Stirling engine are shown in Figure 7.4. As in the molecular gate there are two pistons: a pressure piston and a displacer, in an arrangement that is similar in essence to that shown in Figure 7.2. The working gas is transferred backwards and forward between the "hot space" and the "cold space" by the displacer piston. There is very little difference in pressure between the hot and cold spaces; so the displacer does very little mechanical work. However, the pressure throughout the system varies sinusoidally as a result of the movement of the pressure piston. When the system operates as an engine, gas expands in the hot space and flows into the cold space, driving down the pressure piston. The displacer then moves down, transferring the cold gas at low pressure back to the hot space. The pressure piston is then raised, increasing the pressure in the system, while the gas in the hot space is heated, causing a further rise in pressure, and the cycle is repeated. In this mode of operation the net effect is that heat is transferred by the gas from the hot to cold regions and an equivalent amount of work is delivered to the pressure piston. The system can also be operated in reverse as a heat pump or refrigerator. In that mode, work is done by the pressure piston on the

EXTENSIONS OF THE PSA CONCEPT

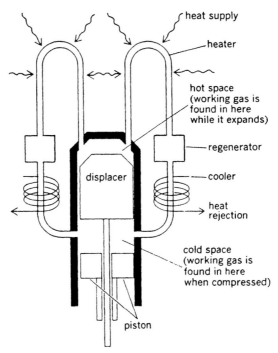

Figure 7.4 The principle of operation of the Stirling engine (displacer type).

working gas, and the equivalent amount of heat is transferred from the cold region to the hot region.

In order to reduce unnecessary heat losses, a regenerative heat exchanger is included between the hot and cold regions. This is essentially a space packed with high-heat-capacity material that picks up heat from the hot gas as it flows to the cold space, stores it, and transfers it to the cold gas returning from the cold space on the next stroke of the displacer.

A thermally coupled PSA (TCPSA) system[4] can be thought of as a Stirling engine in which the regenerator is packed with a selective adsorbent and the gas to be separated is the working fluid. Two possible arrangements are shown in Figure 7.5. The arrangement shown in Figure 7.5(a) is directly analogous to the displacer-type Stirling engine (Figure 7.4). The displacer transfers the cold gas, at high pressure, from the cold space, through the adsorbent bed, where the preferentially adsorbed component is retained. The heat of adsorption raises the temperature of the gas flowing through to the hot space, where it is heated further from an external heat source. The pressure in the system is then decreased and the hot gas is passed back through the adsorbent bed. The preferentially adsorbed species is desorbed and carried down with the gas flow, which is cooled by the heat of desorp-

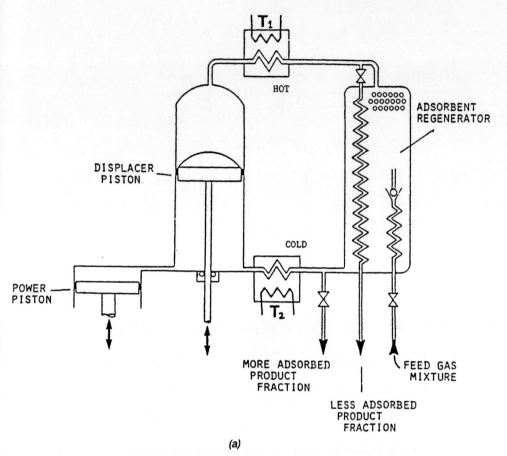

Figure 7.5 Thermally coupled PSA process. (From Keefer[4], with permission.)

tion. The more strongly adsorbed product is removed from the cold end of the system, while the less adsorbed species is removed as product at the hot end. The TCPSA system is thus seen to be similar to the pressure-driven parametric pump but with a temperature gradient across the adsorbent bed. Heat and the more strongly adsorbed component move to the cold end while the less-adsorbed species moves towards the hot end. The same effect can also be accomplished using two pistons operated out of phase [Figure 7.5(b)], as in the "molecular gate."

In addition to the energy recovery that can be achieved through mechanical coupling of the compression and expansion steps, this system also provides for efficient recovery of the heat of adsorption and offers the possibility that the additional energy required to drive the separation can be supplied as

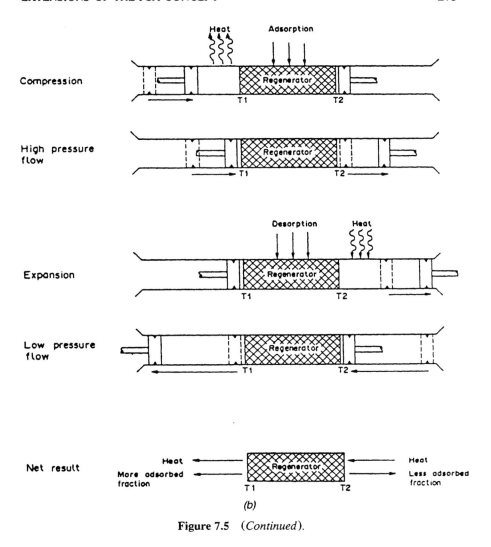

Figure 7.5 (*Continued*).

heat rather than as mechanical energy. This provides the potential for an energy-efficient and cost-effective system powered by relatively low-grade waste heat. The application of a process of this kind to a reacting system to provide continuous separation of the reaction products is an obvious extension of the TCPSA concept.

7.2.1 Test Results

Experimental data obtained for hydrogen purification and air separation (to produce oxygen) in a small laboratory TCPSA unit are summarized in

Figures 7.6–7.8. Even with a modest pressure ratio it is possible to recovery essentially pure hydrogen at fractional recoveries greater than 95% from a feed containing 74% H_2 together with CO and CO_2 (Figure 7.6). Similarly, for air separation (Figure 7.7), essentially complete elimination of nitrogen to produce a product containing about 95% O_2 + 4.5% Ar can be easily accomplished, with a moderate pressure ratio, at a fractional recovery of about 67%. Even at comparatively low cycle speeds the productivity (Figure 7.8) is seen to be far superior to that of conventional PSA systems.

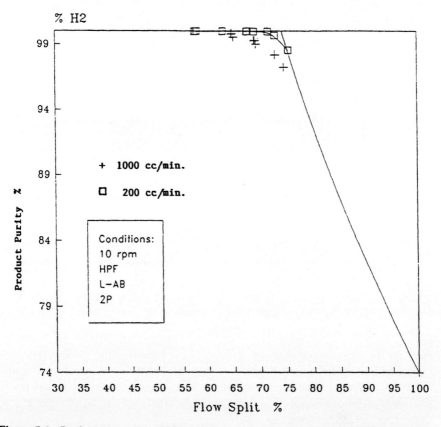

Figure 7.6 Performance of a TCPSA unit in the separation of synthesis gas (74% H_2, 25% CO_2, 1% CO). The hydrogen product purity is plotted against the flow split (i.e., the fraction of feed withdrawn as light product). Adsorbent 13X zeolite ~ 350 cm^3 bed volume. Operating pressure 2.4 atm–1 atm, 10 rpm. Feed 200 cm^3/min, □; 1000 cm^3/min, +. The theoretical line is calculated from the mass balance for 100% pure product. (Courtesy of Highquest Engineering, Inc.)

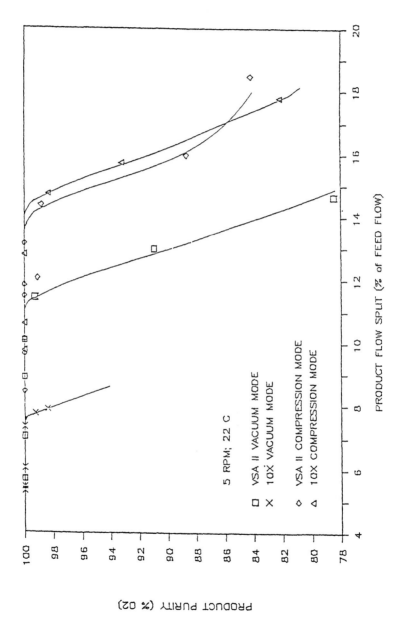

Figure 7.7 Performance of a TCPSA unit in air separation. Adsorbent 10X zeolite. Pressure ratio 2.0, vacuum mode 1.0 atm/0.5 atm; pressure mode 2.0 atm/1 atm. 5 rpm. Product purity means $O_2 + Ar$. (Courtesy of Highquest Engineering, Inc.)

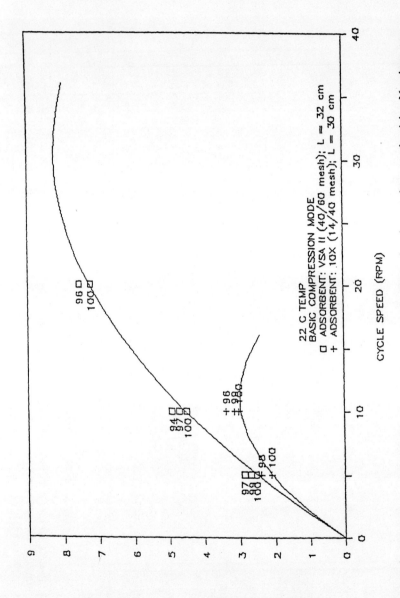

Figure 7.8 Performance of a TCPSA unit in air separation; effect of cycle speed on productivity. Numbers on curves are product purity (%). Comparative data for a conventional PSA process and for the RPSA process are, respectively, 0.5 and 2.3 g/day adsorbent, respectively.[10] (Courtesy of Highquest Engineering, Inc.)

EXTENSIONS OF THE PSA CONCEPT

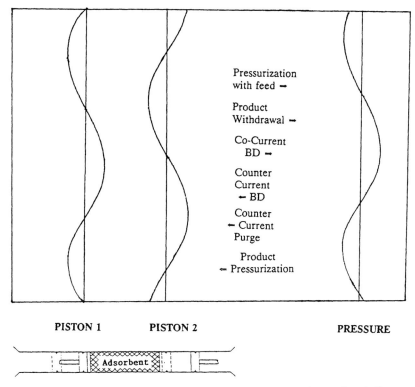

Figure 7.9 TCPSA system showing piston movements and associated flows and pressure changes.

7.2.2 Comparison with Conventional PSA

It is instructive to explore, in greater detail, the relationship between a TCPSA system and a conventional PSA process. Figure 7.9 shows the sequence of the piston movements and the associated pressure changes and gas flows. The corresponding sequence of discrete steps for an equivalent PSA cycle is also indicated:

- Partial pressurization with light product end.
- Pressurization with feed (containing heavy product) from the feed end.
- High-pressure flow from the feed end with removal of a proportion of the light product.
- Cocurrent blowdown (to light product end).
- Countercurrent blowdown (with removal of a proportion of heavy product from feed end).
- Countercurrent purge, at low pressure, with light product.

This is essentially the cycle that would be used, in a conventional PSA process, in order to recover both products. Of course, in a conventional PSA process these steps would be distinct, whereas, in a TCPSA cycle, they are merged, but this does not represent an essential difference.

One may also choose to regard a TCPSA process as analogous to a distillation or countercurrent extraction process in which the light and heavy products are refluxed at each end of the column. The reflux ratio required to produce pure products depends on the separation factor (or relative volatility). Just as in a distillation process it is possible to obtain pure products even when the relative volatility is small, by using a high reflux ratio; so, in a TCPSA process increased reflux may be used to compensate for a low pressure ratio. Whereas conventional PSA processes generally operate at relatively high pressure ratios but with low reflux (in the form of countercurrent purge), TCPSA systems generally operate at much lower pressure ratios but with higher reflux. The tradeoff in terms of power consumption obviously depends on many factors, some of which are system dependent.

7.2.3 Scaleup Considerations

To date only small laboratory-scale versions of a TCPSA system have been built, and, although the viability of the concept has been amply demonstrated, important questions concerning the prospects for scaleup remain to be resolved. The main difficulty is the size of the pistons (and cylinders) (see Section 7.1). In a typical laboratory unit the ratio of cylinder volume to adsorbent volume is about 10, although this figure varies widely depending on the adsorbent and the gas composition. Direct scaleup to a production unit, maintaining this ratio, is obviously unattractive, since the pistons and cylinders become too large and expensive. The most obvious way to avoid this difficulty is to decrease the cycle time. This would give a proportionate increase in throughput for a given size of system. However, mass transfer resistance and pressure drop considerations impose severe restrictions on the cycle time (and the associated gas flow rates). As a result, with a packed adsorbent bed as the mass transfer device the cycle time cannot be reduced below about 2–3 sec (20–30 rpm). This limitation might, however, be overcome by an improved adsorbent configuration. A monolithic adsorbent or a parallel plate contactor with sufficiently small plate spacing and sufficiently uniform gas channels offers the potential for a much faster cycle—up to perhaps 200–300 rpm. At such speeds the cylinder volume per unit throughput becomes much more reasonable; so that intermediate- and large-scale applications appear cost effective relative to conventional PSA systems.

7.3 Single-Column Rapid PSA System

A third PSA variant that may also be regarded as a variant of the parametric pump was suggested by Kadlec and co-workers in the early 1970s.[5,6] This system utilizes a single adsorption column packed with smaller adsorbent particles (100–500 μm). As a consequence of the small particle size, pressure drop through the bed is high (1–2 atm), and the cycle time (typically 3–10 sec) is much shorter than in conventional PSA system; hence the name "rapid pressure swing."

The cycle (Figure 7.10) is very simple, involving in its original conception only two steps of equal duration: a combined pressurization–product withdrawal step and the exhaust step. The RPSA cycle may thus be regarded as a PSA cycle in which the pressurization and feed steps are merged and the purge step is eliminated. Regeneration of the adsorbent occurs only during the countercurrent depressurization step (normally to atmospheric pressure). A large pressure drop in the direction of flow, during the combined pressurization–product withdrawal step, is needed to maintain the required purity of the raffinate product. The pressure gradient between the feed and the product end also allows continuous withdrawal of the raffinate product even during the period in which the bed is being regenerated by depressurization from the feed end.

During the pressurization–product step the more strongly adsorbed species travels less rapidly through the column; so, provided that the duration of the feed step is not too long, the less strongly adsorbed component may be removed at the outlet as a pure raffinate product, just as in a conventional PSA process. However, during the countercurrent depressurization step, withdrawal of the raffinate product continues at the bed outlet while the flow in the inlet region is reversed. The more strongly adsorbed species is thus desorbed and removed as a waste product from the feed end of the column. The result of this pattern of pressure and flow variation is that, in the inlet region of the bed, the concentration front moves alternatively forwards and backwards, but with a net forward bias, as in a parametric pump. Since the wave velocity is higher for the less strongly adsorbed species, the mole fraction of this species increases continuously as the sample of gas progresses towards the outlet of the bed. This mechanism by which the progressive enrichment occurs has been likened to a ratchet.[7]

As a result of the short cycle time the productivity in the type of system is generally much greater than for a conventional PSA process, operating at comparable product purity and recovery. This advantage is, however, offset by the much higher energy requirement. A detailed summary of the earlier experimental studies has been given by Yang.[8]

7.3.1 Modeling and Simulation

The modeling of an RPSA process is similar in principle to that of a conventional PSA system (as discussed in Chapters 4 and 5) except that the assumption of negligible pressure drop, which is generally a valid approximation during the feed and purge steps of a conventional cycle, is no longer valid. The pressure gradient plays a key role in an RPSA process and must

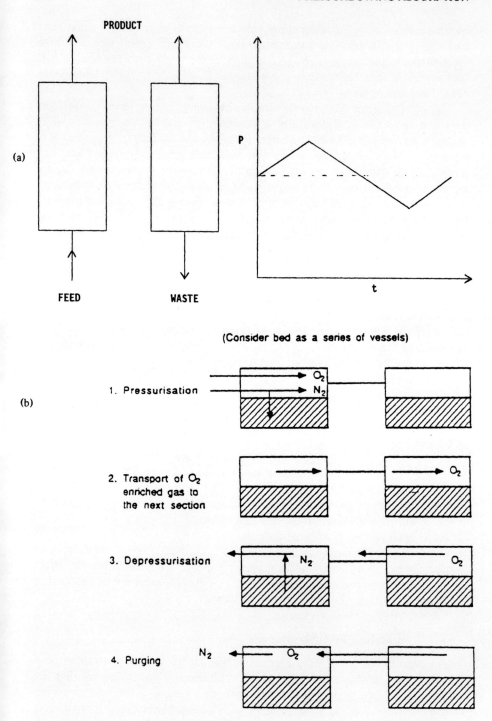

Figure 7.10 The rapid pressure swing process showing (a) pressure variation and (b) the analogy with a ratchet. (From Kenney,[1] with permission.)

EXTENSIONS OF THE PSA CONCEPT

therefore be accounted for explicitly in any model. In general Darcy's Law is used to relate the flow to the pressure drop through the bed:

$$v = -\frac{K}{\varepsilon\mu}\frac{\partial P}{\partial z}; \qquad K = \frac{\varepsilon^3}{(1-\varepsilon)^2}\frac{R_p^2}{38} \qquad (7\text{-}5)$$

Both equilibrium and kinetic models (LDF approximation) have been developed.

7.3.2 Experimental Studies

Turnock and Kadlec[5] studied the separation of a nitrogen–methane mixture by RPSA over 5A zeolite. Representative purity–recovery data and equilibrium theory predictions from their study are shown in Figure 7.11. It is clear that the nitrogen recovery was unacceptably low (< 5% at 90% product

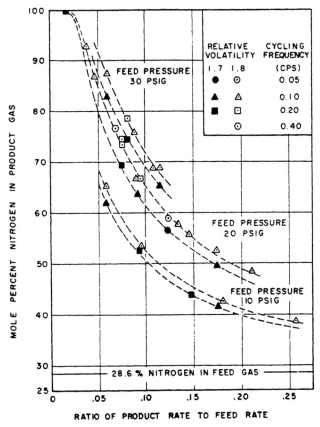

Figure 7.11 Purity of nitrogen product versus product-to-feed ratio for a RPSA process. (From Turnock and Kadlec,[5] with permission.)

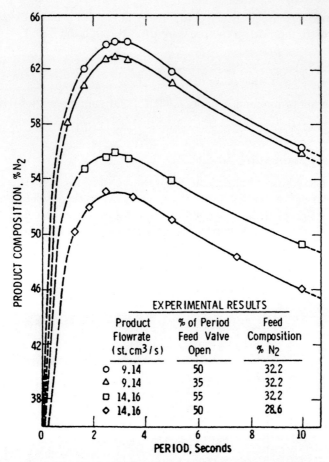

Figure 7.12 Effect of cycle frequency on product purity for separation of methane–nitrogen mixture by RPSA. (From Kowler and Kadlec,[6] with permission.)

purity). In a later study with the same experimental system Kowler and Kadlec[6] identified cycle frequency, duration of the feed step, and product flow rate as the major variables affecting the system performance. The effect of cycling frequency on product purity is shown in Figure 7.12. The optimal frequency of operation appears to be independent of product flow rate. A purity–recovery plot showing the effects of other important process variables, when operating at the optimum frequency, is shown in Figure 7.13. In this study it was also shown that introduction of a no-flow step (both exhaust and product valves closed) did not significantly affect the product purity. However, such a step reduces the exhaust flow rate, thereby increasing the overall recovery of the raffinate product.

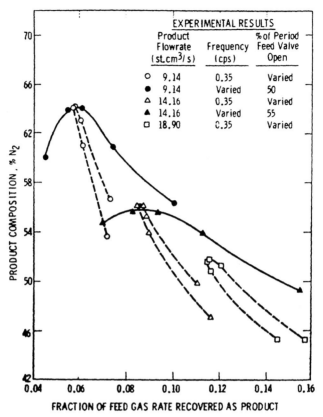

Figure 7.13 Purity–recovery profile for N_2–CH_4 separation by RPSA. (From Kowler and Kadlec,[6] with permission.)

7.3.3 Air Separation by RPSA

A significant improvement in the performance of the RPSA process was achieved by Keller and Jones,[9,11] who introduced a delay step (inlet and outlet valves closed) prior to blowdown together with a shortened pressurization step. The practical feasibility of using such a cycle for air separation (to produce oxygen) was demonstrated. At small scales of operation where power costs are not important, such a process is competitive with the conventional two-column system. Doong and Yang[12] successfully modeled the air separation data of Keller and Jones using a LDF model and showed that mass transfer resistance may become important under rapid cycling conditions. The effects of feed time, delay, and exhaust time on the purity and recovery of oxygen are shown in Figure 7.14.

Very short feed time means inadequate contact time for preferential adsorption of nitrogen and therefore nitrogen appears as contaminant in the

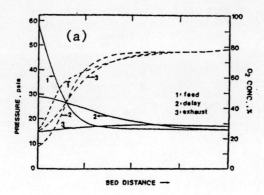

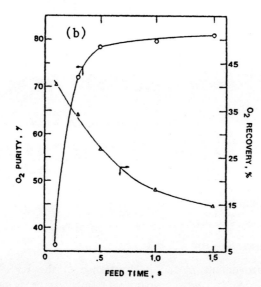

Figure 7.14 Air separation by RPSA over 40–80 mesh Linde 5A zeolite pellets. $P_H = 4.4$ atm, $P_L = 1$ atm, $L = 150$ cm, feed time, 0.5 sec; delay, 2 sec; exhaust, 15 sec. (a) Pressure and O_2 concentration profiles along the column; (b), (c), and (d) effect of varying feed time, delay time, and exhaust time on purity and recovery of O_2 product. Product rate, 48 cc STP/sec. (From Doong and Yang,[12] with permission.)

EXTENSIONS OF THE PSA CONCEPT 285

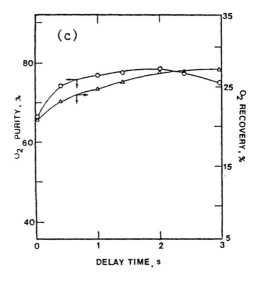

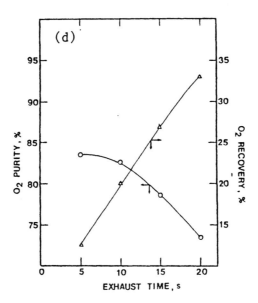

Figure 7.14 (*Continued*).

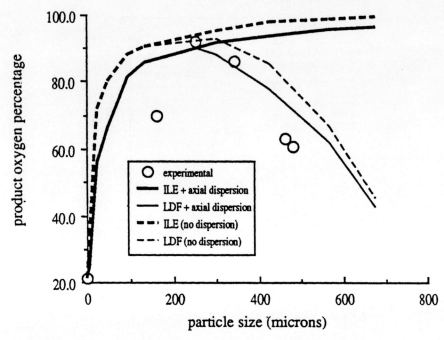

Figure 7.15 Effect of particle size on purity of the oxygen product from a RPSA process showing comparison of experimental results with the profiles calculated from various theoretical models. (From Alpay et al.,[13] with permission.)

product. However, shorter feed time leads to increased oxygen recovery. Energy consumption will therefore depend on the desired product purity. Recovery increases with both increasing delay and exhaust time. The product oxygen concentration shows a broad optimum with increasing delay but decreases monotonically with decreasing exhaust time.

In a more recent study Alpay, Kenney, and Scott[13] investigated the effect of particle size on a RPSA air separation unit using 5A zeolite. The results shown in Figure 7.15 show clearly the existence of an optimum particle size for product enrichment. A simulation including both pressure drop and mass transfer limitations shows that the product enrichment is limited for small particle sizes by axial dispersion and the pressure dynamics of the system, and for larger particles by intraparticle diffusional resistance.

7.4 Future Prospects

None of the systems described in this chapter has yet been developed beyond the laboratory scale; so the question arises as to the prospects for commercialization. The "molecular gate" can operate isothermally only at relatively small scales, at which heat dissipation is rapid relative to the rate of heat generation by adsorption. In order to operate a large-scale version of this process, it would be necessary to introduce heat exchanged at both ends of the adsorbent bed, thus effectively converting the molecular gate to a TCPSA unit.

The prospects for commercialization of the TCPSA process are difficult to assess. The indications are that in terms of energy efficiency and adsorbent productivity the TCPSA process is, for many applications, significantly more economic than a traditional PSA process. The prospects for developing an improved adsorbent contactor appear promising, and this would further enhance the competitive position of TCPSA. The issue of capital cost is more difficult to assess, since no large TCPSA unit has yet been built. However, it seems likely that the remaining economic barriers to the commercialization of TCPSA system will be overcome eventually, and a rapid spread of this technology to a range of commercially important separations may occur within a few years.

The single column RPSA system is, in principle, operable at large scale. However, as the scale of the process increases, the issue of pressure drop (and therefore power consumption) becomes increasingly important. As a result of the high-pressure drop, the power consumption in a RPSA system will always be higher than that of a well-designed two-column process operated under comparable conditions. The economic viability of this type of process is therefore limited to small-scale applications.

Of the three systems described in this chapter only the TCPSA process appears to offer reasonable prospects for large-scale operations. However, the economic operation of such a process at high throughputs depends on the development of a new and more efficient mass transfer device to avoid the inherent limitations of a packed bed contactor.

References

1. N. H. Sweed and R. H. Wilhelm, *Ind. Eng. Chem. Fund.* **8**, 221 (1968). See also N. H. Sweed, *AIChE Symp. Ser.* **80**(233), 44 (1984).
2. R. L. Pigford, B. Baker, and D. E. Blum, *Ind. Eng. Chem. Fund.*, **8**, 848 (1969).
3. G. E. Keller and C. H. A. Kuo, U.S. Patent 4,354,854 (1982), to Union Carbide Corp.
4. B. G. Keefer, U.S. Patents. 4,702,903, 4,816,121, 4,801,308, 4,968,329, 5,096,469, and 5,082,473; Canadian Patent 1256088; European Patent 0143537.
5. P. H. Turnock and R. H. Kadlec, *AIChE J.* **17**, 335 (1971).
6. D. E. Kowler and R. H. Kadlec, *AIChE J.* **18**, 1207 (1972).

7. C. N. Kenney, in "Separation of Gases," *Proceedings of the 5th BOC Priestley Conference*, Birmingham U.K. (1984), Royal Society of Chemistry Special Pub. No. 80, pp. 273–86 (1990).

8. R. Yang, *Gas Separation by Adsorption Processes*, p. 263, Butterworths, Stoneham, MA (1987).

9. R. L. Jones, G. E. Keller, and R. C. Wells, U.S. Patent 4,194,892 (1980), to Union Carbide.

10. G. E. Keller and R. L. Jones, *Am. Chem. Soc. Symp. Ser.* **135**, 275 (1980).

11. R. L. Jones and G. E. Keller, *J. Sep. Process Technology* **2**(3), 17 (1981).

12. S. J. Doong and R. T. Yang, *AIChE Symp. Ser.* **84**(264), 145 (1988).

13. E. Alpay, C. N. Kenney, and D. M. Scott, *Chem. Eng. Sci.* in press.

CHAPTER

8

Membrane Processes: Comparison with PSA

Although pressure swing adsorption and membrane permeation processes operate on quite different principles, they offer economically competitive alternatives for many small- and medium-scale gas separations. The focus of this book is on pressure swing systems, but it seems appropriate to include a brief introduction to membrane processes to provide the reader with the background needed to assess the comparative merits of the membrane alternative. From an overall standpoint pressure swing and membrane processes are similar in that they are both best suited to producing a pure raffinate (retentate) product. Although either process can be adapted to yield a pure extract (permeate) product this cannot be accomplished without a significant loss of efficiency. In both classes of process the main operating cost is the power required to compress the feed stream; so a first estimate of comparative performance can be obtained simply by considering the power requirements.

8.1 Permeability and Separation Factor

The concept of a membrane process is straightforward (Figure 8.1). The separation depends on the difference in permeation rates through a permselective membrane, and the process efficiency is largely dependent on the selectivity and permeability of the membrane material. The permeability (π_i), which provides a quantitative measure of the ease with which a particular

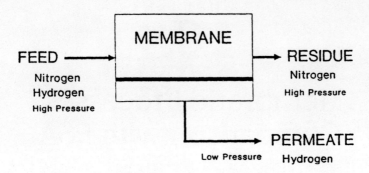

Figure 8.1 Membrane separation process.

component can penetrate the membrane, is defined by:

$$N_i = \pi_i \frac{A}{\delta}(p_{iH} - p_{iL}) \tag{8.1}$$

where $(p_{iH} - p_{iL})$ represents the difference in the partial pressure of component i across the membrane and δ is the thickness.

One may usefully define a local separation factor (α') by reference to the situation sketched in Figure 8.2.:

$$\alpha' = \frac{y/(1-y)}{x/(1-x)} \tag{8.2}$$

The flux ratio is given by:

$$\frac{N_A}{N_B} = \frac{\pi_A}{\pi_B}\left(\frac{\wp x_A - y_A}{\wp x_B - y_B}\right) = \frac{y_A}{y_B} = \frac{y_A}{1 - y_A} \tag{8.3}$$

where $\wp = P_H/P_L$ is the pressure ratio across the membrane. If the back pressure is negligible, $\wp \to \infty$, and the separation factor approaches the permeability ratio (α):

$$\alpha' \equiv \frac{y_A/y_B}{x_A/x_B} \to \frac{\pi_A}{\pi_B} \equiv \alpha \tag{8.4}$$

This is the most favorable situation. Any back pressure will reduce the separation factor to a lower value, and the permeability ratio is therefore sometimes referred to as the "intrinsic separation factor," or the "selectivity."

For an inert (nonadsorbing) microporous solid the permeability ratio is essentially the ratio of Knudsen diffusivities, which is simply the inverse ratio of the square of the molecular weights. Such selectivities are therefore modest and too small to be of much practical interest. If, however, the pores are small enough to offer significant steric hindrance to diffusion or if one or both of the components are adsorbed on the pore wall or dissolved within the solid matrix, much larger separation factors are possible. For such a system the permeability depends on the product of the solubility (or the adsorption

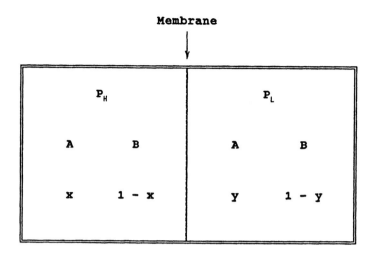

Figure 8.2 Definition of "local" separation factor.

equilibrium constant) and the diffusivity of the adsorbed or dissolved species; so, to a first approximation we may write:

$$\alpha \equiv \frac{\pi_A}{\pi_B} = \frac{K_A D_A}{K_B D_B} \qquad (8.5)$$

This expression is not exact, since it assumes system linearity and ignores the possibility of parallel contributions from other nonselective transparent mechanism such as Poiseuille flow. Nevertheless it serves to delineate the main factors controlling membrane selectivity. Clearly a high selectivity may be achieved from a large difference in either diffusivity or equilibrium constant or from a combination of both these factors. Unfortunately there is

Table 8.1. Compensation of Diffusivity and Solubility for H_2S in Polymer Membranes

Membrane	T (°C)	P (Torr)	π (cm^3 STP cm^{-1} s^{-1} atm^{-1})	D [cm$^2 \cdot$ s^{-1}]	K [cm$^3 \cdot$ cm^{-3}]
Nylon	30	110	2.4×10^{-9}	3×10^{-10}	7.9
	30	621	2.6×10^{-9}	4.9×10^{-10}	5.3
Polyvinyl	30	244	2.2×10^{-9}	5.5×10^{-9}	0.4
trifluoro		453	2.0×10^{-9}	4.9×10^{-9}	0.4
acetate		751	2.0×10^{-9}	6.8×10^{-9}	0.3

often compensation between the kinetic and equilibrium parameters so that the resulting separation factor is smaller than would be predicted by considering either the diffusivity ratio or the equilibrium ratio alone. This is illustrated in Table 8.1.

8.1.1 Nonlinear Equilibrium

Whether the equilibrium is linear or nonlinear, the expression for the flux may be written in the form:

$$N_A = -L_A q_A \frac{d\mu_A}{dz} \tag{8.6}$$

where $\mu_A = \mu_A^0 + RT \ln p_A$, with similar expressions for component B. The flux ratio is given by:

$$\frac{N_A}{N_B} = \frac{L_A}{L_B} \frac{q_A}{q_B} \frac{d\mu_A}{d\mu_B} = \frac{D_{0A}}{D_{0B}} \frac{q_A}{q_B} \frac{p_B}{p_A} \frac{dp_A}{dp_B} \tag{8.7}$$

where $D_{0A} = RTL_A$ is the limiting diffusivity within the Henry's Law region. If the equilibrium isotherm is of binary Langmuir form (Eq. 2.13):

$$\frac{q_A p_B}{q_B p_A} = \frac{b_A}{b_B} \tag{8.8}$$

so that Eq. 8.7 reduces to:

$$\frac{N_A}{N_B} = \frac{D_{0A}}{D_{0B}} \frac{b_A}{b_B} \frac{dp_A}{dp_B} = \alpha \frac{dp_A}{dp_B} \tag{8.9}$$

where α is defined by Eq. 8.5. Integrating across the membrane yields Eq. 8.3, which is thus seen to be applicable even outside the Henry's Law region. It should be noted that this simplification arises only in the special case of an ideal Langmuir system. For other forms of isotherm Eq. 8.5 is not necessarily valid outside the linear region.

8.1.2 Effect of Back Pressure

To quantify the effect of back pressure on the separation factor we may eliminate y between Eqs. 8.2 and 8.3. With $\pi_A/\pi_B = \alpha$ (Eq. 8.4) this yields the Naylor–Backer expression[1]:

$$\alpha' = \left(\frac{1+\alpha}{2}\right) - \frac{1}{2x} - \frac{\alpha-1}{2x\wp}$$

$$+ \left[\left(\frac{\alpha-1}{2}\right)^2 + \frac{r(\alpha-1)-(\alpha^2-1)}{2\wp x} + \left(\frac{\alpha-1+r}{2\wp x}\right)^2\right]^{1/2} \tag{8.10}$$

In the high-pressure limit ($\wp \to \infty$) this reduces simply to $\alpha' = \alpha$. Equation 8.10 provides the most convenient way to account for the effect of back

MEMBRANE PROCESSES

pressure in a system in which the total pressure, on each side of the membrane, is essentially constant but the composition varies with position.

8.1.3 Temperature Dependence

Both equilibrium constants and diffusivities generally vary exponentially with the reciprocal temperature; so a similar form of temperature dependence is to be expected for the permeability ratio.

$$\alpha = \alpha_\infty \exp\left(-\frac{[\delta E + \delta(\Delta H)]}{RT}\right) \qquad (8.11)$$

where $\delta E = E_A - E_B$ and $\delta(\Delta H) = \Delta H_A - \Delta H_B$. Depending on the relative magnitudes of the difference in adsorption energies and in diffusional activation energies, the selectivity may either increase or decrease with temperature. Representative examples are shown in Figure 8.3.[2]

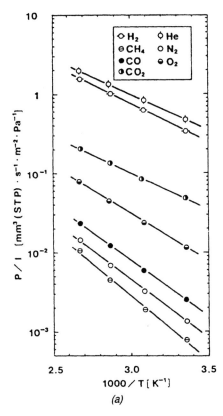

(a)

Figure 8.3(a) Variation of permeation rate with temperture for a polyamide asymmetric hollow fiber membrane ($P_H = 4$ atm, $P_L = 1$ atm). (From Haraya et al.,[2] with permission.)

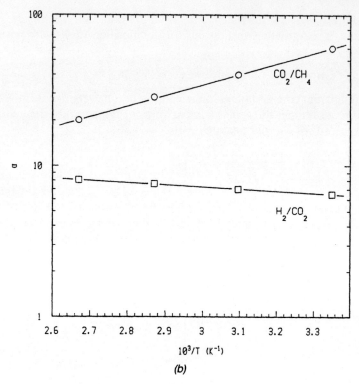

Figure 8.3(b) Variation of selectivity with temperature for a polyamide asymmetric hollow fiber membrane as in Figure 8.3(a).

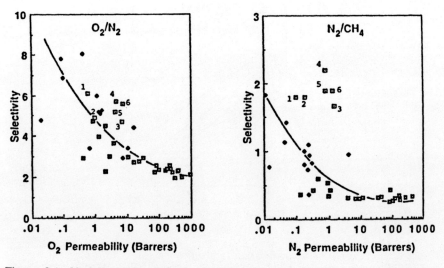

Figure 8.4 Variation of selectivity with permeability for O_2-N_2 separation on polymeric membranes.* (From Koros et al.,[3] with permission.)

*1 Barrer = $10^{-10} \dfrac{cm^3(STP) \cdot cm}{cm^3 \cdot s \cdot cm\ Hg} = 0.335 \dfrac{m \cdot mole}{m \cdot s \cdot TPa}$

MEMBRANE PROCESSES

Table 8.2. Potential of Zeolite Membranes[a]

CO_2/CH_4 separation	CO_2 (A)			CH_4 (B)			
	K	D	$KD = \pi$	K	D	$KD = \pi$	$\alpha = \dfrac{\pi_A}{\pi_B}$
4A zeolite[b]	3.5×10^4	8.6×10^{-10}	3×10^{-5}	30	5×10^{-11}	1.5×10^{-9}	2×10^4
5A zeolite[b]	4500	2×10^{-6}	9×10^{-3}	50	1.4×10^{-6}	7×10^{-5}	130
Rubber	—	—	10^{-6}	—	—	2.3×10^{-7}	4.3
L.D. Polythene	—	—	0.96×10^{-7}	—	—	2.2×10^{-8}	4.4

N_2/CH_4 separation	N_2(A)			CH_4 (B)			
	K	D	$KD = \pi$	K	D	$KD = \pi$	$\alpha = \dfrac{\pi_A}{\pi_B}$
4A zeolite[c]	20	10^{-9}	2×10^{-8}	30	5×10^{-11}	1.5×10^{-9}	13
5A zeolite	24	9×10^{-7}	2.2×10^5	50	1.4×10^{-6}	7×10^{-5}	0.31
Rubber	—	—	6.2×10^{-8}	—	—	2.3×10^{-7}	0.27
L.D. Polythene	—	—	7.4×10^{-9}	—	—	2.2×10^{-8}	0.34

[a] K is in ccSTP/cm^3 atm; D is in cm^2 s^{-1}; π is in cm^3 STP/cm atm sec.
[b] Zeolite has higher permeability *and* selectivity.
[c] 4A zeolite membrane would allow removal of N_2 (minor component) as permeate with permeability comparable with polymeric membrane.

8.1.4 Permeability versus Selectivity

The ideal membrane would have both high selectivity and high absolute permeability (to allow a high throughput per unit area). Unfortunately there is often a high degree of compensation between permeability and selectivity; materials with a high selectivity generally have low permeability and vice versa. Some examples are shown in Figure 8.4 and Table 8.2. The selection of the best material therefore generally involves finding the optimal compromise based on an economic evaluation.

8.2 Membrane Modules

Since the flux varies inversely with the membrane thickness, it is desirable that the active membrane should be as thin as possible. The limitation is of course the physical strength, since the membrane must be strong enough not to rupture under the applied pressure, which is often quite large. For a given pressure difference the throughput is directly proportional to the membrane area. The challenge for the designer is therefore to minimize the membrane thickness and maximize the membrane area per unit of module volume.

The active membrane is generally a thin polymer film supported on a macroporous support that provides physical strength but makes no contribu-

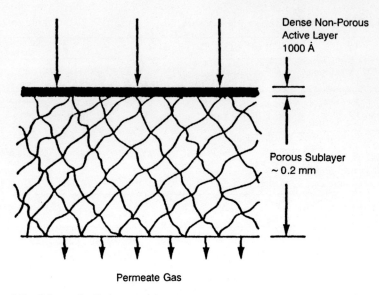

Figure 8.5 Schematic diagram solving construction of an asymmetric membrane.

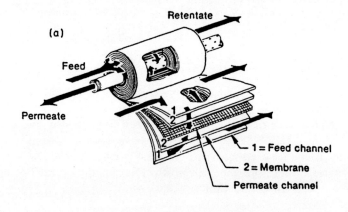

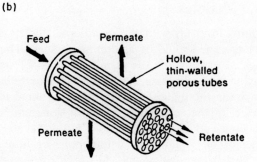

Figure 8.6 Membrane modules: (a) spiral wound, (b) hollow fiber type, and (c) typical small-scale unit. (Courtesy of Dow-Generon Inc.)

MEMBRANE PROCESSES

(c)

Figure 8.6 (*Continued*).

tion to the separation (Figure 8.5). The two most common types of membrane module are shown in Figure 8.6. In the hollow fiber module the fibers are connected between "tube plates", as in a shell and tube heat exchanger, but the fiber bundle is often twisted to promote radial mixing on the shell side. There is some disagreement as to the best flow direction, and commercial systems are available with both inward and outward flow. From the process standpoint outward flow is preferable, since it is easier to maintain a good approximation to plug flow on the high-pressure side if the high-pressure flow goes through the tubes rather than through the shell. This, however, requires that the active membrane be on the interior of the hollow fiber tubes. It is much easier to apply a uniform membrane film to the exterior surface of the tubes, but to take advantage of such an arrangement the feed must be applied to the shell side, on which some deviation from ideal plug flow is inevitable. Such deviations can, however, be minimized by good design; so this arrangement is in fact widely used in commercial systems.

8.2.1 Effect of Flow Pattern

Since the effective separation factor is reduced by back pressure (Eq. 8.10), the flow pattern has a pronounced effect on the performance of a membrane system. This may be clearly shown by calculating the purity–recovery profiles for different flow schemes. As in any mass transfer process, countercurrent flow maximizes the average driving force and therefore provides the most efficient arrangement. It is relatively easy to achieve a reasonable approximation to plug flow on the high-pressure side, but this is much more difficult on the low-pressure side because of the wide variation in the gas velocity (from close to zero at the closed end to a significant value at the permeate exit). If the pressure ratio is large, deviations from plug flow on the low-pressure side have a relatively minor effect on performance, provided that plug flow is maintained on the high-pressure side. The operation of many membrane modules, particularly those of the hollow fiber type, is therefore well represented by the "cross-flow" model, which assumes plug flow on the high-pressure side with perfect mixing on the low-pressure side [Figure 8.7(b)].

The worst case from the point of view of process efficiency is perfect mixing on both sides of the membrane. This provides a useful limiting case

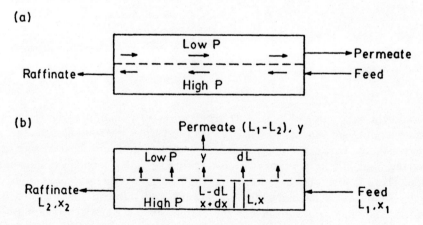

Figure 8.7 (a) Countercurrent and (b) cross-flow membrane elements showing definition of variables used in Eqs. 8.14–8.27.

MEMBRANE PROCESSES

for assessing the effect of flow pattern on performance, but in general one would try to avoid this condition in an operating system.

8.3 Calculation of Recovery – Purity Profiles

8.3.1 Mixed Flow

The fractional recovery (R) is defined simply as the fraction of the less permeable species that emerges in the raffinate product stream:

$$R = \frac{L_2(1 - x_2)}{L_1(1 - x_1)} \tag{8.12}$$

For a well-mixed system the mole fractions x_2 and y_2 in the raffinate and permeate streams are related through Eq. 8.2. The separation factor α' is constant throughout the system and is given by Eq. 8.10 with $x = x_2$. Calculation of the recovery–purity profile is therefore straightforward, requiring only the combination of an overall mass balance for the less permeable species:

$$L_1(1 - x_1) = L_2(1 - x_2) + (L_1 - L_2)(1 - y_2) \tag{8.13}$$

with Eqs. 8.2, 8.10, and 8.12.

8.3.2 Cross-Flow

The calculation is slightly more complex for the cross-flow case, since it is necessary to account for the variation of partial pressure with position on the high-pressure side. For the ideal cross-flow system sketched in Figure 8.6(b), a differential mass balance for the more rapidly diffusing species gives:

$$y\,dL = d(Lx) = L\,dx + x\,dL \tag{8.14}$$

where L is the (local) molar flow rate on the high-pressure side. The local concentrations x and y on the high- and low-pressure sides of the membrane are related by Eq. 8.2. Substitution in Eq. 8.14 and rearranging yields:

$$\frac{dL}{L} = \frac{dx}{(\alpha' - 1)(1 - x)x} + \frac{dx}{(1 - x)} \tag{8.15}$$

which may be integrated from the inlet ($x = x_1$) to any arbitrary exist mole fraction (x_2):

$$\ln\left(\frac{L_2}{L_1}\right) = \ln\left(\frac{1 - x_1}{1 - x_2}\right) + \int_{x_1}^{x_2} \frac{dx}{(\alpha' - 1)(1 - x)x} \tag{8.16}$$

Combining Eqs. 8.12 and 8.16, we obtain:

$$\ln R = \int_{x_1}^{x_2} \frac{dx}{[\alpha'(x) - 1](1 - x)x} \tag{8.17}$$

where $\alpha'(x)$ is given by Eq. 8.10. For any specified feed composition (x_1) and pressure ratio $(\mathscr{P})$ the integration yields directly the relationship between the fractional recovery and purity of the raffinate product.

8.3.3 Counter Current Flow

For the countercurrent flow case[4] the integration is slightly less straightforward, since it is necessary to allow for the variation in composition on both sides of the membrane. A differential balance for each component across the membrane [Figure 8.6(a)] yields:

$$-dL_A = d(Lx) = \pi_A \, dA \, (P_H - P_L y) \tag{8.18}$$

$$-dL_B = -d[L(1-x)] = \pi_B \, dA \, [P_H(1-x) - P_L(1-y)] \tag{8.19}$$

$$d(Lx) = d(L'y); \quad dL = dL' \tag{8.20}$$

Dividing Eqs. 8.18 and 8.19:

$$\frac{-L\,dx - x\,dL}{L\,dx - (1-x)\,dL} = \frac{L + x\,dL/dx}{(1-x)\,dL/dx - L} \tag{8.21}$$

$$= \frac{\alpha(P_H x - P_L y)}{P_H(1-x) - P_L(1-y)}$$

$$-\frac{1}{L}\frac{dL}{dx} = \frac{1}{x + \alpha/\left[(1-\alpha) + (1-\mathscr{P})/(x\mathscr{P} - y)\right]} \tag{8.22}$$

In order to integrate this expression, we must express y in terms of x. This is accomplished by a mass balance over the dotted section in Figure 8.7a.

$$y = \frac{Lx - L_2 x_2}{L'} = \frac{Lx - L_2 x_2}{L - L_2} \tag{8.23}$$

To avoid the need for a trial and error solution, it is easier to change the variable and integrate from the raffinate end:

$$l = \frac{L}{L_2}; \quad y = \frac{Lx - x_2}{l - 1}, \quad l = 1.0, 1 - x = x_2 \tag{8.24}$$

$$P - \frac{1}{l}\frac{dL}{dx} = \frac{1}{x + \alpha/\left[(1-\alpha) + (1-\mathscr{P})/(x\mathscr{P} - y)\right]} \tag{8.25}$$

At the raffinate end y_2 is given by:

$$\frac{N_A}{N_B} = \frac{y_2}{1 - y_2} = \alpha \left(\frac{\mathscr{P} x_2 - y_2}{\mathscr{P}(1 - x_2) - (1 - y_2)}\right) \tag{8.26}$$

which is a simple quadratic equation:

$$y_2^2(1 - 1/\alpha) - y_2\left[(1 - 1/\alpha)(1 + \mathscr{P} x_2) + \mathscr{P}/\alpha\right] + \mathscr{P} x_2 = 0 \tag{8.27}$$

MEMBRANE PROCESSES

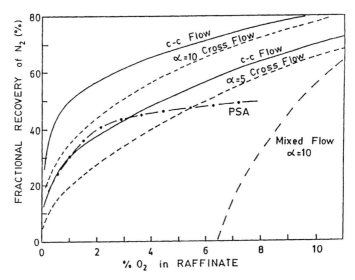

Figure 8.8 Recovery–purity profile for production of nitrogen from air by membrane and PSA processes.[5]

With, α, $\mathscr{P}$, and x_2 specified, the recovery may now be calculated by integration starting from the raffinate product end at which $l = 1.0$, x_2 is fixed, and y_2 is known from Eq. 8.27.

8.3.4 Comparison of Recovery – Purity Profiles

The results of such calculations for a pressure ratio 5.0 and permeability ratios of 5 and 10 are shown in Figure 8.8. These values are typical of the current membrane processes for recovery of nitrogen from air in which the nitrogen is the less permeable species. The strong effects of both permeability ratio and flow pattern on performance are immediately apparent. These effects become most pronounced in the high-purity region, which is generally the region of practical interest.

8.4 Cascades for Membrane Processes

Where a pure product is required it is often advantageous to use more than one membrane element connected in series as a "cascade." The best arrangement depends on several factors, the most important being whether the primary requirement is for a pure raffinate (retentate) product or for a pure permeate. If the requirement is for a pure raffinate product a countercurrent flow system is the best arrangement. If ideal countercurrent flow could be achieved within a membrane element, there would be no advantage to be

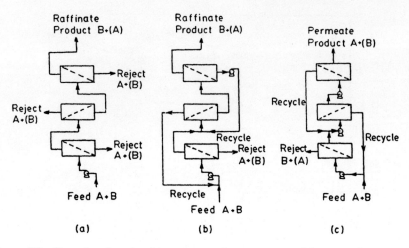

Figure 8.9 Cascades for membrane separation processes: (a) to produce a pure raffinate product with discharge of permeate; (b) to produce a pure raffinate product with recovery of permeate; (c) to produce a pure permeate product. (From Kärger and Ruthven,[6] with permission.)

gained from using more than one stage. However, the deleterious effect of backmixing on the low-pressure side can be largely offset by using elements connected in series [as in Figure 8.9(a)]. A cascade of three or four cross-flow elements connected in this way can achieve a performance approaching that of the ideal countercurrent system; so such schemes are widely used in practice where a pure raffinate is required.

The scheme shown in Figure 8.9(a) is suitable for processes such as air separation (to produce pure nitrogen as the raffinate product), since the permeate (oxygen-enriched air) has little value and can be discharged directly to the atmosphere. Where the permeate is a valuable byproduct or where direct discharge is not allowed, the scheme shown in Figure 8.9(b) is used. In this arrangement the permeate streams from the later stages are recycled; so that there is only a single permeate product stream (from stage 1), and this is at a relatively high concentration. In a well-designed cascade the composition of the recycled streams is matched to the feed composition of the preceding stage so that there is no loss of efficiency by backmixing. The recycle arrangement [Figure 8.9(c)] is of course more expensive than the arrangement of Figure 8.9(a), since recycle compressors are needed.

The arrangement shown in Figure 8.9(c) is used when a pure permeate product is required. Since the permeate is produced at low pressure, interstage compressors are needed, but, by proper choice of the operating pressures for each stage, the same compressors can also handle the raffinate recycle streams. Nevertheless the requirement for additional compressors renders this scheme more expensive in both capital and operating costs than the simple scheme of Figure 8.9(a). For this reason membrane processes are

MEMBRANE PROCESSES   303

at their most economic for production of a pure raffinate product. Although a cascade of several stages can produce a permeate product of high purity, such processes are seldom economical except when the value of the products is unusually high.

8.5 Comparison of PSA and Membrane Processes for Air Separation

8.5.1 Nitrogen Production

The purity–recovery profile for a PSA nitrogen process operating between 5.0 and 1.0 atm on the cycle of Figure 3.17 with the Bergbau–Forschung carbon molecular sieve adsorbent is compared with the corresponding membrane process in Figure 8.7. Brief details of the operating conditions which are typical of such processes, are given in Table 6.1. The profile for a real membrane system will lie between the theoretical curves calculated for the ideal countercurrent and cross-flow cases. For a permeability ratio of 5, which is typical of the present generation of membrane processes, the performance of the membrane system, in the moderate- to high-purity regime, is similar to that of the PSA process. The profiles for the membrane and PSA processes are, however, of different form so that the PSA system gains a marginal advantage in the high-purity region while the membrane system becomes clearly advantageous when product purity requirements are less severe. In both processes the power required to compress the feed air is the main component of the operating cost; so comparing recovery–purity profiles at the same pressure ratio provides a direct comparison of the operating cost component. The overall process economics are, however, modified by differences in capital cost and operational life; so the simple comparison based on recovery–purity profiles provides only a rough guide.

The carbon molecular sieve process has been operational for about 10 years, and the process has therefore been fairly well optimized. The present generation of the CMS adsorbents offers a diffusivity ratio of about 100 but the recovery–purity profile is relatively insensitive to a further increase in this ratio. In contrast, the present generation of membranes have permeability ratios of 5–6, and, although higher selectivity membranes are available, the permeability is generally too low (see Figure 8.4). However, it is clear that a relatively modest increase in membrane selectivity would give the membrane process a clear economic advantage over the competing PSA process for nitrogen production.

Results of a more detailed economic evaluation taking account of both capital and operating costs are shown in Figure 8.10. At sufficiently large scales of operation the cryogenic process is the best choice. PSA and membrane processes are preferred for smaller-scale processes, and when combined with a DEOXO unit (see Section 6.3) both these processes can

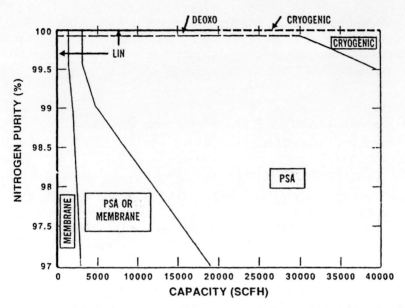

Figure 8.10 Cost effectiveness comparison for nitrogen production by membrane, PSA, and cryogenic processes. (From Thorogood,[7] with permission.)

produce a high-purity product. The overall economic balance between the PSA and membrane systems depends mainly on the scale of operation. Membrane processes offer the best choice at very small scales, PSA processes are most economic at relatively large scales, and there is a significant range of intermediate scales in which there is very little difference in costs between these processes. Although the breakeven points between these different regions are continually changing as the technology evolves, this qualitative pattern is unlikely to change substantially.

8.5.2 Oxygen Production

A similar economic comparison between the PSA and membrane oxygen production processes is shown in Figure 8.11. The membrane process depends on recovering the permeate product, and to recover the permeate in pure form would require two or three stages with intermediate compressors [Figure 8.9(c)]. Such a process is not economically competitive with the corresponding PSA oxygen process with a zeolite adsorbent in which the oxygen is recovered as the raffinate product. The membrane oxygen process is therefore limited by economics to a single stage, and this limits the product purity to about 50% oxygen. Within this restricted range the PSA and membrane processes are competitive, but at higher purities and higher

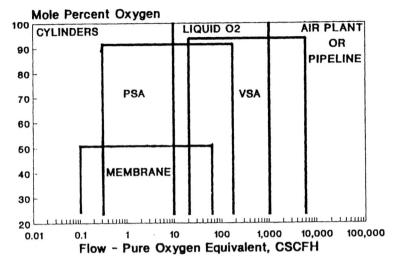

Figure 8.11 Cost effectiveness diagram for oxygen production by membrane, PSA, VSA, and cryogenic processes. (From Spillman,[8] with permission.)

throughputs the balance of economic advantage shifts first to PSA systems, then to vacuum swing systems, and finally to cryogenic distillation.

8.6 Future Prospects

Intensive research programs aimed at developing improved membranes for gas separation are in progress at many academic and industrial laboratories. The development of inorganic membranes formed from a coherently grown layer of zeolite crystals[9] is a particularly promising approach, since such membranes offer substantially higher selectivities and permeabilities compared with polymeric materials (see Table 8.2). They also offer a wide range of thermal stability, making them potentially attractive for membrane reactor applications. The technical challenge is to maintain coherence and physical strength in the scaleup to commercial operations. Nevertheless, the considerations mentioned in Section 8.5 concerning the scaling of the capital costs of membrane processes will probably remain true, regardless of improvements in the selectivity–permeability characteristics. Since the capital costs of both PSA and membrane processes increase almost linearly with throughput while the capital costs for processes such as cryogenic distillation increase less rapidly with increasing throughput (Figure 1.1), one may expect that most future commercial applications of PSA and membrane processes will continue to be at smaller and medium throughputs, rather than at the very largest scales of operation.

References

1. R. W. Naylor and P. O. Backer, *AIChE. J.* **1**, 95 (1955).
2. K. Haraya, T. Hakuta, K. Obata, Y. Shido, N. Itoh, K. Wakabayashi, and H. Yoshitome, *Gas. Sep. and Purif.* **1**, 3(1987).
3. W. J. Koros, G. K. Fleming, S. M. Jordan, and T. H. Kim, and H. H. Hoehn, *Prog. Polymer Sci.* **13**, 339 (1988).
4. G. T. Blaisdell and K. Kammermeyer, *Chem. Eng. Sci.* **28**, 1249 (1973).
5. D. M. Ruthven, *Gas Sep. and Purif.* **5**, 9 (1991).
6. J. Kärger and D. M. Ruthven, *Diffusion in Zeolites and Other Microporous Solids*, John Wiley, New York (1992).
7. R. M. Thorogood, *Gas. Sep. and Purif.* **5**, 83 (1991).
8. R. W. Spillman, *Chem. Eng. Progress* **85**(1), 41 (1989).
9. E. R. Geus, W. J. W. Bakker, P. J. T. Verheijen, M. J. den Exter, J. A. Moulijn, and H. van Bekkum, Ninth International Zeolite Conference, Montreal, July 1992, Proceedings, Vol. 2, p. 371, R. von Ballmsos, J. B. Higgins, and M. M. J. Treacy, eds., Butterworth, Stoveham, MA (1993).

APPENDIX A

The Method of Characteristics

The method of characteristics is a mathematical tool for solving nonlinear, hyperbolic, partial differential equations. The range of potential applications is large and includes such diverse topics as acoustics, catalytic reactors, fluid mechanics, sedimentation, traffic flow, and, of course, adsorption. Further details may be found in the text by Rhee et al.[1] and papers by Acrivos,[2] Bustos and Concha,[3] Dabholkar et al.,[4] Kluwick,[5] and Herman and Prigogine.[6]

The analysis begins with a general quasilinear partial differential equation

$$F(t, z, w)\frac{\partial w}{\partial t} + G(t, z, w)\frac{\partial w}{\partial z} = H(t, z, w) \qquad (A.1)$$

The restrictions on this equation are that F, G, and H are specific, continuously differentiable functions, such that $F^2 + G^2 \neq 0$. A mathematical definition also governs the relation of w to its partial derivatives, viz., the total derivative:

$$dw = \frac{\partial w}{\partial t}\,dt + \frac{\partial w}{\partial z}\,dz \qquad (A.2)$$

These are two independent equations that can be solved for the partial derivatives, as follows.

$$\frac{\partial w}{\partial t} = \frac{\begin{vmatrix} H & G \\ dw & dz \end{vmatrix}}{\begin{vmatrix} F & G \\ dt & dz \end{vmatrix}} \tag{A.3}$$

$$\frac{\partial w}{\partial z} = \frac{\begin{vmatrix} F & H \\ dt & dw \end{vmatrix}}{\begin{vmatrix} F & G \\ dt & dz \end{vmatrix}} \tag{A.4}$$

Numerical values could be found from Eqs. A.3 and A.4 for these partial derivatives, but they are not especially useful in this context. Ironically, their solution becomes meaningless when the denominator vanishes, though the numerator must also vanish for the quotient to remain finite. That property yields expressions among the coefficients that must be valid even though values of the partial derivatives cannot be determined. The following must all be true:

$$F\,dz = G\,dt \tag{A.5}$$
$$F\,dw = H\,dt$$
$$H\,dz = G\,dw$$

Most useful applications of pressure swing adsorption have quasilinear material balance equations, due to the dependence of both velocity and the adsorption isotherm on partial pressure. In fact, Eq. A.1 is equivalent to a continuity equation of component i for a binary mixture in a fixed bed. Furthermore, the appropriate form can be derived from Eq. 4.4, by applying the chain rule for differentiation with some algebraic manipulation. Specifically, that equation is obtained from two independent equations: the first being for component 1 (or 2), and the second representing the sum of components 1 and 2.

In the resulting equation, w is taken to be the mole fraction of component i, and F is adjusted to unity. In so doing, G describes the conveyance due to bulk motion through the fixed bed and the distribution between the fluid and solid phases, while H relates the nature of a composition shift that corresponds to a pressure shift and includes the distribution between the fluid and solid phases. The symbolic definitions are:

$$G = \frac{\beta_A v}{1 + (\beta - 1) y_i}$$

$$H = \frac{(\beta - 1)(1 - y_i) y_i}{1 + (\beta - 1) y_i} \frac{d \ln P}{dt}$$

Even for linear isotherms (where $\beta_i = \{1 + [(1 - \varepsilon)/\varepsilon] k_i\}^{-1}$, and $\beta = \beta_A/\beta_B$), the interstitial fluid velocity depends on the amount of component i

present, which implies that it may be a major constituent and that the adsorbent may have substantial capacity. In these equations, pressure drop in the adsorbent bed is assumed to be negligible. The resulting equalities, corresponding to the first two of Eq. A.5 are:

$$dz = \frac{\beta_A v}{1 + (\beta - 1)y_i} dt$$

$$dy_i = \frac{(\beta - 1)(1 - y_i)y_i}{1 + (\beta - 1)y_i} \frac{d \ln P}{dt}$$

(A.6)

These are equivalent to Eqs. 4.7 and 4.8, and they are called *characteristic equations*. The former defines characteristic trajectories along the bed axis with respect to time. The latter, which must be solved simultaneously with the former, defines the composition variation along each trajectory (e.g., as pressure varies). Although characteristics may appear to be arbitrary lines or curves, they are not. At each position and time there is only one corresponding characteristic. Physically, this is reasonable, because that means that there can only be a single composition at any point and time.

By their nature, however, characteristics that represent different compositions have different slopes. Generally, those having greater amounts of the more strongly adsorbed component have larger slopes, as shown in Eq. A.6. Thus when the influent contains more of the more strongly adsorbed component than the initial column contents, the characteristics of the influent would tend to overlap those of the initial contents. That, as mentioned, would be impossible. The conflict cannot be resolved by merely averaging the compositions or blending the equations. Rather, an overall material balance is performed, looking at the sliver of adsorbent into which passes the high-concentration material, and out of which flows the low-concentration material. In fact, it may be helpful to visualize the composition shift that occurs both in terms of position and time. The first illustration, Figure A.1(a) shows composition profiles at three instants of time, one of which catches the front in the region of interest. The second, Figure A.1(b), shows identical data, but in the form of internal breakthrough curves (i.e., histories at three axial positions). The shapes of the curves do not matter (as long as they are not spreading); all that matters is the composition shift. The fronts in the illustration are sketched as rounded, though the equilibrium theory considers them to be step changes.

The key concept explaining the movement of the front is that there is no accumulation at the front itself. This is equivalent to saying that the adsorbent in the bed is uniform, or that the isotherms are independent of axial position. Following the sketches, it is most useful to consider elements of space and time (e.g., a sliver of adsorbent and a moment of time). For that

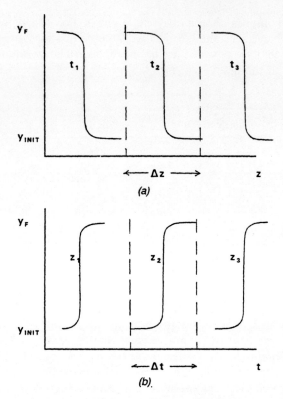

Figure A.1 (a) Composition front for an adsorption step, shown as three instantaneous profiles, as it passes through an element of adsorbent. (b) Composition front for an adsorption step, shown as histories recorded at three axial positions within the adsorbent.

case, a material balance "around the front" is written in the same form as Eq. 4.1

$$\varepsilon\left(\left.\frac{\Delta P y_A}{\Delta t}\right|_z + \left.\frac{\Delta_z v P y_A}{\Delta z}\right|_t\right) + RT(1-\varepsilon)\left.\frac{\Delta n_A}{\Delta t}\right|_z = 0 \tag{A.7}$$

where $\Delta|_t$ represents the shift in a sliver observed at a specific time, and $\Delta|_z$ represents the shift over a moment at a given position, and that $\Delta y_A|_t = -\Delta y_A|_z$, as shown in the figure. When that substitution is made and the definition of θ_A is applied at constant pressure, the following equation for the shock velocity, v_{SH}, is obtained:

$$v_{SH} = \left.\frac{\Delta z}{\Delta t}\right|_{\Delta y} = \theta_A \frac{\Delta v y_A}{\Delta y_A} \tag{A.8}$$

where the shift is taken with respect to position or time, but must be consistent.

APPENDIX A

By writing the same equation for component B, an equivalent but not identical expression is obtained, which can be solved for the interstitial velocity ahead of the shock wave in terms of the velocity behind it. The result is Eq. 4.8.

References

1. H.-K. Rhee, R. Aris, and N. R. Amundson, *First Order Partial Differential Equations*, **1**, Prentice-Hall, Englewood Cliffs, NJ (1986).
2. A. Acrivos, "Method of Characteristics Technique," *Ind. Eng. Chem.* **48**, 703-10 (1956).
3. M. C. Bustos and F. Concha, "Boundary Conditions for the Continuous Sedimentation of Ideal Suspensions," *AIChE J.* **38**, 1135-38 (1992).
4. V. R. Dabholkar, V. Balakotaiah, and D. Luss, "Travelling Waves in Multi-Reaction Systems," *Chem. Eng. Sci.* **43**, 945-55 (1988).
5. A. Kluwick, "The Analytical Method of Characteristics," *Prog. Aerospace Sci.* **19**, 197-313 (1981).
6. R. Herman and I. Prigogine, "A Two Fluid Approach to Town Traffic," *Science* **204**, 148-51 (1979).

APPENDIX B

Collocation Form of the PSA Model Equations

B.1 Dimensionless Form of the LDF Model Equations

The discussion here is restricted to a two-component system (subscripts A and B denote the components). The sum of the mole fractions of the two components in the gas phase at every point in the bed is equal to one. Therefore, solving for only one component in the gas phase is sufficient; the concentration of the other component in the gas phase is obtained by difference. The equations are developed in general terms for a variable pressure step with flow at the column inlet.

Applying ideal gas law $[c_A = C y_A P y_A/(R_g T_0)]$ to the component and overall mass balance equations (Eqs. 1 and 4 in Table 5.2), we obtain:

$$\frac{\partial y_A}{\partial t} = D_L \frac{\partial^2 y_A}{\partial z^2} - y_A \left(\frac{\partial v}{\partial z} + \frac{1}{P} \frac{\partial P}{\partial t} \right) - v \frac{\partial y_A}{\partial z} \quad (B.1)$$

$$- \frac{1-\varepsilon}{\varepsilon} \frac{R_g T_0}{P} \frac{\partial \bar{q}_A}{\partial t}$$

$$\frac{\partial v}{\partial z} + \frac{1}{P} \frac{\partial P}{\partial t} = -\frac{1-\varepsilon}{\varepsilon} \frac{R_g T_0}{P} \left(\frac{\partial \bar{q}_A}{\partial t} + \frac{\partial \bar{q}_B}{\partial t} \right) \quad (B.2)$$

Substituting Eq. B.2 into B.1 yields:

$$\frac{\partial y_A}{\partial t} = D_L \frac{\partial^2 y_A}{\partial z^2} - v \frac{\partial y_A}{\partial z} + \frac{1-\varepsilon}{\varepsilon} \frac{R_g T_0}{P} \left((y_A - 1) \frac{\partial \bar{q}_A}{\partial t} + y_A \frac{\partial \bar{q}_B}{\partial t} \right) \quad (B.3)$$

Equations 5 and 6 in Table 5.2 combined and written in dimensionless form become:

$$\frac{\partial x_A}{\partial \tau} = \alpha_A \left(\frac{\beta_A y_A}{1 + \beta_A y_A + \beta_B(1 - y_A)} - x_A \right);$$
$$\frac{\partial x_B}{\partial \tau} = \alpha_B \left(\frac{\beta_B(1 - y_A)}{1 + \beta_A y_A + \beta_B(1 - y_A)} - x_B \right) \quad \text{(B.4)}$$

Equation B.3 written in dimensionless form and then combined with Eq. B.4 yields:

$$\frac{\partial y_A}{\partial \tau} = \frac{1}{\text{Pe}} \frac{\partial^2 y_A}{\partial Z^2} - \bar{v} \frac{\partial y_A}{\partial Z} \quad \text{(B.5)}$$

$$+ \psi \left[(y_A - 1)\alpha_A \left(\frac{\beta_A y_A}{1 + \beta_A y_A + \beta_B(1 - y_A)} - x_A \right) \right.$$

$$\left. + \gamma_S y_A \alpha_B \left(\frac{\beta_B(1 - y_A)}{1 + \beta_A y_A + \beta_B(1 - y_A)} - x_B \right) \right]$$

The relevant fluid flow boundary conditions (Eq. 7 in Table 5.2) in dimensionless form lead to:

$$\left. \frac{\partial y_A}{\partial Z} \right|_{Z=0} = -\text{Pe}\, \bar{v}|_{Z=0}(y_A|_{Z=0^-} - y_A|_{Z=0});$$
$$\left. \frac{\partial y_A}{\partial Z} \right|_{Z=1} = 0 \quad \text{(B.6)}$$

Equation B.2 written in dimensionless form and then combined with Eq. B.4 takes the form:

$$\frac{\partial \bar{v}}{\partial Z} = -\psi \left[\alpha_A \left(\frac{\beta_A y_A}{1 + \beta_A y_A + \beta_B(1 - y_A)} - x_A \right) \right. \quad \text{(B.7)}$$

$$\left. + \gamma_S \alpha_B \left(\frac{\beta_B(1 - y_A)}{1 + \beta_A y_A + \beta_B(1 - y_A)} - x_B \right) \right] - \frac{1}{P} \frac{\partial P}{\partial \tau}$$

The dimensionless velocity boundary conditions are:

$$\bar{v}|_{Z=0} = \frac{v|_{Z=0}}{v_{\text{OH}}}; \quad \left. \frac{\partial \bar{v}}{\partial Z} \right|_{Z=1} = 0 \quad \text{(B.8)}$$

APPENDIX B

The clean bed initial conditions given by Eq. 12 in Table 5.2 assume the following dimensionless form:

$$y_A(z,0) = 0; \quad y_B(z,0) = 0;$$
$$x_A(z,0) = 0; \quad x_B(z,0) = 0 \tag{B.9}$$

B.2 Collocation Form of the Dimensionless LDF Model Equations

When Eq. B.5 with the boundary conditions given by Eq. B.6 are written in the collocation form based on a Legendre-type polynomial to represent the trial function, the following set of ordinary differential equations is obtained:

$$\frac{dy_A}{d\tau}(j) = \sum_{i=2}^{M+1} [\text{Pm Bx}(j,i) - \bar{v}(j)\text{Ax}(j,i)]y_A(i) \tag{B.10}$$

$$-A_5[\text{Pm Bx}(j,1) - \bar{v}(j)\text{Ax}(j,1)]$$

$$\times \sum_{i=2}^{M+1} [A_3 \text{Ax}(M+2,i) - \text{Ax}(1,i)] y_A(i)$$

$$+A_1[\text{Pm Bx}(j,M+2) - \bar{v}(j)\text{Ax}(j,M+2)]$$

$$\times \sum_{i=2}^{M+1} [A_3 \text{Ax}(M+2,i) - \text{Ax}(1,i)] y_A(i)$$

$$-A_4[\text{Pm Bx}(j,1) - \bar{v}(j)\text{Ax}(j,1)] \sum_{i=2}^{M+1} \text{Ax}(M+2,i) y_A(i)$$

$$+A_5[\text{Pm Bx}(j,1) - \bar{v}(j)\text{Ax}(j,1)]\text{Pe } \bar{v}(1)y_A|_{z=0^-}$$

$$-A_1[\text{Pm Bx}(j,M+2) - \bar{v}(j)\text{Ax}(j,M+2)]\text{Pe } \bar{v}(1)y_A|_{z=0^-}$$

$$+\psi\left[[X_A(j) - 1]\alpha_A\left(\frac{\beta_A y_A(j)}{1 + \beta_A y_A(j) + \beta_B[1 - y_A(j)]} - x_A(j)\right)\right.$$

$$\left.+\gamma_S y_A(j)\alpha_B\left(\frac{\beta_B[1 - y_A(j)]}{1 + \beta_A y_A(j) + \beta_B[1 - y_A(j)]} - x_B(j)\right)\right],$$

$$j = 2, \ldots, M+1$$

Equation B.4 now becomes:

$$\frac{dx_A}{d\tau}(j) = \alpha_A\left(\frac{\beta_A y_A(j)}{1 + \beta_A y_A(j) + \beta_B[1 - y_A(j)]} - x_A(j)\right); \quad \text{(B.11)}$$

$$\frac{dx_B}{d\tau}(j) = \alpha_B\left[\frac{\beta_B[1 - y_A(j)]}{1 + \beta_A y_A(j) + \beta_B[1 - y_A(j)]} - x_B(j)\right],$$

$$j = 2, \ldots, M + 1$$

The following set of linear algebraic equations is obtained when the dimensionless overall material balance equation (Eq. B.7) and the velocity boundary conditions (Eq. B.8) are combined and written in collocation form:

$$\sum_{i=2}^{M+1}\left(\text{Ax}(j,i) - \frac{\text{Ax}(M+2,i)\text{Ax}(j,M+2)}{\text{Ax}(M+2,M+2)}\right)\bar{v}(i) \quad \text{(B.12)}$$

$$= -\psi\left[\alpha_A\left(\frac{\beta_A y_A(j)}{1 + \beta_A y_A(j) + \beta_B[1 - y_A(j)]} - x_A(j)\right)\right.$$

$$\left.+ \gamma_S \alpha_B\left(\frac{\beta_B[1 - y_A(j)]}{1 + \beta_A y_A(j) + \beta_B[1 - y_A(j)]} - x_B(j)\right)\right]$$

$$- \left[\text{Ax}(j,1) - \frac{\text{Ax}(j,M+2)\text{Ax}(M+2,1)}{\text{Ax}(M+2,M+2)}\right]\bar{v}(1) - \frac{1}{P}\frac{\partial P}{\partial \tau},$$

$$j = 2, \ldots, M + 1$$

The following equations, which are derived from the boundary conditions, give the values at $j = 1$ and $M + 2$:

$$y_A(1) = -A_5 \sum_{i=2}^{M+1} [A_3 \text{Ax}(M+2,i) - \text{Ax}(1,i)]y_A(i) \quad \text{(B.13)}$$

$$- A_4 \sum_{i=2}^{M+1} \text{Ax}(M+2,i)y_A(i) + A_5 \text{Pe}\, \bar{v}(1) y_A|_{z=0^-}$$

$$y_A(M+2) = A_1 \sum_{i=2}^{M+1} [A_3 \text{Ax}(M+2,i) - \text{Ax}(1,i)]y_A(i) \quad \text{(B.14)}$$

$$- A_1 \text{Pe}\, \bar{v}(1) y_A|_{z=0^-}$$

$$\bar{v}(M+2) = -\frac{\text{Ax}(M+2,1)}{\text{Ax}(M+2,M+2)}\bar{v}(1) \quad \text{(B.15)}$$

$$- \left(\sum_{i=2}^{M+1} \text{Ax}(M+2,i)\bar{v}(i)\right)\Big/\text{Ax}(M+2,M+2)$$

Note: Here j refers to the axial location in the bed and is different from the

APPENDIX B

Table B.1. Summary of the Changes to Eqs. B.10–B.15 Necessary for Describing the Individual Steps in a Skarstrom Cycle

Step	Operation	j^a	$\dfrac{1}{P}\dfrac{\partial P}{\partial \tau}$	$y_A\|_{z=0^-}$	$\bar{v}(1)$	Parameters $\psi, \alpha_A, \alpha_B,$ β_A, β_B
1	Pressurization of bed 2 (pressure changing linearly from P_L to P_H in time t_P)	2	$\dfrac{P_H - P_L}{P_2 \tau_P}$	Mole fraction of A in feed	$\dfrac{P_H}{P_2}$	Change with changing column pressure[b]
	Blowdown of bed 1 (pressure changing linearly from P_H to P_B in time t_B)	1	$\dfrac{P_B - P_H}{P_1 \tau_P}$	$y_{A1}(1)$	0	Change with changing column pressure[c]
	High-pressure flow in bed 2 (constant pressure)	2	0	Mole fraction of A in feed	1	Constant
	Purge flow in bed 1 (constant pressure)	1	0	Mole fraction of A in product from bed 2	G	Constant (values for high- and low-pressure steps are different)

[a] The subscript j ($= 1$ for bed 1 and 2 for bed 2), which should properly appear with all the dependent variables and the parameters $P, \psi, \alpha_A, \alpha_B, \beta_A, \beta_B$, is omitted from the equations for simplicity.

[b] The change in column pressure is defined by $P_2 = [(P_H - P_L)/t_P] + P_L$.

[c] The change in column pressure is defined by $P_1 = [(P_B - P_H)/t_B] + P_H$.

subscript j mentioned in Table B.1. In the preceding equations: M is the number of internal collocation points. In Eq. B.10: Pm = $1/\text{Pe}$. In Eqs. B.10, B.13, and B.14:

$$A_1 = 1/\{\text{Ax}(1, M + 2) - [A_3 \text{Ax}(M + 2, M + 2)]\}$$

$$A_2 = \text{Ax}(M + 2, M + 2)/\{\text{Ax}(1, M + 2) - [A_3 \text{Ax}(M + 2, M + 2)]\}$$

$$A_3 = [\text{Ax}(1, 1) - \text{Pe}\,\bar{v}(1)]/\text{Ax}(M + 2, 1)$$

$$A_4 = 1/\text{Ax}(M + 2, 1)$$

$$A_5 = A_2 A_4$$

Equations B.10–B.15 are the collocation forms of the LDF model equations describing a variable pressure step with flow at the column inlet. The appropriate changes to these general equations necessary for describing the individual steps in a two-bed process operated on a Skarstrom cycle are summarized in Table B.1. In a Skarstrom cycle (see Figure 3.4) steps 1 and 2 differ from steps 3 and 4 only in the direction of flow. Following the same procedure discussed here, a similar set of collocation equations was derived for steps 3 and 4. The set of coupled algebraic and ordinary differential equations thus obtained describing steps 1–4 in the two beds was solved by Gaussian elimination and numerical integration, respectively. For numerical integration the Adam's variable-step integration algorithm as provided in the FORSIM package (Ref. 49 in Chapter 5) was used.

B.3 Dimensionless Form of the Pore Diffusion Model Equations (Table 5.6)

The discussion here is also restricted to a two-component system, but the equations are developed in general terms for a constant-pressure step with flow at the column inlet. The variable diffusivity case is considered. The following equations are taken from Table 5.2. Fluid phase mass balance:

$$-D_L \frac{\partial^2 c_A}{\partial z^2} + v \frac{\partial c_A}{\partial z} + c_A \frac{\partial v}{\partial z} + \frac{\partial c_A}{\partial t} + \frac{1-\varepsilon}{\varepsilon} \frac{\partial \bar{q}_A}{\partial t} = 0 \tag{B.16}$$

Continuity condition:

$$c_A + c_B = C \text{ (constant)} \tag{B.17}$$

Overall mass balance:

$$C \frac{\partial v}{\partial z} + \frac{1-\varepsilon}{\varepsilon}\left(\frac{\partial \bar{q}_A}{\partial t} + \frac{\partial \bar{q}_B}{\partial t}\right) = 0 \tag{B.18}$$

APPENDIX B

Boundary conditions for fluid flow:

$$D_L \frac{\partial c_A}{\partial z}\bigg|_{z=0} = -v|_{z=0}(c_A|_{z=0^-} - c_A|_{z=0});$$

$$\frac{\partial c_A}{\partial z}\bigg|_{z=L} = 0 \tag{B.19}$$

The velocity boundary conditions for the steps other than pressurization:
For $v|_{z=0}$ see Eqs. 10b–d in Table 5.2;

$$(\partial v/\partial z)|_{z=L} = 0 \tag{B.20}$$

In the dimensionless form Eq. 1 in Table 5.6 becomes:

$$\frac{\partial y_A}{\partial \tau} = \Gamma[y_A - y_{AP}|_{\eta=1}];$$

$$\frac{\partial Y_B}{\partial \tau} = \frac{\Gamma}{\gamma_S}[1 - y_A - y_{BP}|_{\eta=1}] \tag{B.21}$$

Substituting Eq. B.21 in the dimensionless form of Eq. B.18, we obtain:

$$\frac{\partial \bar{v}}{\partial z} = -\frac{\psi \Gamma}{W}\left[(y_A - y_{AP}|_{\eta=1}) + (1 - y_A - y_{BP}|_{\eta=1})\right] \tag{B.22}$$

Equation B.16 written in dimensionless form and combined with Eq. B.22 yields:

$$\frac{\partial y_A}{\partial \tau} = \frac{1}{\text{Pe}} W \frac{\partial^2 y_A}{\partial Z^2} - \bar{v}W \frac{\partial y_A}{\partial Z} + \psi\Gamma[(1 - y_A)(y_A - y_{AP}|_{\eta=1}) \tag{B.23}$$

$$+ y_A(1 - y_A - y_{BP}|_{\eta=1})]$$

The fluid flow boundary conditions (Eq. B.19) assume the following dimensionless form:

$$\frac{\partial y_A}{\partial z}\bigg|_{Z=0} = -\text{Pe } \bar{v}|_{Z=0}(y_A|_{Z=0^-} - y_A|_{Z=0});$$

$$\frac{\partial y_A}{\partial z}\bigg|_{Z=1} = 0 \tag{B.24}$$

The velocity boundary conditions given by Eq. B.20 take the following dimensionless form:

$$\bar{v}|_{Z=0} = \frac{v|_{Z=0}}{v_{\text{OH}}}; \qquad \frac{\partial \bar{v}}{\partial Z}\bigg|_{Z=1} = 0 \tag{B.25}$$

The dimensionless form of the velocity boundary conditions for pressurization (given by Eq. 2 in Table 5.6) is:

$$\bar{v}|_{Z=1} = 0 \tag{B.26}$$

The following dimensionless equations are obtained from the particle balance equations (Eqs. 11 and 12 in Table 5.6) and the related boundary conditions (Eqs. 4, 13, and 14):

$$\frac{\partial x_{CA}}{\partial \tau} = \frac{1}{1 - x_{CA} - x_{CB}} \left[(1 - x_{CB}) \left(\frac{\partial^2 x_{CA}}{\partial \eta^2} + \frac{2}{\eta} \frac{\partial x_{CA}}{\partial \eta} \right) \right. \quad (B.27)$$

$$\left. + x_{CA} \left(\frac{\partial^2 x_{CB}}{\partial \eta^2} + \frac{2}{\eta} \frac{\partial x_{CB}}{\partial \eta} \right) \right] + \frac{1}{(1 - x_{CA} - x_{CB})^2}$$

$$\times \left((1 - x_{CB}) \frac{\partial x_{CA}}{\partial \eta} + x_{CA} \frac{\partial x_{CB}}{\partial \eta} \right) \left(\frac{\partial x_{CA}}{\partial \eta} + \frac{\partial x_{CB}}{\partial \eta} \right)$$

$$\frac{\partial x_{CB}}{\partial \tau} = \frac{\gamma_K}{1 - x_{CA} - x_{CB}} \left[(1 - x_{CA}) \left(\frac{\partial^2 x_{CB}}{\partial \eta^2} + \frac{2}{\eta} \frac{\partial x_{CB}}{\partial \eta} \right) \right. \quad (B.28)$$

$$\left. + x_{CB} \left(\frac{\partial^2 x_{CA}}{\partial \eta^2} + \frac{2}{\eta} \frac{\partial x_{CA}}{\partial \eta} \right) \right] + \frac{\gamma_K}{(1 - x_{CA} - x_{CB})^2}$$

$$\times \left((1 - x_{CA}) \frac{\partial x_{CB}}{\partial \eta} + x_{CB} \frac{\partial x_{CA}}{\partial \eta} \right) \left(\frac{\partial x_{CB}}{\partial \eta} + \frac{\partial x_{CA}}{\partial \eta} \right)$$

$$\left. \frac{\partial x_{CA}}{\partial \eta} \right|_{\eta=0} = 0; \quad \left. \frac{\partial x_{CB}}{\partial \eta} \right|_{\eta=0} = 0 \quad (B.29)$$

$$\left. \frac{\partial x_{CA}}{\partial \eta} \right|_{\eta=1} = \Gamma'(1 - x_{CA})(y_A - y_{AP}|_{\eta=1}) \quad (B.30)$$

$$- \frac{\Gamma'}{\gamma_K \gamma_S} x_{CA}(1 - y_A - y_{BP}|_{\eta=1})$$

$$\left. \frac{\partial x_{CB}}{\partial \eta} \right|_{\eta=1} = \frac{\Gamma'}{\gamma_K \gamma_S}(1 - x_{CB})(1 - y_A - y_{BP}|_{\eta=1}) \quad (B.31)$$

$$- \Gamma' x_{CB}(y_A - y_{AP}|_{\eta=1})$$

The equilibrium isotherms (Eq. 7 in Table 5.6) assume the following dimensionless form:

$$y_{AP}|_{\eta=1} = \frac{1}{\beta_A} \frac{x_{CA}}{1 - x_{CA} - x_{CB}};$$

$$y_{BP}|_{\eta=1} = \frac{1}{\beta_A \gamma_E} \frac{x_{CB}}{1 - x_{CA} - x_{CB}} \quad (B.32)$$

B.4 Collocation Form of the Dimensionless Pore Diffusion Model Equations

When Eq. B.23 with the boundary conditions given by Eq. B.24 is written in the collocation form, the following set of ordinary differential equations is

APPENDIX B   321

obtained:

$$\frac{dy_A}{d\tau}(j) = \sum_{i=2}^{M+1} [\text{Qm Bx}(j,i) - W\bar{v}(j)\text{Ax}(j,i)] y_A(i) \tag{B.33}$$

$$-A_5[\text{Qm Bx}(j,1) - W\bar{v}(j)\text{Ax}(j,1)]$$

$$\times \sum_{i=2}^{M+1} [A_3 \text{Ax}(M+2,i) - \text{Ax}(1,i)] y_A(i)$$

$$+A_1[\text{Qm Bx}(j,M+2) - W\bar{v}(j)\text{Ax}(j,M+2)]$$

$$\times \sum_{i=2}^{M+1} [A_3 \text{Ax}(M+2,i) - \text{Ax}(1,i)] y_A(i)$$

$$-A_4[\text{Qm Bx}(j,1) - W\bar{v}(j)\text{Ax}(j,1)] \sum_{i=2}^{M+1} \text{Ax}(M+2,i) y_A(i)$$

$$+A_5[\text{Qm Bx}(j,1) - W\bar{v}(j)\text{Ax}(j,1)] \text{Pe } \bar{v}(1) y_A|_{z=0^-}$$

$$-A_1[\text{Qm Bx}(j,M+2) - W\bar{v}(j)\text{Ax}(j,M+2)] \text{Pe } \bar{v}(1) y_A|_{z=0^-}$$

$$+ \psi\Gamma\{[y_A(j) - 1][y_A(j) - y_{AP}|_{\eta=1}(j)]$$

$$+y_A(j)[1 - y_A(j) - y_{BP}|_{\eta=1}(j)]\},$$

$$j = 2, \ldots, M+1$$

$y_{AP}|_{\eta=1}(j)$ and $y_{BP}|_{\eta=1}(j)$ are obtained from Eq. B.41. $y_A(1)$ and $y_A(M+2)$ are given by Eqs. B.13 and B.14, respectively.

B.4.1 Pressurization

The following set of linear algebraic equations is obtained when the dimensionless overall material balance equations (Eq. B.22) and the velocity boundary condition (Eq. B.26) are combined and written in the collocation form:

$$\sum_{i=2}^{M+1} \text{AH}(j,i)\bar{v}(i) = -\frac{\psi\Gamma}{W}\{[y_A(j) - y_{AP}|_{\eta=1}(j)] \tag{B.34}$$

$$+ [1 - y_A(j) - y_{BP}|_{\eta=1}(j)]\},$$

$$j = 2, \ldots, M+1$$

The trial function chosen to satisfy the zero exist velocity boundary condition is:

$$\bar{v} = (1-z)\sum_{i=1}^{M} a_i P_{i-1}(z) = \sum_{i=1}^{M+1} \xi_i z^{i-1} \tag{B.35}$$

$$\bar{v}|_{z=0} = \bar{v}(1) = \xi_1 \tag{B.36}$$

Solution of Eq. B.34 gives velocities along the column at the internal collocation points $Z_2, Z_3, \ldots, Z_{M+1}$. The exit velocity is known from the specified boundary condition. The velocity at the column inlet may be obtained by solving the following matrix constructed from Eq. B.35:

$$\begin{bmatrix} \bar{v}(2) \\ \bar{v}(3) \\ \vdots \\ \dot{\bar{v}}(M+2) \end{bmatrix} = \begin{bmatrix} 1 & Z_2 & Z_2^2 & \cdots & Z_2^M \\ 1 & Z_3 & Z_3^2 & \cdots & Z_3^M \\ \vdots & & & & \vdots \\ 1 & 1 & 1 & \cdots & 1 \end{bmatrix} \begin{bmatrix} \xi_1 \\ \xi_2 \\ \vdots \\ \bar{\xi}_{M+1} \end{bmatrix} \quad (B.37)$$

B.4.2 Other Steps

The following set of linear algebraic equations is obtained when Eq. B.22 and the velocity boundary conditions given by Eq. B.25 are combined and written in collocation form:

$$\sum_{i=2}^{M+1} \left[\mathrm{Ax}(j,i) - \frac{\mathrm{Ax}(M+2,i)\mathrm{Ax}(j,M+2)}{\mathrm{Ax}(M+2,M+2)} \right] \bar{v}(i) = \quad (B.38)$$

$$-\psi \left[\alpha_A \left(\frac{\beta_A y_A(j)}{1 + \beta_A y_A(j) + \beta_B [1 - y_A(j)]} - x_A(j) \right) \right.$$

$$\left. + \gamma_S \alpha_B \left(\frac{\beta_B [1 - y_A(j)]}{1 + \beta_A y_A(j) + \beta_B [1 - y_A(j)]} - x_B(j) \right) \right]$$

$$- \left(\mathrm{Ax}(j,1) - \frac{\mathrm{Ax}(j,M+2)\mathrm{Ax}(M+2,1)}{\mathrm{Ax}(M+2,M+2)} \right) \bar{v}(1) - \frac{1}{P} \cdot \frac{\partial P}{\partial \tau}$$

$$j = 2, \ldots, M+1$$

$\bar{v}(M+2)$ is given by Eq. B.15. In Eq. B.33:

$A_1 = 1/\{\mathrm{Ax}(1, M+2) - [A_3 \mathrm{Ax}(M+2, M+2)]\}$

$A_2 = \mathrm{Ax}(M+2, M+2)/\{\mathrm{Ax}(1, M+2) - [A_3 \mathrm{Ax}(M+2, M+2)]\}$

$A_3 = [\mathrm{Ax}(1,1) - \mathrm{Pe}\, \bar{v}(1)]/\mathrm{Ax}(M+2, 1)$

$A_4 = 1/\mathrm{Ax}(M+2, 1)$

$A_5 = A_2 A_4$

$\mathrm{Qm} = W/\mathrm{Pe}$

APPENDIX B

The collocation form of the particle balance equations is:

$$\frac{\partial x_{CA}}{\partial \tau}(j,k) = \frac{1}{1 - x_{CA}(j,k) - x_{CB}(j,k)} \tag{B.39}$$

$$\times \left([1 - x_{CB}(j,k)] \sum_{i=1}^{N+1} B(k,i) x_{CA}(j,i) \right.$$

$$\left. + x_{CA}(j,k) \sum_{i=1}^{N+1} B(k,i) x_{CB}(j,i) \right)$$

$$+ \frac{1}{[1 - x_{CA}(j,k) - x_{CB}(j,k)]^2}$$

$$\times \left([1 - x_{CB}(j,k)] \sum_{i=1}^{N+1} A(k,i) x_{CA}(j,i) \right.$$

$$\left. + x_{CA}(j,k) \sum_{i=1}^{N+1} A(k,i) x_{CB}(j,i) \right)$$

$$\times \left(\sum_{i=1}^{N+1} A(k,i) x_{CA}(j,i) \right.$$

$$\left. + \sum_{i=1}^{N+1} A(k,i) x_{CB}(j,i) \right)$$

$$\frac{\partial x_{CB}}{\partial \tau}(j,k) = \frac{\gamma_K}{1 - x_{CA}(j,k) - x_{CB}(j,k)} \tag{B.40}$$

$$\times \left([1 - x_{CA}(j,k)] \sum_{i=1}^{N+1} B(k,i) x_{CB}(j,i) \right.$$

$$\left. + x_{CB}(j,k) \sum_{i=1}^{N+1} B(k,i) x_{CA}(j,i) \right)$$

$$+ \frac{\gamma_K}{[1 - x_{CA}(j,k) - x_{CB}(j,k)]^2}$$

$$\times \left([1 - x_{CA}(j,k)] \sum_{i=1}^{N+1} A(k,i) x_{CB}(j,i) \right.$$

$$\left. + x_{CB}(j,k) \sum_{i=1}^{N+1} A(k,i) x_{CA}(j,i) \right)$$

$$\times \left(\sum_{i=1}^{N+1} A(k,i) x_{CB}(j,i) \right.$$

$$\left. + \sum_{i=1}^{N+1} A(k,i) x_{CA}(j,i) \right),$$

$$j = 2,\ldots, M+1;\ k = 1,\ldots, N$$

$x_{CA}(j, N+1)$ and $x_{CB}(j, N+1)$ are obtained from Eqs. B.42 and B.43.

The collocation forms of the equilibrium isotherms are:

$$y_{AP}|_{\eta=1}(j) = \frac{1}{\beta_A} \frac{x_{CA}(j, N+1)}{1 - x_{CA}(j, N+1) - x_{CB}(j, N+1)}; \qquad (B.41)$$

$$y_{BP}|_{\eta=1}(j) = \frac{1}{\beta_A \gamma_E} \frac{x_{CB}(j, N+1)}{1 - x_{CA}(j, N+1) - x_{CB}(j, N+1)},$$

$$j = 2, \ldots, M+1$$

$x_{CA}(j, N+1)$ and $x_{CB}(j, N+1)$ are obtained from Eqs. B.42 and B.43.

The boundary conditions at the particle surface (Eqs. B.30 and B.31) written in the collocation form and then combined with Eq. B.41 lead to a set of coupled nonlinear algebraic equations:

$$\sum_{i=1}^{N} A(N+1, i) x_{CA}(j, i) + A(N+1, N+1) x_{CA}(j, N+1) \qquad (B.42)$$

$$- \frac{\Gamma'}{\gamma_K \gamma_S} [1 - x_{CA}(j, N+1)]$$

$$\times \left(y_A(j) - \frac{1}{\beta_A} \frac{x_{CA}(j, N+1)}{1 - x_{CA}(j, N+1) - x_{CB}(j, N+1)} \right)$$

$$+ \frac{\Gamma'}{\gamma_K \gamma_S} x_{CA}(j, N+1)$$

$$\times \left(1 - y_A(j) - \frac{1}{\beta_A \gamma_E} \frac{x_{CB}(j, N+1)}{1 - x_{CA}(j, N+1) - x_{CB}(j, N+1)} \right) = 0$$

$$\sum_{i=1}^{N} A(N+1, i) x_{CB}(j, i) + A(N+1, N+1) x_{CB}(j, N+1) \qquad (B.43)$$

$$- \frac{\Gamma'}{\gamma_K \gamma_S} [1 - x_{CB}(j, N+1)]$$

$$\times \left(1 - y_A(j) - \frac{1}{\beta_A \gamma_E} \frac{x_{CB}(j, N+1)}{1 - x_{CA}(j, N+1) - x_{CB}(j, N+1)} \right)$$

$$+ \Gamma' x_{CB}(j, N+1)$$

$$\times \left(y_A(j) - \frac{1}{\beta_A} \frac{x_{CA}(j, N+1)}{1 - x_{CA}(j, N+1) - x_{CB}(j, N+1)} \right) = 0,$$

$$j = 2, \ldots, M+1$$

The solution of this set of coupled nonlinear algebraic equations gives $x_{CA}(j, N+1)$ and $x_{CB}(j, N+1)$.

In these equations M is the number of internal collocation points along the column axis (j refers to axial location). (Note: j used in these equations is different from the subscript j used in Table B.2.) N is the number of internal collocation points along the radius of the adsorbent particle (k refers to location inside a microparticle).

APPENDIX B

Table B.2. Summary of the Changes to Eqs. B.33–B.43 and Eqs. B.13–B.15 Necessary for Describing the Individual Steps of a Modified Skarstrom Cycle with Pressure Equalization and No Purge

Operation	j^a	$\bar{v}(1)$	Velocity profile along the column
Pressurization of bed 2 (square wave change in pressure)	2	Given by Eq. B.37	From Eq. B.34
Blowdown of bed 1 (square wave change in pressure)	1	0	From Eqs. B.38 and B.15
High-pressure flow in bed 2 (constant pressure)	2	1	From Eqs. B.38 and B.15
"Self-purging" of bed 1	1	0	From Eqs. B.38 and B.15

^{a}The subscript j (= 1 for bed 1 and 2 for bed 2), which should properly appear with all the dependent variables and the parameters ψ, β_A, Γ, and Γ', is omitted from the equations for simplicity. During pressurization and high-pressure adsorption in bed 2, $y_A|_{z=0}$ is the mole fraction of component A in the feed gas. For blowdown and self-purging steps in bed 1, $y_A|_{z=0} = y_{A1}(1)$. ψ, β_A, Γ, and Γ' have different values for low- and high-pressure steps.

Equations B.33–B.43 are the collocation forms of the (variable-diffusivity) pore diffusion model equations describing a constant-pressure step with flow at the column inlet. The appropriate changes to these general equations necessary for describing the individual steps in a two-bed process operated on a modified Skarstrom cycle with pressure equalization and no purge are summarized in Table B.2. In a modified Skarstrom cycle with pressure equalization and no purge (see Figure 3.16) steps 3 and 6 are the pressure equalization steps. The pressure equalization step is difficult to handle in a rigorous manner. The approximate representation of this step is discussed in Section 5.2. Steps 1 and 2 differ from steps 4 and 5 only in the direction of fluid flow. Following the same procedure discussed here, a similar set of collocation equations was derived for steps 4 and 5. A set of coupled algebraic (linear and nonlinear) and ordinary differential equations describes the operations in steps 1, 2, 4, and 5 in the two beds. The nonlinear algebraic equations were solved by the IMSL routine NEQNF (Ref. 50 in Chapter 5). The linear algebraic equations were solved by Gaussian elimination. The ordinary differential equations were integrated in the time domain using Gear's stiff (variable-step) integration algorithm as provided in the FORSIM package (Ref. 49 in Chapter 5).

APPENDIX

C

Synopsis of PSA Patent Literature

C.1 Introduction

This appendix reviews some PSA developments that appeared as patents, a few of which were cited in Chapter 1. In addition, it covers the origins of many cycles that are mentioned throughout the book, especially in Chapters 4 and 6. The subject has been set aside here because there was no ideal place for it in the body, yet it seemed too important to omit.

Although individual patents are commonly filled with new technology, their focus is generally very narrow. Sometimes, if there are profound new ideas in a patent, instead of subtle modifications of existing know-how, the ideas are muddled because of the verbose legalistic jargon and baffling drawings. Sometimes they are merely austere announcements that technical milestones have been passed, and not necessarily that commercialization efforts are underway. In fact, the earliest known patents, awarded to Hasche and Dargan in 1931[1] and Finlayson and Sharp in 1932,[2] seem to have been largely ignored. Perhaps their ideas were made to work on a small scale but not on a large scale, and consequently, they were viewed by those few who became aware of them as laboratory curiosities and were quickly forgotten. Nevertheless, they explained the basic ideas of PSA.

Patents awarded since the 1930s have also frequently escaped notice, and early accomplishments and ideas have been overlooked or forgotten. Perhaps this has occurred because keeping track of patented technology is extremely time consuming, and reading patents is tedious and confusing. An unfortunate outcome has been that some of the basic ideas of PSA have been claimed an innovations over and over again. For example, several patents that were issued in the 1930s, 1940s, and 1950s described the principles of PSA, yet Skarstrom frequently receives credit for inventing PSA, apparently because of the thoroughness of his first patent, which was issued in 1960.[3]

This appendix covers a variety of PSA patents, emphasizing cycles and key concepts and the rich diversity of ideas that led to success. There are hundreds from which to choose, so the coverage presented here is by no means comprehensive; it is regrettable but unavoidable that some seminal contributions have been overlooked. That the material discussed here is predominately drawn from U.S. patents is not meant to slight the development of PSA technology by inventors in other nations. For those who seek additional information, Tondeur and Wankat[4] reviewed the field of PSA technology by categorizing patents and publications according to cycle attributes (such as the number of steps and number of columns), separation applications, and corporations to which patent rights had been assigned. They cited well over 100 sources, many of which were patents. Similarly, Ball[5] reviewed many U.S. patents and compiled an annotated bibliography. In addition, rather than restricting this review to very recent patents, the emphasis is on earlier patents because they contain most of the fundamental ideas that have proved to be widely applicable to various adsorbents and gas mixtures. In fact, one might observe that most recent patents tend to be permutations of the fundamental ideas introduced in the early patents. Considering that, one cynical reaction might be that, by now, all the patentable ideas in the field of PSA have already been claimed. There is a strong case for that opinion, but it is a complex issue that should be left to patent attorneys to argue.

Timing is critical for patents, not only for legal reasons, but also because one of the few rewards an inventor receives is recognition. An unfortunate experience for some patents is that they have languished, some for more than five years, while being reviewed in the Patent Office. So, from that point of view, the filing date often reveals more about the context of an invention (i.e., what other ideas were accessible in the public domain) than the date of issue. Thus, both dates are often mentioned here. In fact, patents are introduced roughly in chronological order, with cross references to more recent patents that have used some of the same approaches.

C.2 Inventors and Patents

C.2.1 Hasche and Dargan

In their patent application, filed in 1927 and approved in 1931, Hasche and Dargan[1] described a pressure swing process for recovering sulfur dioxide from smelter gases using silica gel. In that process, the feed was compressed

APPENDIX C

so that the partial pressure of sulfur dioxide was above atmospheric but below its critical pressure. The compressed gas was then cooled and admitted to the adsorbent bed, while purified gas was withdrawn. Upon imminent breakthrough, the column was blown down, and the energy of compression was recovered. They stated that "The process may, therefore, be described as substantially isothermal and utilizes the property of the adsorptive material whereby the amount of gases held therein is substantially proportional to the gaseous pressure." They did not, however, include a process flowsheet, and no specific performance was claimed, for example, in terms of purity, recovery, or energy consumption.

C.2.2 Finlayson and Sharp

The patent by Finlayson and Sharp was filed in 1931 and approved in 1932,[2] and it covered more of the basic concepts of PSA. For example, they explained several specific principles that are fundamental to the operation of PSA cycles and gave a detailed example of a single PSA system being used to alter the ratio of hydrogen to carbon monoxide in water gas. Most notably, they described a pressurization step followed by a production step in which "the first fraction or fractions being richer in the less easily adsorbed component or components... and the final fraction or fractions being richer in the more easily adsorbed component or components." Furthermore, they noted that the "pressure of the adsorbed gas may be released for instance to a limited extent or substantially to atmospheric pressure or even a vacuum may be applied." This idea was embellished later by Skarstrom and Heilman.[6] Even their simple idea of using subatmospheric pressure for desorption has been listed among the claims of several other patents,[7-9] and recently it has been popular to tout this version of PSA as VSA, as discussed in Sections 3.2 and 6.2. They did not include a schematic diagram of their process or any information describing performance or timing.

C.2.3 Perley

A patent by Perley[10] happened to overlap, not only with the intended application, but also with the dates of the patent by Finlayson and Sharp[2] (it was submitted in 1928, and awarded in 1933). It described a process for adsorbing carbon monoxide and carbon dioxide from a water gas mixture, to yield technically pure hydrogen. A schematic diagram is shown in Figure C.1. The adsorbent was regenerated by reducing the pressure and, optionally, heating the adsorbent via purge gas. Interestingly, the idea of combining pressure swings with temperature swings is still considered novel, and has been covered in recent patents.[11-13]

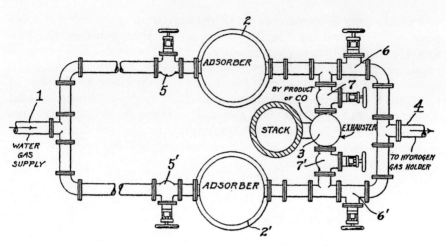

Figure C.1 Flowsheet of the PSA apparatus suggested by Perley in 1928, which appeared in U.S. Patent No. 1,896,916 in 1933. This equipment was apparently operated manually, and was designed for splitting hydrogen from water gas.

C.2.4 Erdmann

Another early patent that was essentially ignored was awarded to Erdmann in 1941.[14] His process seemed to be a reduction to practice of some of the ideas suggested by Finlayson and Sharp[2] and Perley.[10] Specifically, he used activated carbon and a feed of 93% hydrogen and the balance carbon monoxide to produce a pure product of 98 to 99% hydrogen, and an equimolar byproduct. The operating temperature was $-50°C$ (to exploit higher adsorption capacities at such a low temperature), and the pressure range was apparently 1 to 0.13 atm.

C.2.5 Guerin de Montgareuil and Domine

A patent that has received some recognition was awarded in 1957 to Guerin de Montgareuil and Domine.[15] It employed pressurization by feed, followed by cocurrent blowdown, exploiting a type of kinetic effect, to yield the light product, and finally complete blowdown, yielding an enriched heavy component. One advantage it offered was simplicity, due to the absence of flow reversal, but that was met by the disadvantage of obtaining both products at relatively low pressures. Despite that, it exploited, to a greater degree than previous patents, the coupling of pressure and flow to achieve PSA separations.

C.2.6 Skarstrom

Skarstrom's first PSA patent[3] was filed in 1958 and approved in 1960. He described PSA systems for gas drying and carbon dioxide removal, for oxygen

APPENDIX C

enrichment from air (up to about 85% at nearly nil recovery), and for nitrogen enrichment from air (up to 99% at nearly nil recovery). The latter was the first kinetics-based PSA separation (in fact, the equilibrium selectivity is in opposition to the kinetic selectivity), while the other applications he cited exploited equilibrium selectivity. The adsorbents he cited were silica gel and activated alumina, zeolite 5A, and zeolite 4A, respectively. He showed how hydrocarbons, including paraffins, isoparaffins, olefins, di-olefins, and aromatics could be split via three zeolites. He also explained high-pressure feed, blowdown, purge, and pressurization steps. A flowsheet of his basic cycle is shown in Figure C.2. Finally, Skarstrom suggested linking oxygen- and nitrogen-producing systems by recycling the secondary products "to provide at least a portion of the feed for the other adsorber concentration system." This combined cycle is shown in Figure C.3. That idea, (i.e., using

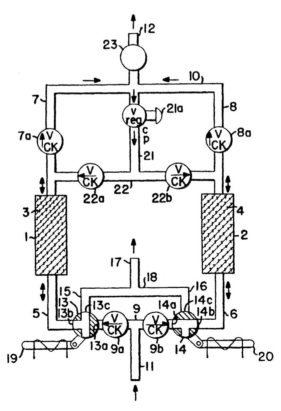

Figure C.2 Flowsheet of the basic PSA apparatus suggested by Skarstrom in 1958, which appeared in U.S. Patent No. 2,944,627 in 1960. This equipment was apparently designed for general purposes, from air drying to splitting nitrogen or oxygen from air.

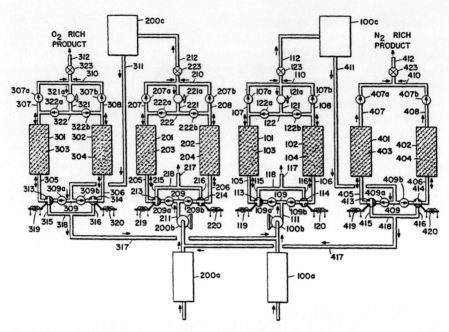

Figure C.3 Flowsheet of a second PSA apparatus suggested by Skarstrom in 1958, which appeared in U.S. Patent No. 2,944,627 in 1960. This equipment was apparently designed for extracting two nearly pure products from a single feed, such as splitting nitrogen and oxygen from air.

two or more adsorbents and coupling PSA systems) has been reinvented (with subtle twists) or borrowed several times.[16-24]

C.2.7 Skarstrom

Some early PSA patents concerned with hydrogen purification, which followed up on much earlier suggestions by Perley[10] and Erdmann,[14] were filed by Skarstrom in 1960 and awarded in 1963–64.[25-27] The first patent[25] described upgrading reformer recycle gases from various types of processes, in the range of 50 to 80% hydrogen, with an ultimate purity of greater than 95% hydrogen, using activated carbon as the adsorbent. The second described specific operating conditions, but did not specify the resulting performance. The third revealed more process details, and mentioned a potential recovery of 70%, but did not state the corresponding purity.

C.2.8 Hoke, Marsh, et al.

Within a year of filing of Skarstrom's PSA patents related to hydrogen purification, his colleagues, Hoke, Marsh, Bernstein, Pramuk, and he filed

APPENDIX C 333

related disclosures that were approved in 1964.[28,29] Their first process was also used for upgrading reformer recycle gas, in the range of 50% hydrogen, with an ultimate purity of greater than 99% hydrogen, using activated carbon and activated alumina as the adsorbents. The operating conditions included a pressure range of from 200 to 650 psig for adsorption, and 8 to 12 psia for desorption and purging. The cycle included a monthly thermal regeneration at about 600°F, via an inert gas. The second patented process was apparently the first that explicitly mentioned a cocurrent blowdown step that was used directly partially to pressurize a parallel column. That idea, which is often referred to as *pressure equalization*, has been claimed as a vital part of *many* other patents, of which only a few are cited here as examples.[30-34]

C.2.9 Wilson

Another early patent application concerning the actual (versus hypothetical) separation of oxygen and nitrogen from air in a single system was filed by Wilson in 1959, although it was not approved until 1965.[7] It contained several novel ideas that recurred in later patents. For example, a three-step cycle was used: pressurization (with predried air), cocurrent blowdown (producing enriched oxygen, at 70 to 90%), and countercurrent blowdown (producing enriched nitrogen, at subatmospheric pressure). He suggested an option of using heat to achieve a higher degree of desorption than by pressure alone. He also suggested using an oscillating piston to drive the pressure swing, and that idea has been borrowed, slightly modified, or reinvented several times.[35-39] He suggested that the same ideas could be used to remove carbon dioxide from air (e.g., in submarines) and to separate helium from natural gas.

C.2.10 Stark

About 2 years after the flurry of patent applications for PSA hydrogen purification were filed by Skarstrom and his colleagues (see C.2.7), another patent that extended the separation capability was filed by the same company (Stark, filed in 1963 and awarded in 1966[16]). The application was also the upgrading of reformer recycle gas, in the range of 80% hydrogen, to attain an ultimate purity of 99.9% hydrogen at a recovery of 75%, using silica gel as the adsorbent. The operating conditions included a pressure range of from 500 psig for adsorption and 0 psig for desorption and purging.

C.2.11 Basmadjian and Pogorski

An early PSA cycle that was apparently the first to use a rinse step (see Section 4.4.5) was invented by Basmadjian and Pogorski. Their application was filed in 1963 and the patent issued in 1966.[40] This cycle was intended to recover helium and possibly other light components from nitrogen or natural gas. The basic cycle consisted of pressurization with feed, rinse and production of enriched light product, blowdown and recovery of the heavy component. They improved the purity of the light component by conducting the pressurization and rinse in discrete steps in sequential columns. More recent patents usually employ a different set of steps. That might make the resemblance difficult to recognize. Nevertheless, a wide variety of other patents have borrowed, subtly modified, or reinvented the idea of rinse.[9,41-48]

C.2.12 Berlin

The first high-performance PSA oxygen generator was disclosed by Berlin in two applications filed in 1963 and approved in 1966–1967.[8,49] He claimed oxygen purities up to 93% (for which the balance was claimed to be 7% argon), at 53% recovery, but at a pressure ratio of 50 to 1000, and he suggested a strontium-exchanged X zeolite as the adsorbent.

C.2.13 Feldbauer

Shortly after the patent application for PSA hydrogen purification was filed by Stark (see C.2.10), another was filed from the same company that further extended the separation capability. In particular, Feldbauer filed in 1964 and the patent was awarded in 1967.[50] The concepts also applied to upgrading of reformer recycle gas, in the range of 40% hydrogen, to attain an ultimate purity of 96% hydrogen and a recovery of 94%, using 5A zeolite, activated carbon, and/or activated alumina as the adsorbent. The operating conditions included a pressure range of from 35 bar (500 psig) for adsorption and 1.4 bar (5 psig) for desorption and purging.

C.2.14 Wagner

Shortly before the patent for PSA hydrogen purification was awarded to Stark in 1967, another patent was filed by Wagner (awarded in 1969[31]) that significantly extended the separation capability. The concepts also applied to upgrading of reformer recycle gas, using a four-bed process, with a feed of about 77.1% hydrogen and 22.5% CO_2 with traces of other components. The process attained an ultimate purity of 99.9999 + % hydrogen and a recovery of 76.5%, using activated carbon and 5A zeolite as the adsorbents in a compound bed. The operating conditions included a pressure range of from 13.4 bar (200 psia) for adsorption and 15 psia for desorption and purging.

C.2.15 Batta

After the patent application was filed by Wagner in 1967 for PSA hydrogen purification, another was filed by Batta (awarded in 1971[51]) that slightly improved the inherent recovery. Among the patented concepts were: upgrad-

APPENDIX C 335

ing of reformer recycle gas containing about 75% hydrogen, attaining an ultimate purity of 99.999% hydrogen and a recovery of 80%, and using a four-bed process with 5A zeolite or activated carbon as the adsorbent. The operating conditions included a pressure range of from 9.7 bar (145 psia) for adsorption and 1 bar (15 psia) for desorption and purging. A patent having similar chains was awarded to Shell et al. in 1974.[52]

C.2.16 Batta

Following up on Berlin's patent[8] for oxygen separation from air, another application was filed by Batta that slightly improved the oxygen purity. It did not quite match the former patent's recovery, but it employed a pressure ratio of only about 3 rather than 50 (or greater). It was awarded in 1973.[53] The general concepts claimed had been discovered earlier for other applications. For example, he suggested a four-bed process, cocurrent blowdown, and pressure equalization. These ideas allowed an ultimate purity of 95 to 96% oxygen and a recovery of 40 to 45% to be achieved, using 5A or 13X zeolite as the adsorbent. The operating conditions included pressures of 3.2 bar (47 psia) for adsorption to 1 bar (15 psia) for desorption and purging.

C.2.17 Fuderer and Rudelstorfer

Until the mid-1970s, PSA systems had generally consisted of four parallel beds or fewer. Fuderer and Rudelstorfer, in a patent awarded in 1976,[54] changed all that with a system of as many as ten parallel beds. For that embodiment, the cycle comprised twenty steps, employing 54 timed valves, although any individual column only underwent 11 separate steps. Their application was splitting hydrogen from impurities such as carbon dioxide and nitrogen. They gave specific pressures, step times, pressure ratios, and other process details.

C.2.18 Munzner, Jüntgen, et al.

One of the most significant patents in the past 2 decades disclosed a process for separating nearly pure nitrogen from air by exploiting a carbon molecular sieve (CMS) adsorbent in which the diffusion rate differences of oxygen and nitrogen were dramatic. The original patent was granted in Germany in 1976.[55] Patents were awarded to Munzner et al. in 1977[56,57] and to Jüntgen et al. 1981,[58] which were assigned to Bergwerksverband GmbH. The idea and technology were not radically new, since CMS had been introduced separately in 1971,[59] and Skarstrom[3] had already shown that kinetic separations were possible. The patent by Jüntgen et al., however, showed significantly better performance than the system of Skarstrom: 99.9% nitrogen at a recovery of about 40% versus 99% nitrogen at essentially nil recovery. A schematic diagram of the process is shown in Figure C.4. It is instructive to

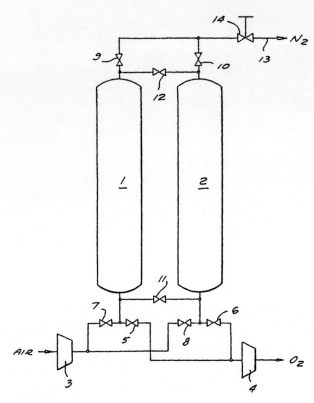

Figure C.4 Flowsheet of the kinetics-based PSA apparatus suggested by Juntgen et al. in 1979, which appeared in U.S. Patent No. 4,264,339 in 1981. This equipment was designed for splitting nitrogen from air, using a carbon molecular sieve.

notice the minor differences between this process and the one suggested by Perley.

C.2.19 Collins

One of the less conventional patent applications was filed by Collins in 1975 for PSA oxygen separation from air. The patent, awarded in 1977,[34] sheds light on the deviations from isothermal behavior of relatively large diameter beds of zeolite (e.g., 30 cm diameter or larger). It shows that expected recoveries often exceed those achieved in such large columns due to depressed temperatures within the bed. Accordingly, it shows actual temperature profiles along the bed axis during the PSA cycle. The primary contribution was a means for suppressing the temperature deviations by inserting aluminum plates or rods in the bed to conduct heat axially, thus reducing or eliminating temperature gradients.

C.2.20 Sircar and Co-Workers

Several patents were awarded to Sircar and his colleagues in the span of 1977–1988 that applied to hydrogen purification,[60] air separation,[9,23] air purification,[61] splitting reformer off-gas to get hydrogen and carbon dioxide, and recovering hydrogen and methane from hydrodesulfurization plant effluent.[62,63] The key unifying feature of these patents was: two or more sets of parallel beds were connected sequentially, and the sets usually contained different adsorbents in order to isolate different components of the feed mixture. They were also orchestrated to ensure that each product was obtained with as little power input as practical, and that each byproduct was fully exploited before it was released.

C.2.21 Jones, Keller, and Wells

The patent by Jones, Keller, and Wells, which issued in 1980,[64] pushed the limits of fast cycling close to the limits allowed by fluid mechanics and valve dynamics (see Figure C.5). That is, it employed feed step times of about 0.5 to 2.0 s, and exhaust times of 0.5 to 20 s. It also systematically explored the relations between bed length, particle size, feed pressure, and adsorbent type on product purity and recovery. It covered several applications, including air separation, splitting nitrogen and ethylene, and splitting hydrogen from methane, carbon monoxide, and carbon dioxide.

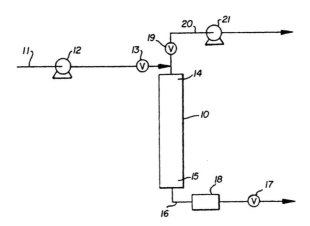

Figure C.5 Flowsheet of the rapid PSA process suggested by Jones et al. in 1979, which appeared in U.S. Patent No. 4,264,339 in 1981. This equipment was designed for splitting oxygen from air, using a zeolite molecular sieve. It employed smaller particles and exploited pressure waves that propagate along the axis by coupling flow and pressure shifts in a synchronous way. Two relatively pure products are produced, and the adsorbent productivity is high.

C.2.22 Kumar et al.

One of the more difficult separations among atmospheric gases is to split argon from oxygen. This is due to nearly identical adsorption isotherms that these components exhibit with most zeolites. Kumar et al., who filed in 1982 and received a patent in 1984,[19] approached the problem of purifying argon containing minor amounts (e.g., less than 10% each) of nitrogen and oxygen by first removing the nitrogen via zeolite, then removing oxygen using carbon molecular sieve. Essentially, they used loosely coupled PSA systems in which the product of the first was the feed to the second. Following similar reasoning, Hayashi et al., who filed in 1983 and were awarded a patent in 1985,[20] were able to split air to obtain both high-purity oxygen and argon.

C.2.23 Wiessner and Bolkart

A subtle variation on the technique for recovering the most strongly adsorbed components from a mixture was suggested by Wiessner and Bolkart in 1988.[65] They allowed blowdown to an intermediate pressure, then purged the residual heavy component. Subsequently, they completed blowdown, and as an option purged again. This allows a valuable component to be recovered at a moderate pressure, above the minimum pressure in the PSA cycle, allowing higher overall recovery with minimal recompression cost. It could also permit a moderately adsorbed component to be removed prior to dropping the pressure sufficiently to desorb a more strongly adsorbed component.

C.2.24 Tagawa et al.

An example of the refinement that is prevalent in recent PSA patents was given by Tagawa et al. in 1988.[66] They compared three different methods for pressure equalization, in a process to split oxygen from air using 5A zeolite. Their options included: (1) from the product end of one column to the feed end of a parallel column, (2) from the product end of one column to the product end of a parallel column, and (3) from the feed end of one column to the feed end of a parallel column. They state that the first is in the prior art. They also claim that (2) and (3) can be conducted simultaneously.

C.3 Concluding Remarks

To close this appendix, some comments are appropriate. First, a few overall impressions are given about specific patents and phases that they seem to fall into. Then a few comments are made about the relation of this appendix to other parts of this book.

First, it is remarkable that the basic flowsheet for PSA has been in the public domain for over sixty years (cf. Figure C.1), yet some aspects are still

APPENDIX C 339

not perfectly understood. In addition, it has been possible to secure patents with equivalent flowsheets even in the 1980s (cf. Figure C.3). Those facts speak volumes about the subtleties involved in PSA.

In one of the very first disclosures of PSA concepts, Finlayson and Sharp[2] made some suggestions that *now* might be thought of as mundane, but were truly farsighted at the time. For example, they gave a detailed example of a single PSA system being used to alter the ratio of hydrogen and carbon monoxide in water gas from about 1 : 1 to 9 : 1, 2 : 1 and 1 : 8, respectively, for subsequent synthesis of ammonia, methanol, and aliphatic acids. They also recognized activated carbon and silica gel as promising adsorbents. Finally, they suggested that, "nitrogen, hydrogen, oxygen, carbon monoxide, and methane" were candidates for purification by PSA. Even though these applications were suggested over *60 years ago*, the PSA hydrogen purification (the easiest of the group) did not become practical on a commercial scale until the late 1960s. Similarly, separation of either relatively pure oxygen or nitrogen from air did not become commercially viable until the late 1970s, and splitting carbon monoxide and methane from contaminants only became moderately successful in the late 1980s. Despite the achievements of modern technology, improvements in all of the applications are still being sought. One wonders what, if Finlayson and Sharp were with us today, they would expect to be possible via PSA 60 years from now.

In view of the incremental nature of Batta's patent in 1971,[51] it could be though of as a sort of turning point, at least for hydrogen production. It seemed to mark the passage from dramatic innovations in which substantial improvements over prior art were claimed, to an onslaught of minor refinements of existing cycles. A similar turning point occurred for PSA oxygen production in the early 1980s. That is not to say that innovations ceased. It was just that most of the "obvious" modes of PSA operation had already been discovered, if not perfected.

Most of the topics covered in this book are based on mathematical models, although in some instances experimental verification is mentioned. Conversely, this appendix covers the main repository of original ideas related to pressure swing adsorption cycles, and apparently none of them depended on mathematical models to reach fruition. This is not to say that some may not have achieved more resounding success if an adequate mathematical model had been available. It may be worthwhile to consider, for a moment, the reinforcing aspects of the patent literature and the conclusions obtained independently, though much later, via mathematical models.

For example, there are significant distinctions in the ways in which even the simplest steps can be accomplished. From a theoretical point of view, Section 4.6 shows that pressurization with product is virtually always superior to pressurization with feed. That point is made as strongly, if not as succinctly, in patents by different methods of pressure equalization, as illustrated by the references listed in Sections C.2.8 and C.2.24. Specifically, when the purified product from one column is used partially to pressurize a parallel column by interconnecting the product ends, it is possible to obtain more high-quality product than when the feed ends are interconnected. Another point that was made in Section 4.4.4 was that simultaneous depressurization and production always improves recovery over isobaric production or simultaneous pressurization and production. Accordingly, in patents the concepts of cocurrent blowdown either into a parallel bed (in tandem with the concept of pressurization with product and pressure equalization topic mentioned previously) or to obtain the light product has also appeared in several forms, as discussed in Sections C.2.5, C.2.9, and C.2.16, to name a few. Finally, the idea of a rinse step was discussed in Section 4.4.5 from a theoretical perspective. That section showed the advantages in terms of recovery that could be realized by adding that step. Long before that theory was conceived, however, the idea of the rinse step had been invented and reduced to practice by Basmadjian and Pogorski[40] (cf. Section C.2.11). They and several others have used it as a means for recovering the heavy component(s) or recycling unadsorbed feed. In view of all that, it is clear that the results of mathematical models reinforce what has already been learned in practice. With that in mind, it is promising that such models could provide guidance for improving performance of existing PSA cycles, and for moving efficiently towards nearly optimum conditions for new cycles.

References

1. R. L. Hasche and W. N. Dargan, "Separation of Gases," U.S. Patent No. 1,794,377 (1931).

2. D. Finlayson and A. J. Sharp, "Improvements in or Relating to the Treatment of Gaseous Mixtures for the Purpose of Separating Them into Their Components or Enriching Them with Respect to One or More of Their Components," British Patent No. 365,092 (1932).

3. C. W. Skarstrom, "Method and Apparatus for Fractionating Gaseous Mixtures by Adsorption," U.S. Patent No. 2,944,627 (1960).

4. D. Tondeur and P. C. Wankat, "Gas Purification by Pressure Swing Adsorption," *Separ. and Purif. Methods* **14**, 157–212 (1985).

5. D. J. Ball, "Patent Search of Pressure Swing Adsorption Related Processes," unpublished (1985).

6. C. W. Skarstrom and W. O. Heilman, "Technique with the Fractionation of Separation of Components in a Gaseous Feed Stream," U.S. Patent No. 3,086,339 (1963).

7. E. M. Wilson, "Method of Separating Oxygen from Air," U.S. Patent 3,164,454 (1965).

8. N. H. Berlin, "Vacuum Cycle Adsorption," U.S. Patent No. 3,313,091 (1967).

9. S. Sircar and T. R. White, "Vacuum Swing Adsorption for Air Separation," U.S. Patent 4,264,340 (1981).

10. G. A. Perley, "Method of Making Commercial Hydrogen," U.S. Patent No. 1,896,916 (1933).

11. R. Kumar, S. Sircar, and W. C. Kratz, "Adsorptive Process for the Removal of Carbon Dioxide from a Gas," U.S. Patent No. 4,472,178 (1984).

12. S. Sircar, "Regeneration of Adsorbents," U.S. Patent No. 4,784,672 (1988).
13. S. Sircar, "Closed-Loop Regeneration of Adsorbents Containing Reactive Adsorbates," U.S. Patent No. 4,971,606 (1990).
14. K. Erdmann, "Process of the Removal of Carbon Monoxide from Mixtures thereof with Hydrogen," U.S. Patent No. 2,254,799 (1941).
15. P. Guerin de Montgareuil and D. Domine, French Patent No. 1,233,261 (1957).
16. T. M. Stark, "Gas Separation by Adsorption Process," U.S. Patent No. 3,252,268 (1966).
17. D. Domine and L. Hay, "Gas Separation by Adsorption," U.S. Patent No. 3,619,984 (1971).
18. P. J. Gardner, "Process and Compound Bed Means for Evolving a First Component Enriched Gas," U.S. Patent No. 4,386,945 (1983).
19. R. Kumar, S. Sircar, T. R. White, and E. J. Greskovitch, "Argon Purification, U.S. Patent No. 4,477,265 (1984).
20. S. Hayashi, H. Tsuchiya, and K. Haruna, "Process for Obtaining High Concentration Argon by Pressure Swing Adsorption," U.S. Patent No. 4,529,412 (1985).
21. G. S. Glenn, V. K. Rajpaul, and R. F. Yurczyk, "Integrated System for Generating Inert Gas and Breathing Gas on Aircraft," U.S. Patent No. 4,681,602 (1987).
22. K. S. Knaebel, "Complementary Pressure Swing Adsorption," U.S. Patent No. 4,744,803 (1988).
23. S. Sircar, "Preparation of High Purity Oxygen," U.S. Patent No. 4,756,723 (1988).
24. K. Haruna, K. Ueda, M. Inoue, and H. Someda, "Process for Producing High Purity Oxygen Gas from Air," U.S. Patent No. 4,985,052 (1991).
25. C. W. Skarstrom, "Process for the Recovery of Hydrogen from Hydrocarbon Gas Streams," U.S. Patent No. 3,101,261 (1963).
26. C. W. Skarstrom, "Timing Cycle for Improved Heatless Fractionation of Gaseous Materials," U.S. Patent No. 3,104,162 (1963).
27. C. W. Skarstrom, "Apparatus and Process for Heatless Fractionation of Gaseous Constituents," U.S. Patent No. 3,138,439 (1964).
28. R. C. Hoke, W. D. Marsh, J. Bernstein, and F. S. Pramuk, "Hydrogen Purification Process," U.S. Patent No. 3,141,748 (1964).
29. W. D. Marsh, F. S. Pramuk, R. C. Hoke, and C. W. Skarstrom, "Pressure Equalization Depressuring in Heatless Adsorption," U.S. Patent No. 3,142,547 (1964)
30. C. W. Skarstrom, "Oxygen Concentration Process," U.S. Patent No. 3,237,377 (1966).
31. J. L. Wagner, "Selective Adsorption Process," U.S. Patent No. 3,430,418 (1969).
32. N. R. McCombs, "Selective Adsorption Gas Separation Process," U.S. Patent No. 3,738,087 (1973).
33. H. Lee and D. E. Stahl, "Pressure Equalization and Purging System for Heatless Adsorption," U.S. Patent No. 3,788,036 (1974).
34. J. J. Collins, "Air Separation by Adsorption," U.S. Patent No. 4,026,680 (1977).
35. D. B. Broughton, "Adsorptive Separation of Gas Mixtures," U.S. Patent No. 3,121,625 (1964).
36. G. A. Rutan, "Apparatus and Method for Drying a Gaseous Medium," U.S. Patent No. 3,236,028 (1966).

37. W. Betteridge and J. Hope, "Separation of Hydrogen from other Gases," U.S. Patent No. 3,438,178 (1969).

38. R. Eriksson, "Fractionating Apparatus," U.S. Patent No. 4,169,715 (1979).

39. G. E. Keller II and C.-H. A. Kuo, "Enhanced Gas Separation by Selective Adsorption," U.S. Patent No. 4,354,859 (1982).

40. D. Basmadjian and L. A. Pogorski, "Process for the Separation of Gases by Adsorption," U.S. Patent No. 3,279,153 (1966).

41. T. Tamura, "Absorption Process for Gas Separation," U.S. Patent No. 3,797,201 (1974).

42. P. Wilson, "Inverted Pressure Swing Adsorption Process," U.S. Patent No. 4,359,328 (1982).

43. S. Matsui, Y. Tukahara, S. Hayashi, and M. Kumagai, "Process for Removing a Nitrogen Gas from Mixture Comprising N_2 and CO or N_2, CO_2, and CO," U.S. Patent No. 4,468,238 (1984).

44. M. Whysall, "Pressure Swing Adsorption Process," U.S. Patent No. 4,482,361 (1984).

45. T. Inoue and K. Miwa, "Separation Process for a Gas Mixture," U.S. Patent No. 4,515,605 (1985).

46. E. Richter, W. Korbacher, K. Knoblauch, K. Giessler, and K. Harder, "Method of Separating Highly Adsorbable Components in a Gas Stream in a Pressure-Sensing Adsorber System," U.S. Patent No. 4,578,089 (1988).

47. M. Yamano, T. Aono, and M. Uno, "Process for Separation of High Purity Gas from Mixed Gas," U.S. Patent No. 4,775,394 (1988).

48. K. S. Knaebel, "Pressure Swing Adsorption," U.S. Patent No. 5,032,150 (1991).

49. N. H. Berlin, "Method for Proving an Oxygen-Enriched Environment," U.S. Patent No. 3,280,536 (1966).

50. G. F. Feldbauer, "Depressuring Technique for ΔP Adsorption Process," U.S. Patent No. 3,338,030 (1967).

51. L. B. Batta, "Selective Adsorption Process," U.S. Patent No. 3,564,816 (1971).

52. D. C. Shell, D. A. Tanner, and R. D. Brazzel, "Separation Process," U.S. Patent No. 3,788,037 (1974).

53. L. B. Batta, "Selective Adsorption Process for Air Separation," U.S. Patent No. 3,717,974 (1973).

54. A. Fuderer and E. Rudelstorfer, "Selective Adsorption Process," U.S. Patent No. 3,986,849 (1976).

55. H. Jüntgen et al., F.R.G. Patent No. 2,652,486 (1976).

56. H. Munzner, U.S. Patent No. 4,011,065 (1977).

57. H. Munzner, U.S. Patent No. 4,015,956 (1977).

58. H. Jüntgen, K. Knoblauch, J. Reichenberger, H. Heimbach, and F. Tarnow, "Process for the Recovery of Nitrogen-Rich Gases from Gases Containing at Least Oxygen as Other Component," U.S. Patent No. 4,264,339 (1981).

59. H. Munzner et al., F.R.G. Patent No. 2,119,829 (1971).

60. S. Sircar and J. W. Zondlo, "Hydrogen Purification by Selective Adsorption," U.S. Patent No. 4,077,779 (1978).

61. S. Sircar and W. C. Kratz, "Removal of Water and Carbon Dioxide from Air," U.S. Patent No. 4,249,915 (1981).
62. S. Sircar, "Separation of Multicomponent Gas Mixtures," U.S. Patent No. 4,171,206 (1979).
63. S. Sircar, "Separation of Multicomponent Gas Mixtures by Pressure Swing Adsorption," U.S. Patent No. 4,171,207 (1979).
64. R. L. Jones, G. E. Keller II, and R. C. Wells, "Rapid Pressure Swing Adsorption Process with High Enrichment Factor," U.S. Patent No. 4,194,892 (1980).
65. F. Wiessner and A. Bolkart, "Adsorbate Recovery in PSA Process," U.S. Patent No. 4,717,397 (1988).
66. T. Tagawa, Y. Suzu, S. Hayashi, and Y. Mizuguchi, "Enrichment in Oxygen Gas," U.S. Patent No. 4,781,735 (1988).

Author Index

Ackley, M. W., 180, 184
Acrivos, A., 307
Alexis, R. W., 238
Alpay, E., 181, 286
Anzelius, A., 224

Backer, P. O., 292
Ball, D. J., 104, 328
Banerjee, R., 259–61
Batta, L. B., 79, 228, 238, 334, 335
Basmadjian, D., 333, 340
Berlin, N. H., 76, 77, 334
Bernstein, J., 332
Betteridge, W., 333
Bolkart, A., 338
Boniface, H., 186
Breck, D. W., 22
Broughton, D. B., 333
Brunauer, S., 25, 28
Bustos, M. C., 307

Campbell, M. J., 227
Carter, J. W., 176, 202
Cassidy, R. T., 5, 80, 238, 245

Cen, P. L., 85, 177, 191, 208, 211
Chan, Y. N. I., 99, 107, 110, 138
Chen, Y. D., 180
Cheng, H., 156
Chihara, K., 19, 49, 176, 181, 202, 208
Chlendi, M., 166
Coe, C. G., 34
Collins, J. J., 146, 336

Dabholkar, V., 307
Danner, R. P., 33, 34
Dargan, W. N., 328–29
Davis, J. C., 79
de Montgareuil, P. G., 4, 5, 72, 82, 330
Derrah, R. I., 32
Desai, R., 28
Doetsch, I. H., 181
Domine, D., 4, 5, 72, 82, 330
Dominguez, J. A., 49
Doong, S. J., 114, 156, 173, 177, 180, 208, 211, 284
Dorfman, L. R., 33
Doshi, K. J., 238, 240
Dow-Generon, 297

Note: References are generally indexed by first author only.

Eagan, J. D., 33
Erdman, K., 330–32
Eriksson, R., 333
Espitalier-Noel, D. M., 213

Farooq, S., 52, 74, 86, 177, 180, 181, 183, 185, 187, 188, 190, 192, 198, 203, 206, 208, 213, 215, 254
Feldbauer, G. F., 334
Fernandez, G. F., 71
Finlayson, B. A., 4, 5, 184, 329
Flores-Fernandez, G., 133, 151, 165
Fuderer, A., 238, 335

Gardner, P. J., 332
Garg, D. R., 147
Glenn, G. S., 332
Glueckauf, E., 58, 181
Grebbell, J., 246
Greskovitch, E. J., 332
Guerin de Montgarcuil, 4, 5, 72, 82, 330

Haas, O. W., 86
Habgood, H. W., 195
Haq, N., 33, 185
Haraya, K., 293
Harrison, I. D., 23
Hart, J., 156, 160, 161
Haruna, K., 332, 338
Hasche, R. L., 5, 328–29
Hassan, M. M., 180, 185, 194, 196, 202
Hayashi, S., 332, 338
Heilman, W. O., 329
Heimbach, H., 335
Herman, R., 307
Hill, F. B., 111, 138, 156
Hoke, R. C., 332
Hope, J., 333
Horvath, G., 34
Huang, J. T., 33

Izumi, J., 244

Jones, R. L., 283, 337
Juntgen, H., 18, 230, 243, 244, 335

Kadlec, R. H., 95, 278, 281, 282, 283
Kahle, 4, 5
Kapoor, A., 85, 86, 91, 177, 180, 182, 189, 192

Karger, J., 36, 43, 302
Kawazoe, K., 34, 181
Kayser, J. C., 98, 106, 117, 119, 133, 138, 165
Keefer, B. G., 271–76
Keller, G. E., 238, 267, 283, 333, 337
Kenney, C. N., 71, 114, 133, 151, 279, 280, 286
Kidnay, A. J., 33
Kirkby, N., 4, 114
Kluwick, A., 307
Knaebel, K. S., 85, 89, 111, 134, 138, 165, 177, 180
Knoblauch, K., 85–87, 230, 241, 244, 335
Kolliopoulos, K. P., 126
Koresh, J., 18
Koros, W. J., 294
Kowler, D. E., 278, 282, 283
Kratz, W. C., 248, 329, 343
Kumar, R., 33, 152, 246, 248, 329, 332, 338
Kuo, C. H. A., 267, 333

Lapidus, L., 184
Lederman, P. B., 33
Lee, H., 333
Lee, L.-K., 65
Le Van, 156, 161, 213
Liapis, A. I., 62
Loureiro, J. M., 156, 161
Lu, Z. P., 156, 161

Marsh, W. D., 78, 332
Matsumura, Y., 14
Matz, M. J., 73, 106, 114, 134, 150
McCombs, N. R., 333–34
Meunier, F., 184
Miller, G. W., 33, 186
Mitchell, J. E., 95, 99, 110, 133, 137, 176, 184
Mizuguchi, Y., 338
Munkvold, G., 174, 176
Munzner, H., 335
Myers, A. L., 31, 34

Nakao, S., 181, 182, 197, 206
Naylor, R. W., 292
Nolan, J. T., 34

AUTHOR INDEX

Perley, G. A., 4, 5, 329, 332
Pigford, R. L., 162
Pilarczyk, E., 230, 241, 243, 244
Pogorski, L. A., 333, 340
Pramuk, F. S., 332

Raghavan, N. S., 176, 184, 189, 197, 202
Rajpaul, V. K., 332
Reichenberger, J., 335
Reinhold, H., 231
Rhee, H. K., 307
Rikkinides, E. S., 257
Ritter, J. A., 202, 215, 257
Roberts, C. W., 17
Rodriguez, A. E., 156, 161
Round, G. F., 43, 195
Rousar, I., 106, 114, 152
Rudelstorfer, E., 238, 335
Rutan, G. A., 333
Ruthven, D. M., 11, 24, 27, 32-34, 39-40, 44, 46, 49, 59, 86, 180-81, 197, 223, 254, 301

Schroter, H. J., 40, 230, 243, 244
Scott, D. M., 156, 159-61, 181, 286
Seinfeld, J. H., 184
Sharp, A. J., 329
Shell, D. C., 335
Shendalman, L. H., 95, 99, 110, 137, 176
Shin, H. S., 85, 88-90, 177, 180
Sircar, S., 3, 135, 137, 234, 235, 246, 256, 329, 332, 333, 335, 337
Skarstrom, C. W., 4, 5, 72, 74, 75, 78, 106, 221, 226, 328
Smith, O. J., 240
Smolarek, J., 227
Sood, S. K., 255, 257
Sorial, G. A., 30, 33
Spillman, R. W., 305
Springer, C., 33
Stahl, S. E., 333
Stakebake, J. L., 33
Stark, J. M., 332, 333

Stewart, H. A., 238
Suh, S. S., 83, 106
Sun, L. M., 184
Sundaram, N., 156
Suzu, Y., 338
Suzuki, M., 11, 181, 182, 197, 202, 206, 208, 236
Sweed, N. H., 265

Tagawa, T., 338
Tarnow, F., 335
Thorogood, R. M., 304
Tomita, T., 237, 238
Tondeur, D., 328
Tsuchiya, H., 332, 338
Turnock, P. H., 95, 278, 281

Valenzuela, D. P., 34
van der Vlist, 33
Vansaut, E. F., 16
Vereslt, H., 33
Villadsen, J. V., 184
von Rosenburg, D. U., 184

Wagner, J. L., 81, 114, 238, 334
Wakasugi, J., 33
Wankat, P. C., 106, 114, 328
Wells, R. C., 337
Westerberg, A. W., 240, 244
White, D. H., 222-25
White, T. R., 332, 333, 337
Wiessner, F., 338
Wilhelm, R. H., 265
Wilson, E. M., 333

Xu, Z., 49

Yang, R. T., 11, 84, 85, 114, 173, 177, 182, 208, 213, 257, 279, 284
Yon, C. M., 147
Yucel, H., 38, 44
Yurczyk, R. F., 332

Zondlo, J. W., 337

Subject Index

Activated alumina, 21
Activated carbon, 17–20
Adsorbents, 11–23
  physical properties, 19
  pore size distribution, 13
  structure, 35
Adsorption equilibrium, 23
Adsorption step, 68
Air drier, 73, 75, 148, 211–26, 331
Air Liquide cycle, 82
Air separation, 70, 71, 74, 78, 80, 87, 90, 93, 187, 196, 226–35, 281, 330–34, 337
  Bergbau-Forschung process, 87, 230
  equilibrium theory, 133–42
  Lindox process, 228
  nitrogen production, 195–201, 230, 331, 335
  oxygen and nitrogen, 232, 256, 275
  oxygen production, 187, 190, 226, 331–38
  RPSA, 283–86
  TCPSA, 276
  vacuum swing process, 230
Applications of PSA, 7
Argon recovery, 338
Atmospheric gases
  adsorption data, 32–34

Blowdown step, 68, 88, 151–62
  concentration of heavy component, 253
  losses, 76
Burnauer's classification, 25
Bulk separation, 173

Carbon dioxide
  recovery, 242–44
  separation (from methane), 91, 192
Carbon molecular sieve, 17–20, 86–88, 196–201
Characteristics, method of, 100, 184, 307–11
Co-current depressurization, 88
Coke oven gas, 242
Collocation, 184, 313–25
Column pressure,
  constant, 173
  variable, 174
Compensation, diffusivity/solubility, 291
Concentration (of trace component), 157, 251
Concentration profiles (during pressurization), 71
Constant pattern, 55, 101
Continuous counter-current model, 201
Corrected diffusivity, 40
Cryogenic PSA, 255

Cycles (for PSA), 67–94
  analysis of, 105–33
  five-step, 122
  four-step, 98, 107
  self-purging, 88
  Skarstrom, 106

Dead volume, 125
Depressurization, 83
Desiccants, 20
Design example, 143–46
Desorption, 68
Diffusion,
  Fickian, 40
  Knudsen, 37
  macropore, 36
  micropore, 43
  model, 191–201, 318–24
  molecular, 37
  Poiseuille flow, 37
Diffusivity, 40
  corrected, 40
  effective, 46
Dispersed plug flow, 174
Dispersive front, 101
Dynamic modelling, 57, 165–219
Dynamic models, summary of, 167
Dynamics (adsorption column), 52

Economics of PSA (N_2), 231
Efficiency (of PSA process), 258–63
Elementary steps (of PSA process), 67
Enrichment, 108
Equilibrium controlled separations, 71, 95–162
Equilibrium/kinetic effects reinforce, 90
Equilibrium selectivity, 50
Equilibrium theory, 53, 95–163
  data for atmospheric gases, 32–34
  experimental validation, 133–37
Energy analysis, 258–62
Experimental PSA system, 89
  verification of equilibrium theory, 133–37
Extensions of PSA concept, 265–87
Extract, 1

Favorable and unfavorable isotherms, 25
Five-step cycle, 84

Flow models, 172
Fluid film resistance, 42
Forces of adsorption, 11
Four-step cycle, 107, 137–42

Gemini-8 process, 248
Gemini-9 process, 246

Heat effects in PSA processes, 146–51, 207–17
Heat transfer, 47
Heats of sorption, 13
Helium recovery, 333
Henry constants, 23
  for atmospheric gases, 32
Henry's law, 23
History of PSA, 4, 95, 327–40
Hydrocarbon separations, 246
Hydrogen and carbon dioxide, PSA process, 246
Hydrogen recovery, 235–42, 274, 330, 332, 334, 335, 337
  Bergbau-Forschung process, 242
  four-bed process, 236
  polybed process, 239
Hydrogen-helium separation, 254
Hydrophilic/hydrophobic properties, 12

Ideal adsorbed solution theory, 31
Isosiv process, 245
Isotherm, 25–32, 177–80
  BET, 28
  binary and multicomponent, 29
  carbon sieve, 91
  favorable/unfavorable, 25
  Freundlich, 27
  Gibbs, 28, 31
  Langmuir, 25
  Sipps, 30

Kinetic selectivity, 48, 52, 87
Kinetics (of sorption), 34–61
Knudsen diffusion, 290

Langmuir isotherm, 25, 179
LDF model, 180–83, 185–93, 209, 313–18
  rate expression, 57
Lindox process, 228
Losses, blowdown, 76

SUBJECT INDEX

Macropore diffusion, 36, 45
Mass transfer models, 180
Mass transfer resistance, 55
   effect on productivity, 190
Membrane, 289
   cascades, 301
   modules, 295
   process, cascades for, 301, 302
   process, comparison with PSA, 289–305
   process, effect of back pressure, 292
   process, effect of flow pattern, 298
   process for N_2, 303
   process for O_2, 304
   zeolite, 295
Mesopores, diffusion in, 36
Methane recovery, 244
Methane-carbon dioxide separation, 192
Micropore diffusion, 37, 43
Model,
   continuous countercurrent, 201–7
   diffusion, 191–201, 318–24
   LDF, 184–91, 313–18
Molecular diameter, effect on diffusion, 19, 23, 41
Molecular gate, 267, 287
Molecular sieve carbon, 17, 49
Moments (chromatographic), 60
Multicomponent systems, 61
Multiple bed systems, 79, 239, 335
Multiple cyclic states, 213–17, 230–32

Nitrogen PSA, 196–200, 205, 230–32, 331
Non-isothermal systems, 61, 146–51, 207–13
Numerical methods, 183, 313–25

Oxygen production, 74, 187, 226–30, 304, 331
   and nitrogen production, 232–35

Parametric pump, 265
Patents (PSA), 3, 327–43
Permeability, 289, 294
Physical strength (of adsorbents), 17
Polybed process, 239, 335
Pore diffusion, 36
   model, 318
Pore size distribution, 13

Pressure,
   equalization, 69, 76, 193, 333, 338
   profile, 151–61
   ratio, effect of, 90, 108–32
   variation, effect of, 118, 158
Pressurization, 68
   feed, 110, 112, 141
   product, 107, 137, 139
   step, 151–62
Proportionate pattern, 55, 101
Publications (PSA), 3
Purge, 68, 88, 224
   incomplete, 73, 114, 118
   pressure, effect of, 90
   with strong adsorptive, 84

Raffinate, 1
   product, 71
Rapid PSA. *See* RPSA
Recovery. *See also* Individual gases
   effect of dead composition, 109, 112
   effect of dead volume, 131
   effect of pressure, 109–32, 142
   light product, 85, 118
   purity profile, 301
   rapidly diffusing species, 93
   strong adsorptive, 83, 251–57
Reformer gases (H_2 recovery), 246, 330, 332, 335
Resistances to mass transfer,
   combination of, 60
   diffusional, 36–45
   effect of, 55
   fluid film, 42
   surface, 42, 49
Rinse, 69, 122
RPSA, 278–86, 337
   air separation, 283
   future prospects, 286
   modelling, 279
   performance, 281

Selectivity
   effect of, 104
   equilibrium, 12, 50, 90–93
   kinetic, 12, 52, 90–93
   membrane, 290
   permeability, 294
   temperature dependence, 293

Self-purging cycles, 85, 87, 88, 230–32
Self-sharpening profile, 101
Separation factor, 30, 51, 289, 290
  effect of back pressure, 292
  effect of temperature, 293
Shock wave, 55, 101
Silica gel, 20
Simple wave, 101
Simulation of PSA processes, 165–219
  continuous countercurrent model, 201–7
  LDF model, 184–91
  pore diffusion model, 191–201
Size selectivity, 23
Skarstrom cycle, 72, 106, 221–26
Smelter gases (purification), 328
Spreading pressure, 28
Stirling engine, 271
Sulfur dioxide (recovery), 328
Surface resistance, 42

TCPSA, 270–77
  air separation, 275
  comparison with PSA, 277
  scale-up, 278
  synthesis gas separation, 274
Temperature variation, 146–51, 207–17
Thermal wave, 63
Thermally coupled PSA. See TCPSA
Trace component concentration, 251–57
Trace systems, 172

Uptake rates, 41–50

Vacuum swing, See VSA
Voidage, 99
VSA, 7, 329
  cycle, 81
  economics, 230
  process, 229

Water gas, 330
Wave velocity, 266

Zeolites, 21–23

Printed in the United States
54023LVS00002B/17